WINNING SPACE

WINNING
SPACE

HOW AMERICA REMAINS
A SUPERPOWER

BRANDON J. WEICHERT

REPUBLIC
BOOK PUBLISHERS

ISBN 9781645720119 (Hardcover)   9781645720126 (ebook)

For inquiries about volume orders, please contact:

Republic Book Publishers

501 Slaters Lane #206

Alexandria VA 22314

editor@republicbookpublishers.com

Published in the United States by Republic Book Publishers

Distributed by Independent Publishers Group

www.ipgbook.com

Book designed by Mark Karis

Printed in the United States of America

For Ashley, Charlotte, and Madison—my three loves. To God, for giving me the passion to do this. To my parents, who have always supported me in my endeavors (including this one). To my IWP family for giving me the tools to succeed. And to Frank and Esther for letting me stay at their beautiful home for all of my lectures in Washington, D.C. over the years. Thank you for everything.

CONTENTS

FOREWORD

MACKUBIN THOMAS OWENS

BRANDON WEICHERT has written a lively and informative book on the evolution of US space policy and its future. He argues that we are at a crossroads: if we continue on the path we have tread in recent decades, we could face a "space Pearl Harbor" or worse. Not only China and Russia but also Iran and North Korea are developing the capabilities to attack strategic US space assets.

Our use of space is pervasive, so much so that Americans don't even recognize how dependent we are on space, not only for national security but also for civilian communications and commerce, which rely at a minimum on GPS and satellite-based telecommunications.

The fact is that the United States has the most to lose if space based assets are ever attacked.

From a national security standpoint, space is a "strategic domain." Weichert observes that US space policy has been shaped by three "schools:" the "sanctuary" school, which fears the militarization of space and seeks to apply international cooperation and arms control measures to counter a space arms race; the "survivability" school, which seeks to make the United States less reliant on satellites while enhancing the survivability of existing satellite constellations to withstand attacks; and "space control" or "space superiority," a defensive posture relies on deterrence. This is the school favored by most military officers.

Weichert lays out the shortcomings of each of the schools and argues on behalf of a more radical posture: "space dominance." This approach treats space as the strategic "high ground," giving the United States an advantage in its pursuit of its interests.

Although space as a strategic domain shares many characteristics of air power, it is more analogous to sea power. As the legendary John Collins observed in a chapter of his watershed book, *Military Geography*, space can be divided into four regions: Earth and its atmosphere; circumterrestrial space; the Moon and its environs; and the outer envelope. Here we can find analogs to choke points and sea lines of communications. Controlling these features can have strategic effects. As the First Lord of the Admiralty, Admiral Sir Jackie Fisher remarked in 1904, "Five keys lock up the world! Singapore. The Cape. Alexandria. Gibraltar. Dover." And England controlled them all by means of the Royal Navy.

But sea power works indirectly over time. Although Trafalgar was a decisive battle that ultimately sealed the fate of Napoleonic France, it would be ten more years before Napoleon was finally defeated. As Alfred Thayer Mahan wrote in *The Influence of Sea Power upon the French Revolution and Empire, 1793–1812*, "Those far distant, storm-beaten ships, upon which the Grand Army never looked, stood between it and the dominion of the world." In contrast, space power can have an immediate strategic impact.

A problem that transcends a philosophy of space is that the assets necessary to ensure space dominance must compete against other claimants of limited resources. And space planning to counter something like a space Pearl Harbor is an example of what Frank Hoffman calls a "pink flamingo": a predictable event that is ignored due to cognitive biases of a senior leader or a group of leaders trapped by powerful institutional forces. These are the cases which are "known knowns," often brightly lit, but remaining studiously ignored by policymakers.

In this book, Weichert sounds a clarion call for taking space dominance seriously. Whether one agrees with him or not, his argument is one with which national security experts must contend.

INTRODUCTION

MAKE SPACE GREAT AGAIN

IN 2013, as a young congressional staffer working for a Tea Party Republican—a real "Mr. Smith Goes to Washington"—I was asked to attend a meeting on his behalf in my first week. It was a seemingly innocuous meeting where a group of space advocates gave a rundown on some of the critical issues regarding satellite management. During that meeting, though, someone raised the point that the satellites the US military relied on were getting old and that they were increasingly hard to maintain because of their age. Some of the space policy people cautioned that these aging satellite systems were susceptible to attack from rival states.

In fact, another added, America's enemies were actively developing counterspace weapons—systems that could be used to deny the United States military access to the strategic domain of space in the event of a crisis. Satellites are important because they provide bandwidth to a modern society such as ours. Without that bandwidth everything would grind to a halt. The most amazing part of the meeting, at least to me, was the ambivalent reaction from most of the lawmakers, Republican and Democrats alike. Some of them did take note of what was said. Ultimately, however, nothing came of it. After all, space was too expensive. Better to let that sleeping dog lie. And the longer we refused to allow the space dog to hunt, the more numerous the prey in the field became. The prey even started evolving into predators themselves.

That meeting stuck with me for the next several years. I soon found myself enrolled in a master's program at the Institute of World Politics in Washington, DC. At that school, all of the students were challenged on a routine basis to think as unconventionally as possible. Most of the students were mid-career civil servants from the executive and legislative branch. I was one of the youngest students there. Ultimately, I found several professors who encouraged me to pursue my own research interests. Given that I was so interested in space policy, professors there, like Dr. John Lenczowski, the founder of IWP, as well as my adviser, John Tsagronis, and Dr. Marek Chodakiewicz, all compelled me to continue expanding my research in this critical issue that no one was talking about at that time. Finally, Dr. Mackubin Owens gave me the added push I needed to keep up my research, even when all seemed lost (being mocked as a "space cadet" would grow tiresome over the years).

For Dr. Owens's class, I was required to write a final paper on any topic we covered in his military theory course. We had covered much military history—all of it very important (although Jomini is still perplexing to me). Every fiber of my being told me to stop talking about space, to focus on something earthlier, and to simply follow the bureaucratic consensus. And I certainly did not want to risk my high GPA by messing up a class that I did exceedingly well in. But I was never a consensus builder, which is why I left Congress after just three years, and I rarely played it safe. So, rather than write something on the importance of sea power today (a very important topic, but one that bores me somewhat—sorry to any sailors out there), I chose to stick with my obsession. I handed in a thirty-two-page manuscript with pages of citations and maps of various orbits, when Dr. Owens had asked for a meager ten to twelve pages. I thought for certain he would fail me on the sheer size of the piece. Yet, he encouraged me. Dr. Owens enjoyed the piece so much that he converted it into an article for *Orbis: The Foreign Policy Research Institute's Journal of World Affairs,* for which he is the longtime editor.

Thanks to these great teachers, I continued to research and constantly question my assumptions about this significant and often overlooked area of research. Soon thereafter, I was published in a variety of online news outlets, ranging from *American Greatness*, the *American Spectator*, *Real Clear World*, *Real Clear Defense*, *Real Clear Politics*; and *Real Clear Policy*. Much of what I wrote online pertained to various topics touched on in this book. I usually focused on the confluence of emerging technology and national security, by relating it to the strategic domain of space (although I do write on other topics related to national security as well). After my media profile was established, I was soon asked by interested parties in both the academic and defense communities to lecture on this subject. All of these things have allowed me to continue to ponder what many believe to be imponderable: Is the United States losing the second space race? If so, what are the implications of this? And how might the United States arrest this perceived decline in such a critical domain? This book endeavors to answer these questions while issuing a dire, near-term warning: that America's enemies know that the United States is paradoxically dependent on satellites in space yet is also oblivious to the vulnerabilities of those systems. As such, the country is likely in store for a "space Pearl Harbor." This book is the apotheosis of my professional and academic work over the last six years.

Space had been a minor policy concern since the end of the Cold

War. Even when America had the space shuttle, few Americans cared. Congress was hard-pressed to either fund a replacement for the aging system or to extend its life span. As for the military's space policy, the situation was bleaker. The basic way that the Pentagon purchased, produced, and placed satellites into orbit had remained the same for far too long. In subsequent chapters you will see how the stifling red tape of the Pentagon's bureaucracy, coupled with the corruption of some of America's top defense contractors, along with congressional ambivalence, has led to a crisis situation in America's military space policy. The satellites on which our military depends are more vulnerable than at any other time in history. This is a critical national security concern that should be on the front pages of every newspaper in America. Yet, it is often relegated to the back page, left to languish with the obituaries (the bureaucratic graveyard being exactly where US space policy has been headed).

Donald J. Trump changed all of that. The controversial billionaire real estate tycoon from Manhattan who promised to put "America First" in foreign policy, to "Make America Great Again" by building a wall to stop illegal immigration—as well as to repair America's dying economy—also spoke about space in ways few others dared. For many in the Western political establishment, Trump was a troglodyte who was unworthy of any office, be it a small-town dogcatcher, let alone the president of the United States. Of course, Trump persisted. He bested not one but *two* political dynasties—the nominally Republican Bush clan and the mostly neoliberal Clinton cabal. Trump did this by spending less money than his opponents, keeping overhead costs low for his spit-and-polish-style campaign, and focusing on the major topics that Americans cared most for. Yes, much of his campaign was predicated on earthly issues, like immigration, the economy, and health care. Yet, Trump had the vision and gumption to talk about the future as well.

"We must have American dominance in space," Trump declared during a speech at the National Space Council, which the Trump administration reconstituted after the Obama administration had unceremoniously disbanded the organization. "I'm hereby directing the Department of Defense to immediately begin the process to establish a space force as the sixth branch of the armed forces," Trump said, further explaining to his audience that "we are going to have the air force, and we are going to have the space force. Separate but equal. It's going to be something so important."[1] Trump had openly campaigned on it throughout the contentious 2016 presidential election. What few knew is that these words would kick off a firestorm, both within the entrenched military bureaucracy, which did not want to fight for their slice of the budgetary pie with yet another branch of the military, and between the "Never Trump" elites in Washington and the Trump administration. Whoever won the bureaucratic tug-of-war over US space policy would determine whether the United States could remain a dominant space power or not. Thus far, despite claims that the space force is proceeding, the entrenched bureaucracy is winning the space force fight. Thankfully, though, the president has not given up on his ambitious space policies. In fact, on December 20, 2019, the United States Space Force officially became the sixth branch of the United States military. While an independent branch, at least for now, the relatively small space force will exist under the Department of the Air Force; the Marine Corps, too, has a similar arrangement with the Department of the Navy.[2]

Trump's various remarks since 2016 in support of a more robust American presence in space are heartening. It indicates a strategic seriousness that the last four US presidents have lacked. After all, the only reason that the United States has been faced with the prospect of having its vital satellite constellations attacked by a rival is because the last four presidents have failed to adequately strengthen America's dominance in orbit. This

---

1 Oriana Pawlyk, "It's Official: Trump Announces Space Force as 6th Military Branch," Military. com, 18 June 2018. https://www.military.com/daily-news/2018/06/18/its-official-trump-announces-space-force-6th-military-branch.html.

2 Meghann Myers, "The Space Force Is Officially the Sixth Branch. Here's What That Means." The *Air Force Times.* 20 December 2019. https://www.airforcetimes.com/news/your-military/2019/12/21/the-space-force-is-officially-the-sixth-military-branch-heres-what-that-means/.

is what many analysts have referred to as a "space Pearl Harbor." Such a catastrophic surprise attack could emanate from China, the United States' number one, long-term strategic competitor; it could come from Russia, whose leadership understands that they have, at most, a decade to fundamentally alter the geopolitical chessboard in their favor; or a "space Pearl Harbor" might derive from one of the two nuclear-arming rogue states, North Korea or Iran. Specifically, North Korea—and, to a lesser extent, Iran—possess the means to threaten the United States with orbital electromagnetic pulse (EMP) weapons. As you will see in succeeding chapters, the EMP threat to the United States is significant. And such a destructive weapon, when married to a space program, such as North Korea's, could be devastating to the United States.

Trump is the only American leader who not only recognized these threats but was willing to do something about it. Then, of course, the public face of America's space policy, the National Aeronautics and Space administration (NASA), has sadly become a parody of itself. An organization that once routinely placed men on the moon has not returned astronauts to the moon since the 1970s (or sent astronauts beyond the lunar surface, to Mars). And what few successes NASA has enjoyed since those times, while beneficial to society, have been miniscule and unworthy of the cost of NASA. Whereas NASA was once a key contributor to national prestige, the nation's high-tech economy, and US national security, NASA has become a national embarrassment. In 2017, President Trump announced his administration's intention to return Americans to the moon by 2024. Almost immediately, forces inimical to the president's agenda *from within* NASA (you know: the country's top "space exploration" group) as well as his congressional enemies have worked to prevent the completion of this goal. Meanwhile, China intends to have their astronauts on the moon's south pole, establishing a permanent colony there by 2024. While many academicians chortle at the suggestion that China's space threat is serious, I have seen little reason over the last two decades to doubt China's intentions or capabilities to achieve their national goal.

The private space sector in space is also booming. It is the private space sector that offers the United States its greatest hope since NASA seems impervious to change. Unlike NASA and the Pentagon bureaucracy, the private space sector is innovative. And thanks to competition within that sector, most private space companies help to lower the overall cost to spaceflight operations. As you will see throughout this work, cost is the greatest hindrance to space operations for either the private or public sectors. The private space sector, unfortunately, remains restrained by onerous regulations imposed by the various international agreements governing the use of space. The United States must work on either rewriting these rules or simply abandoning them. There is much economic opportunity to be enjoyed, not only in the space launch service sector, but also in the budding space tourism industry and, more important, in the nascent space mining economy. America's private space sector must be given more latitude and resources to fully exploit space economically.

America is a juggernaut. When we get moving as a country, we are unstoppable. Taking the initial steps, however, is always the hardest for our country. Today, the United States faces a space Pearl Harbor—and everyone in Washington knows it.[3] Similarly, the United States risks falling behind in both the new space economy and in the national prestige and scientific research categories. All of these redound to the benefit of US rivals. In this book, you will read about the extent of China's, Russia's, North Korea's, and Iran's space capabilities and how their newfound space capabilities threaten the United States. You will also see how other countries, like France, Israel, Brazil, and India, are all rising to counteract the apparent decline of the United States in space.

Fundamentally, the United States is in a second space race. It managed to win the first space race in the Cold War through commitment

---

3 Baker Spring, "U.S. National Security Policy Toward Space: The Debate That Should End but Won't," *Space News,* 10 October 2005. http://spacenews.com/us-national-security-policy-toward-space-debate-should-end-wont/

and a sound strategy. Right now, America's strategy in space is confusing. Everyone knows that there are threats to critical American systems in orbit. The president has wanted to protect against these threats. Yet, few in the bureaucracy or in Congress are willing to enable the president to fully achieve his goal, if only because they personally—vehemently—dislike the president. It's a case of "Orange Man Bad" becoming official national security and space policy. It threatens the entire country, as the United States depends heavily on space systems for its very survival.

When the president told audiences in 2018 that he sought "space dominance," he also likely set much of the space policy community against him. For the American space policy community, I believe, is generally divided into two groups that often overlap. One group is the utopians, who believe that space should be a weapons-free sanctuary, like Antarctica. The other group is the naysayers. These are usually engineers or scientists, with a liberal bent, who simply want to tell you how something *cannot* work. We all know people like this. Nassim Taleb refers to such people as the "Intellectual-Yet-Idiot." At each turn, though, whenever the naysayers have explained how a new or proposed piece of technology, such as the EmDrive, a propellantless propulsion system for space, cannot work, China proves them wrong. For the utopian crowd, space dominance is simply too patriarchal and too pro-American and would inevitably lead to the weaponization of space. The naysayers are technocrats at heart who simply disagree that the United States should spend the money or has the technological capacity to produce the equipment needed to achieve space dominance, whereas the utopians are ideologically opposed to the full development of space.

Space dominance is exactly what it sounds like: it is a national plan for being the hegemon in space. Such a hegemon would determine who would have access to critical orbits in space, or even who got into space. That power would also have the ability to protect all of its space assets, thereby denying rivals the capability to hold its sensitive systems hostage or to simply destroy them. Space dominance is militaristic and would ultimately allow for the weaponization of space, which slaughters many sacred cows in the peacenik space policy community. But any space force that did not seek space dominance would be a useless entity. The reason one would invest such massive amounts of taxpayer dollars in the creation of a new military branch dedicated to space would be precisely because the United States wanted to dominate such an important—possibly the most important—strategic domain.

Other concepts presented as "rivals" to space dominance are the aforementioned sanctuary view, which as you will see throughout this work, is totally impractical and no longer applicable today. The other is the survivability school, which simply seeks to make the United States less reliant on satellites while enhancing the ability of existing satellite constellations to withstand anti-satellite (ASAT) attacks. The third school is space control, or "space superiority." This is the concept that most military officers, particularly during the Obama era, tended to favor. The space control/space superiority model embraces the need for a military role in space, yet it is entirely defensive and is based on deterrence—as in, if a rival attacked US satellites, the US military would retaliate in kind. Of course, in the next chapters, you will see how the United States relies disproportionately more on satellites than any other country in the world. Therefore, deterrence will not work, since there is no balance to maintain. The last school is the space dominance model, which calls not only for the militarization of space—which has already happened—but for the weaponization of space. Further, the space dominance school, because of the imbalance of forces, insists on an offensive, compellent, as opposed to deterrent, capability to tempt American rivals into attacking the United States in orbit, so as to allow for the US to destroy that foe from space using the offensive space weapons that I advocate for throughout this book.

With the exception of the sanctuary school of thought, though, the other three (survivability, space control/space superiority, and space dominance) can actually exist together. Rather than thinking of these concepts as mutually opposed, think of them as comingling on a spectrum. They do not separate ideas but are rather merely steps on a strategic ladder ascending to the total dominance of the United States from space. The United States must make its existing satellite constellations, military and civilian alike, more survivable; this appeals to the survival school. The US must then threaten rival forces with certain doom if they attack US space assets; this is a deterrent policy and, therefore, falls under the rubric of space control/space superiority. Once these things are accomplished, though, Washington must redefine its strategic parameters from simply being equal to any power in space to becoming the undisputed master of space, as it was before the 2000s. Once the United States loses its vaunted position as the preeminent space power, which it very much is in danger of today, all will be lost.

It will require far more effort, with far greater risks, to reacquire from a determined enemy than most in the United States will be willing to support. From a cultural perspective, as you will see, since the 1960s, the martial prowess of the United States has been severely reduced by sociopolitical developments. In essence, America's culture has been gutted since the 1960s to the point that no one knows who we are anymore. Such a culture will be unable to take back something as important as space once it is lost to a more determined foe. It is, therefore, best to simply prevent a determined foe from displacing the American position in space in the first place.

President Trump is the only politician seeing clearly on the matter of space policy. It is high time that Washington worked to fully implement his plan—before it is too late. The creation of the US Space Force is not the end of the plan. It is the beginning. Presently, US rivals are working hard to gain superiority over the United States. Washington will prevent them from doing so by completely changing the way it views space and behaves in that vital strategic domain.

1

2022: THE YEAR "SPACE PEARL HARBOR" HAPPENS

THE RUSSIAN CO-ORBITAL SATELLITE, known as a "space stalker," is a devilish little machine. It is smaller and much less advanced than an ordinary satellite, designed to lurk in the darkness of space and tailgate quietly behind a large, undefended, and critical US military satellite in geosynchronous orbit, the highest orbit away from Earth's equator, and the orbit in which most of America's sensitive military satellites operate. Yet, the small size and fast movement of the space stalker would make it hard for the US military to track. Unlike other satellites, the space stalker is a cannibal, a devourer of other satellites. What's more, they are very cheap to produce and launch, meaning that many space stalkers can easily be placed in Earth's orbit.

The space stalker is much like the old German U-boat from the First and Second World Wars. It is a prowler, designed to lurk in distant depths and strike out against critical infrastructure in stealthy and devastating ways. The U-boats of old hid beneath the cold depths of the North Atlantic and stalked important Allied transports, sinking them to harm the Allied war effort in Europe. Similarly, the space stalkers of today are designed to hunt critical US satellites and disable them to grant the forces of a country, like Russia or China, key advantages over the technological wizardry of the US military. The US military overwhelmingly depends on satellite constellations to function. For America's various, lower-tech enemies to have a chance at beating US forces in combat, any first strike against the potent US military would have to be sudden. For such a lightning attack to succeed, though, an attacking force would have to reduce the ability for the US military to quickly recover from such a surprise attack.

Russian space stalkers will have spent years trailing behind one of the satellites that composed the US military's highly sensitive Wideband Global Satcom (WGS) satellite constellation. The WGS constellation is a collection of highly expensive and hard-to-replace satellites that formed the backbone of the US military's global communications network. Beginning in 2019, the United States Congress had authorized the funding of additional WGS satellites to augment the existing constellation, but they would not be deployed until 2024.[1] The vaunted US military superiority in places such as Europe, therefore, was a technological Potemkin Village: it was outwardly menacing but, remove enough of its vulnerable satellite linkages, and the entire force was rendered ineffective. Despite necessary legislation having been passed allowing for the creation of a US Space Force as far back as 2020, many of the necessary reforms to America's satellite constellations will not

---

1 "New WGS-11 Satellite to Offer Greater Coverage, Efficiency Than Predecessors," *Aerotech News*, 27 December 2019. https://www.aerotechnews.com/blog/2019/12/27/new-wgs-11-satellite-to-offer-greater-coverage-efficiency-than-predecessors/

be enacted before 2030. It is more than likely that the Russians, given their increasing hostility toward the West, would not simply wait for the United States to plug the strategic gaps in its space defenses. The year 2022, when there is sure to be another contentious midterm election in the United States—when sociopolitical divisions inside the United States would be high—might be a pristine moment for Moscow to attack Europe and therefore US forces charged with helping to defend vulnerable European states along the Russian border.

US-Russian relations are at crisis levels. There are strategic priorities that Moscow has, and Washington has simply abandoned any pretense of working diplomatically to ameliorate Russia's concerns. Russia's autocratic president, Vladimir Putin, had hoped to get a comprehensive deal with President Donald Trump over key issues, ranging from Ukraine to overall NATO expansion to sanctions relief. Despite President Trump's more sympathetic stance toward Putin's Russia, though, he will have been dogged by domestic American politics even in his second term, which will prove to be insurmountable. Thus, Trump will never fully realize his desire for the United States to enjoy healthier relations with Russia. Meanwhile, Moscow simply kept pushing its luck in its dealings with the West, causing levels of tension so great that it was only a matter of time before a war erupted. Both Trump and Putin were in a prickly situation: neither really wanted to go to war with each other, yet both were locked into conflict. After all, neither the Russian nor the American government could back down from their decades of increased hostility toward each other without one appearing weak to the rest of the world. And for a country like Russia—with a nationalist-imperialist leader like Vladimir Putin—appearing weak was more dangerous, at times, than courting war with the West.

Many words have been used to describe Vladimir Putin. "Stupid" is not one of them. "Desperate" is also not a word often associated with the Russian strongman, who loves bearing his exposed chest to all while on staged hunting expeditions in the Russian wilderness. Yet, Putin *is* desperate. In fact, he's one of the most desperate world leaders there is.

On top of being desperate, Putin has a relatively well-armed military with tons of nukes and an axe to grind with Russia's European neighbors. What's more, Russia's population is shrinking.[2] The Russian economy has also been impacted by American sanctions, though it is likely that the impact of Western sanctions has been to further militate Russia against the United States rather than to change Russian misbehavior.[3] Even without sanctions having been imposed, though, Russia's GDP is roughly equivalent to that of Italy and it is entirely subject to the unpredictable swings of the global price of oil, since fossil fuel is a key driver of its entire economy. Overall, Russia lacks the prestige it once had—a fact that Moscow's leaders, specifically Vladimir Putin, are all too aware of, which is a primary motivator for Russia's continued aggression against the West.[4] As the largest country in the world, in terms of sheer landmass, with a rapidly shrinking population and turgid economy, Russia *cannot* maintain its current defensive perimeter over the long run.[5] If Russia is to ensure its territorial integrity, it must redraw its borders—particularly to its west, in nearby Europe. Putin must act within the next decade to decisively rejigger the European map to better favor his designs—namely, to more easily defend Moscow and the Russian "core" that sits adjacent to Europe's borders.[6] If Moscow cannot accomplish these ends peacefully, it will resort

2 "Russia's Natural Population Declines for 4th Straight Year—Audit Chamber," *Moscow Times,* 7 November 2019. https://www.themoscowtimes.com/2019/11/07/russias-natural-population-declines-for-4th-straight-year-audit-chamber-a68066

3 Andrew Chatzky, "Have Sanctions on Russia Changed Putin's Calculus?" Council on Foreign Relations. 2 May 2019. https://www.cfr.org/in-brief/have-sanctions-russia-changed-putins-calculus

4 Oliver Bullough, "Vladimir Putin: The Rebuilding of 'Soviet' Russia," BBC. 28 March 2014. https://www.bbc.com/news/magazine-26769481

5 Peter Zeihan, "The Russian Grab," YouTube, 21 April 2017. https://www.youtube.com/watch?v=rkuhWA9GdCo

6 Boris Toucas, "Russia's Design in the Black Sea: Extending the Buffer Zone," Center for Strategic and International Studies, 28 June 2017. https://www.csis.org/analysis/russias-design-black-sea-extending-buffer-zone

to force. And since Putin is staring down another round of Russian general decline over the next decade, Russia's grand strategy will have a play clock rapidly counting down on it, much as a quarterback in American-style football has. Either Putin makes a bold play soon to move the strategic ball down the field or he cedes the initiative—and, therefore, the future—to his Western rivals. A more apt comparison for Putin would be judo, a sport that Putin is a master of. In that sport, two fighters wrestle and wrangle with each other; each making minute changes to his position, until the one side exposes a weakness—allowing for the other (in this case, Putin) to exploit that weakness. Once properly exploited, the attacker can then flip his opponent and pin him to the ground, beating him. Since at least 2007, when Vladimir Putin went to Munich and gave a speech in which he accused Europe of being a vassal to an American imperium bent on global domination, Russia and the United States have been engaged in an intense judo match. Washington has failed to exploit inherent Russian weaknesses, whereas Moscow has exploited American vulnerabilities with maximal effect. That trend in the 2020s will become more pronounced and will ultimately reach its inevitable, violent end.

Under present conditions, though, no amount of Russian military spending or reform could defeat the United States military in combat. To achieve a herculean task, like retaking Russia's lost European territories, Russian forces would first need to defeat the US military stationed in Europe. Short of an all-out nuclear exchange, Russian forces would be unable to accomplish such a feat. Russian forces could, however, potentially beat an American force that was deprived of its technological accoutrements by first debilitating the US military in space—which is something the Russians have been strategizing about for years.[7]

Without warning, in 2022, Russian space stalkers tailgating behind

---

7 Kyle Mizokami, "Trump's Space Force Isn't the Only Military Space Program: Here's What China and Russia Are Up To," *Jalopnik,* 25 February 2019. https://foxtrotalpha.jalopnik.com/as-trump-s-space-force-ramps-up-what-are-russia-and-ch-1832772367

sensitive US military satellites, such as those belonging to the WGS, receive coded signals from their controllers on Earth. The engines of the small space stalkers pulse to life. The tiny space devils in the cold, black, soundless void of space menacingly extend their strong robotic claws outward, vectoring in toward the American satellites they had been tailgating behind, in some cases for years—at full speed—and smashing into dozens of those sensitive US satellites, sending the cumbersome American satellites out of their orbits, and crashing them toward Earth's surface. For years, Moscow has been signaling their intentions to place space stalkers in orbit and to make ready for a devastating attack on American systems in orbit at a time of their choosing.[8] The Americans, despite knowing how critical access to satellites in a time of military crisis would be, did little to prepare for such an attack. Bureaucratic inertia and the banality of domestic US politics prevented timely defenses from being erected.

Within minutes of Russia's space attack, US military units globally would not be able to coordinate or communicate with their combatant commands—or each other. Everything would have slowed down for US military forces worldwide. Slowing America's military down is critical for an American adversary. Once the American military defenders could not coordinate with each other, much less effectively respond to a Russian invasion, the Russian victory on the ground would be assured. After such a thrashing, with America's military capabilities shattered, Moscow would assume that Washington and Brussels would negotiate with the Kremlin. Rather than risk escalating into nuclear war, Moscow believes Washington and Brussels would instead negotiate and agree to create a new political order in eastern Europe that favored the Russians. After all, to restore the status quo in Europe, Washington would have to wage another world war that it is hardly prepared to fight. And

---

8 "Mysterious Russian Space Object Could Be the Return of Istrebitel Sputnikov—the 'Satellite Killer,'" *National Post,* 19 November 2014. https://nationalpost.com/news/mysterious-russian-space-object-could-be-the-return-of-istrebitel-sputnikov-the-satellite-killer

according to Russia's nationalist-imperialist leaders, the West, as led by the United States, was "declining and decadent, though still hegemonically inclined." Putin's aim, therefore, would be to disabuse the West of its hegemonic inclinations. As Vladimir Putin himself has argued (and he is not entirely incorrect in this assessment), "We can see how many of the Euro-Atlantic countries are actually rejecting their roots, including the Christian values that constitute the basis of Western civilization. They are denying moral principles and all traditional identities: national, cultural, religious, and even sexual."[9]

Setting the cultural arguments aside, from a strategic standpoint, the United States might not win such a war against a foe like Russia. In fact, "after about nine months of intense peer conflict, attrition would grind the US armed forces down to something resembling the military of a regional power."[10] This is precisely what the strategists in Russia are hoping for, since this set of circumstances would allow for them to rewrite the geopolitical boundaries in what they view as their spheres of influence—while avoiding a total war with the United States, though such a gambit on the part of Russia would be risky. But fortune favors the bold, and such a move on the part of Putin would only be proof of how desperate his current geopolitical situation is.

WHEN RUSSIA STRIKES AMERICAN SATELLITES

It would not only be the Wideband Global Satcom constellation of satellites that Russia would target. Russia would use a variety of weapons to debilitate US space systems, though space stalkers are probably the most efficient mode of surprise attack, since they are already near their targets and can avoid causing collateral damage to other satellites

---

9 Marc Champion, "Putin Is Trump's Brother from Another Motherland," *Bloomberg Businessweek*, 9 July 2018. https://www.bloomberg.com/news/articles/2018-07-09/putin-is-trump-s-brother-from-another-motherland

10 Mark Cancian, "Long Wars and Industrial Mobilization: It Won't Be World War II Again," *War on the Rocks*, 8 August 2017. https://warontherocks.com/2017/08/long-wars-and-industrial-mobilization-it-wont-be-world-war-ii-again/

operating in nearby orbits. As you will see later in this book, Russian forces have perfected their ability to jam America's Global Positioning System (GPS). Also, a Russian first strike against US forces would not only occur in space. Russian forces are trained to use cyberspace as a zone of attack against the United States as well as the electromagnetic (EM) spectrum. The point of the Russian attacks would be to *degrade* US military capabilities to such a point that they are rendered combat ineffective—which, given the systems Russia is targeting and the methods they plan to use against American forces, they likely will be able to do to any US or NATO force arrayed against them.

For example, most US military weapons and forces rely on GPS to navigate. Russian forces might also target US early warning missile launch satellites in geosynchronous orbit. Further, Russian attacks on American nuclear command and control satellites would effectively degrade America's vaunted nuclear counterstrike capabilities. This would be a key Russian move to better protect invading Russian forces in the aftermath of a Russian surprise attack against American forces in Europe. Ultimately, both Russia and America possess ungodly levels of nuclear weapons. Should things deteriorate too much between the two sides, the nuclear genie might just be let out of the bottle. Yet, removing American command and control capabilities—as well as their early missile warning abilities—will give Russia key advantages over their American rivals. Another critical piece in the US military technological house of cards will have been removed, causing more of it to collapse. In effect, US forces could not communicate with each other, could not see any incoming attacks, and could not operate the lethal speeds at which they were trained to operate. This would allow for the larger, though less sophisticated, Russian forces to have decisive advantages over the Americans and their European allies.

The Russians have long employed misdirection in their military operations. Russia desires to take dominance over the Baltic states. Yet, Russia also seeks greater influence over northwestern Europe as much as they seek dominion over central and eastern Europe. While not as strategically important for Moscow, Russia and Finland do share a long history of conflict with each other. And in times of larger conflict, such as during the Second World War, the Soviets under Joseph Stalin's leadership conspired to not only take the Baltic states from the Nazi forces but to keep their German foes off-balance by first striking hard into Finland. In 1944, German intelligence had determined that Stalin was readying to attack German positions in Belarus. Yet, Hitler disagreed. He believed that Stalin would first strike to the south, taking the rich oil fields of Romania. Both German intelligence and Hitler were ultimately wrong. Stalin instead chose to nab the low-hanging fruit that was Finland, keeping his rival uncertain as to his wider strategic goals while forcing Germany to spend more time and precious resources—with far fewer troops than the Soviets had—trying to defend a large area of Europe. After taking key parts of Finland, Stalin would turn the bulk of his forces against the more important Baltic states.[11]

Something similar may be at play today. No, the United States and its NATO allies are not the Nazis (neither is Putin's Russia, for that matter). But the West is in a strategic position similar to what Germany was in by 1944 on the Eastern Front. Instead of the Soviet Union, the Russian Federation today is pressing hard against European territories to Moscow's west. And there are insufficient levels of Western forces arrayed against the hulking Russians. There is a handful of places where the Russians might attempt to attack. Just as during the Second World War, the Baltic states are the most obvious targets. Yet, as recent history has proven, the Russians do not like to take the most obvious path of attack. While coveting the Baltic states more than any other part of Europe, Russia might decide to initiate an attack on Finland just as they did in 1944. Since Finland today is not a NATO member, any Russian move against Finland (or, at least parts of Finland) would likely have the same effect that Russian moves against Georgia in 2008 and Ukraine

---

11 Jonathan Martin, *World War II In Colour,* episode 8, December 31, 2010, IMDB, https://www.imdb.com/title/tt2101277/

in 2014 had: they would not provoke the Western allies to direct military action against Russia. Moscow's moves would so greatly confound NATO and exacerbate its internal divisions that it would inherently weaken the alliance at a time when Russia was yet again on the march. Additionally, an attack on Finland would ultimately enhance Russian power in Europe relative to that of the United States.[12] Each Russian slice of the European salami fulfills Russia and starves the United States.

American and NATO forces would be reeling from the dazzling attacks on US satellite constellations as Russian tanks rolled into Finland. US, Finnish, and other NATO elements have been deployed to Finland. But their numbers are far smaller than what the Russians will send against them. The Western defenders have long relied on finesse and technological wizardry to offset any of their numerical disadvantages. With Russia having degraded or entirely removed the technological advantage with a space attack, though, finesse would be insufficient to deter the far more numerous Russian heavy infantry that would bombard Finland. More dauntingly, Russia's Finland attack would not be their main thrust into Europe. Inevitably, the Russians would push deeper into the Baltics—even risking a wider war as they targeted NATO members, the more emboldened Russia was by Western weakness. With the reliability of the US security umbrella in doubt, the idea of NATO would have been destroyed while the real impact would be that Moscow would have defeated the West. Whatever happened next, Russia would have shown the world that the United States can be beaten, that the technology its military has relied upon for decades—what the Chinese refer to as "American magic"—is vulnerable, and exploiting that vulnerability is the key to victory against US forces.

The Russian attack in space will be the most devastating attack conducted on American forces since either Pearl Harbor or 9/11. Just like Pearl Harbor, the attackers will have enjoyed a series of early victories

---

12 Kevin Ponniah, "How Pragmatic Finland Deals with Its Russian Neighbour," BBC, 27 July 2017. https://www.bbc.com/news/world-europe-40731415

that would fundamentally reorient the geopolitical map in the attackers' favor. If, as Carl von Clausewitz once argued, war is the extension of politics through other means, Russia will have decisively beaten their technologically superior American rivals in Europe. Returning things to the status quo thereafter would mean a costly American effort and would risk nuclear confrontation with the Russians. Many Russian leaders believe the Americans have an irrational fear of nuclear war. With America's nuclear weapons command and control capabilities degraded by Russian attacks in space, the Russian forces might gain "escalation dominance" over the American forces.[13] For their part, the Americans would not merely sit on their proverbial hands. Even as the United States lost its satellite capabilities, it would have retained some ability to respond in kind to the Russian space attacks. Parts of America's force that were not totally impacted by the Russian space Pearl Harbor would target Russian satellite constellations and destroy them. At that point, both sides might risk an endless escalation in space where, inevitably, the two sides end up targeting civilian systems. Soon thereafter, the entire global economy and communications network will have been thrown into disarray. Yet, should the combatants stop just shy of causing such unmitigated global chaos, any in-kind American attack on Russian military satellite constellations would not be as devastating to Russian forces. After all, the Russian military is not as reliant on satellites as the American forces are. And the Russian government has been endeavoring to condition their citizens and soldiers to exist—and survive—in an environment of long-term technological degradation. In 2020, for example, the Russian government successfully tested their ability to disconnect the entirety of Russia from the American-dominated internet.

---

13 Aaron Miles, "Escalation Dominance in America's Oldest New Nuclear Strategy," *War on the Rocks,* 12 September 2018. https://warontherocks.com/2018/09/escalation-dominance-in-americas-oldest-new-nuclear-strategy/

Instead, Russia became home to the world's largest intranet.[14] While a successful Russian disconnection from the American-dominated global internet will remove key opportunities for Russia's civilian economy, in a time of warfare, it would further degrade America's ability to retaliate against a Russian attack (US strategists have viewed cyberspace as a key area to retaliate against Russia for any potential attack on the West).[15]

Of course, the Trump administration has consistently stood firm against Russian aggression. So, even if the odds were slim for America to truly win a war against Russia, it is unknowable whether President Trump would endure the sheer humiliation that Putin's surprise attack in space would visit upon his presidency. Perceived slights against Trump's ego are dangerous. After all, the president infamously questioned the manhood of a Republican presidential challenger in 2016 by mocking the size of his challenger's hands. It, therefore, remains unclear whether Trump would allow himself—or his country—to be humiliated by a Russian attack. What's more, the president has consistently proven himself to care for US forces. When Russian mercenaries threatened a group of US Special Forces operators in eastern Syria, Trump ordered the destruction of the Russian mercenary force. As a result, nearly two hundred Russians were killed. Should American forces in Europe be threatened by Russian attack—or should some American troops be killed or captured by an invading Russian force—Trump just might rally the country and give the Russians a fight, no matter the cost.

At the same time, though, the president has proven himself to be unwilling to risk a nuclear war. In his previous dealings with the nuclear-arming North Korean regime, President Trump shocked the world by embracing a more accommodating stance toward the nuclear

---

14 Catalin Cimpanu, "Russia Successfully Disconnected from the Internet," *ZD Net,* 23 December 2019. https://www.zdnet.com/article/russia-successfully-disconnected-from-the-internet/

15 Ryan Johnston, "Report: Obama Admin Planted Cyber 'Bombs' Inside Russian Infrastructure," *Cyber Scoop,* 23 July 2017. https://www.cyberscoop.com/report-obama-planted-cyber-bombs-inside-russian-infrastructure/

rogue state. This occurred even after Kim Jong-un, North Korea's leader, publicly mocked and embarrassed the president. The president gave peace a chance rather than risk a nuclear war, however limited, with North Korea. Given Russia's much larger nuclear arsenal, and the fact that Russia will have likely achieved a major victory and taken much of its desired territories in Europe following an effective surprise attack on American satellites by the time US forces could conceivably counterattack, it is possible that President Trump might sue for peace with Russia. By that point, President Trump's business background would enter into his strategic calculus. He might conduct a simple cost-benefit analysis and determine that the short-term buy-in costs to restoring the status quo in Europe after a successful Russian strike would simply be too high for whatever long-term benefits existed. At the very least, this is the Trump that Putin is likely hoping to deal with, although Putin should remember that Trump is also a successful entrepreneur who took great risks throughout his business career. Should the cost-benefit Trump go up against the judo-playing Putin, though, Russia will likely get what it wants in Europe.

Thus, a new paradigm would have been birthed in Europe. With the termination of America's once-dominant position in Europe, the US dominant place in space would also be removed. This would have profoundly negative long-term implications for the United States globally, as other actors—including Russia—would rush to replace the American position in orbit. Meanwhile, Americans would have to reel from the fact that their undefeated military would have been thoroughly routed by a less advanced, though more strategically innovative, Russian foe.

WHY WOULD RUSSIA EVER DO SUCH A THING?

The Kremlin believes that Russia, not the United States, should be the dominant force in Europe. What's more, Russia's leaders believe that American power in Europe is declining. Russia's leaders do not think it would take much to sweep away the detritus of the post–Cold War,

US-led Liberal International Order in Europe.[16] To achieve this ultimate goal of Russian strategic dominance over Europe, though, the Kremlin must break apart NATO and fundamentally weaken the European Union. Moscow believes that once these organizations are gone, the influence and power of the United States in Europe will be removed also.[17]

To convince the Americans and their European allies that NATO has died and the EU is useless, Russia must deal a humiliating blow to the Americans. If the myth of American military superiority is shattered, Moscow believes that European states will then seek an accommodation with Russia. This will allow for Moscow to fundamentally rewrite the territorial boundaries of the region in Russia's favor. Given Russia's incredibly shrinking population and military force, a new defensive perimeter in Europe that incorporates natural barriers, such as the Baltic Sea and the Carpathian Mountains, will allow for that shrinking Russian force to better defend Russia's European border, as there will be fewer gaps for Russian forces to plug.

Before Russia would opt to conduct its devastating surprise attack on US satellite constellations in 2022, though, Moscow will have likely engaged in a series of escalations with most of their neighbors. For example, it is likely some terrifying war of words between Putin's regime and Poland's leaders would occur at some point, as Russian forces expanded on the other side of Poland's border. There's a strong possibility that Moscow would engineer a coup in Belarus in order to justify a permanent increase of Russian forces in that country. The Belarusian coup could be conducted by Russian intelligence operatives who would impersonate "'Belarusian' ultranationalists to stoke the fires of radicalism and compromise nonviolent resistance against the

---

16 Arktos, "Alexander Dugin: 'Eurasian Mission' (Arktos, 2014)," YouTube, 13 April 2015. https://www.youtube.com/watch?v=xuVcwNUWncQ

17 Strategy Stuff, "The Strategy of Eurasianism," YouTube, 15 February 2019. https://www.youtube.com/watch?v=5Z98moTOa7Y

dictatorship."[18] This manufactured Russian ultranationalist resistance would see the autocratic ruler of Belarus call on his bigger Russian ally for help in snuffing out the attempted coup.[19] For the last decade, Russian leaders have spoken of their need to reabsorb Belarus. Given the cultural linkages and the geographical proximity, Russia's reclamation of Belarus is a question of *when* rather than *if.* Soon thereafter, Russian forces could more reliably use Belarus as a highway into Europe, just as Soviet forces had done during the twentieth century—ultimately building up along the borders of the Baltic states and/or Finland.

Debates would soon divide NATO about whether both Finland and neighboring Sweden should be brought into the NATO alliance. Moldova, another Russian target in the Baltics, is also not in NATO, but its neutrality is enshrined in its constitution. Plus, the presence of large numbers of Russian-speaking Moldovan citizens would likely prevent Moldova from joining NATO, even if it wanted to. The eastern Europeans would think the inclusion of Finland and NATO was a no-brainer—especially since Finland is set to join the alliance by 2025. Sweden would likely be interested in joining NATO, as they have spent the last several years dealing with Russia's renewed naval threat in the North Sea.[20] Finland might balk at joining the alliance before 2025, though. Finnish leaders have tried to keep tensions with Russia relatively low over the years, although Finland has developed a potent defensive capability that would be able to inflict maximum damage on any invading Russian force. Despite these Finnish military preparations,

---

18 Paul Goble, "There Is an 'Operation Trust' in Belarus—but Not What Regnum Editor Describes," *Window on Eurasia—New Series,* 25 March 2017. http://windowoneurasia2.blogspot.com/2017/03/there-is-operation-trust-in-belarus-but.html

19 Marek Jan Chodakiewicz, "Business as Usual in Belarus: The Alternance Dance," *The Weichert Report,* 5 April 2017. https://theweichertreport.com/2017/04/05/business-as-usual-in-belarus-the-alternance-dance/

20 David Crouch, "Swedish Navy Returns to Vast Underground HQ Amid Russia Fears," *Guardian,* 30 March 2019. https://www.theguardian.com/world/2019/sep/30/swedish-navy-returns-to-vast-underground-hq-amid-russia-fears

however, Finland does not have the numbers or strategic depth to withstand the kind of pounding that Russia was planning to inflict on them. NATO has a wide range of territory to defend and nowhere near enough troops or political unity to defend it all. Moscow has preferred to keep their NATO opponents off-kilter by pushing tensions to their boiling point with more than one of Russia's European neighbors. This has forced NATO to spread its forces along a wide territory, thinning the numbers of those defensive forces. Many Western analysts have assuaged concerns about Western defensive vulnerabilities in Europe by proclaiming the glories of the magical elixir that is American high technology. But not even the hallowed American *wünderwaffe* would be effective against the kind of attack Russia was planning to conduct—any more than the US military was prepared to prevent either the Japanese attack on Pearl Harbor or al-Qaeda's attacks on September 11, 2001.

Russia has the capability to rapidly winnow down NATO forces charged with defending Europe once the American technological advantages have been neutralized. Plus, the Russians may have quietly reconstituted their *limited* nuclear first-strike policy from the Soviet era.[21] Essentially, this policy was not a total nuclear first-strike policy. The Russians, after all, have never shared the American view of nuclear weapons. Since the 1960s, American leaders from both political parties have viewed atomic weapons in apocalyptic terms.[22] These weapons are nation destroyers. Russian military planners, though, have viewed nuclear weapons—particularly, smaller, tactical nukes—as nothing more than big artillery pieces to soften up the NATO defensive lines at the start of a major Russian offensive into Europe.[23]

---

21 Brandon J. Weichert, "Russia's Preemptive Nuclear War Doctrine," *The Weichert Report,* 14 October 2016. https://theweichertreport.com/2016/10/14/russias-preemptive-nuclear-war-doctrine/

22 "Russian/Soviet Doctrine," Federation of American Scientists, 4 September 2000. https://fas.org/nuke/guide/russia/doctrine/intro.htm

23 Mark B. Schneider, "The Russian Nuclear Threat," *Real Clear Defense,* 28 May 2019. https://www.realcleardefense.com/articles/2019/05/28/the_russian_nuclear_threat_114457.html

What's more, under these conditions, the US forces deployed to defend Europe are too small. The NATO forces that would stand against a Russian invasion would either be quickly overrun or encircled by larger, armored Russian forces. US and NATO forces deployed to several Baltic states and Finland are nothing more than so-called tripwires, intended to prevent the Russians from attacking those lands. The concept of a tripwire goes back to the Cold War and was something that NATO used to keep the Russians from attacking them.[24] Yet, any American decision to increase its tripwires in Europe will not deter Russia in 2022, as similar American actions had throughout the Cold War. Moscow's strategic calculus is simply different today from what it was during the Cold War. Russian leaders believe they are vulnerable along their border with Europe—and that their current boundary is entirely indefensible compared to what it was during the Cold War. Therefore, Russia intends to secure the area where Moscow believes it is most vulnerable at all costs—while Russia still has the capabilities to do so. Washington simply does not understand the strategic importance of a European buffer zone to Russian leaders.

SATELLITES: AMERICA'S ACHILLES' HEEL

By crippling the Americans' military technology in space, Russia will have thrown the Americans back to a pre-1970s era of warfare. Russian troops could fight and win with pre-1970s-era technology because that was precisely the level of advancement that most of Russia's forces operated at. Could the Americans? *Likely not.*

Lacking the advantages that satellites provided, the relatively small American forces in Finland and the Baltics could not call in for effective airstrikes or long-range artillery strikes. Their numbers were too small to offer direct, effective resistance against the Russians. And what US

---

24 Dianne Pfundstein Chamberlain, "NATO's Baltic Tripwire Forces Won't Stop Russia," *The National Interest,* 21 July 2016. https://nationalinterest.org/blog/the-skeptics/natos-baltic-tripwire-forces-wont-stop-russia-17074

forces were in Finland and the Baltics would prove unable to mass into an effective counterattacking force against the Russians. Writing of his experiences in the Napoleonic Wars, Carl von Clausewitz, one of the fathers of the modern, Western way of warfare, said, "To achieve victory we must mass our forces at the hub of all power and movement. The enemy's 'center of gravity.'" This center of gravity "is always found where the mass is concentrated most densely."[25] American forces being so relatively small must first mass together, using technology to amplify their strength, and punch through larger, though less advanced, forces. The Russians, as well as America's other rivals, like the Chinese, have used Clausewitz for their own purposes. They have identified the satellite linkages that make the massing process of US Armed Forces possible.

In 2019, the Mitchell Institute for Aerospace Studies and the MITRE Corporation issued a report claiming that "when it comes to nuclear modernization, NC3 [nuclear command, control, and communications] is the least expensive, yet perhaps the most critical." Should this NC3 capability be degraded or destroyed, then Washington could not "provide convincing nuclear deterrence." Two of the critical elements for America's nuclear response and defense are the Space Based Infrared System (SBIRS) and the Air Force's Advanced Extremely High Frequency (AEHF) satellites. These two satellite constellations, which support America's NC3 capabilities, are highly vulnerable to attack. They are also very expensive and unwieldy. Should any of them go down in an anti-satellite attack, replacing them in a timely fashion would be onerous. During the time these systems were off-line, an American rival, like Russia, would have an unprecedented advantage over the United States. While the US military is already planning for a more survivable SBIRS, for example, this upgraded and better-defended system will not be available until 2029 or 2030. In effect, then, the United States'

---

25 Joseph L. Strange and Richard Iron, "Center of Gravity: What Clausewitz Really Meant," *Joint Forces Quarterly,* no. 35 (2004): p. 22. https://theforge.defence.gov.au/sites/default/files/adfwtc06_strange_and_iron_-_clausewitz.pdf

critical NC3 functions will be seriously vulnerable to foreign attack throughout the 2020s.[26] This was why I chose 2022 for the hypothetical Russian space attack and invasion of Europe. Moscow would wait to see what President Trump might do at the start of his second term. Any failure to reset the US-Russia relationship, thereby assuaging Russian concerns about their vulnerability in Europe, would set Putin on this most destructive course. And there are limits to what deals the United States can make with Moscow before it loses all credibility globally.

The scenario detailed above is the nightmare scenario that so many defense experts, going back to the now-forgotten Report of the Commission to Assess United States National Security, Space Management, and Organization that was led by Donald Rumsfeld in 2000, had warned a succession of American policymakers about.[27] Yet, no one in power took this threat seriously. It was all just too theoretical— especially after the horrendous 9/11 attacks occurred, forcing Washington to spend an inordinate amount of time focusing on global terrorism and the threat of nuclear-arming rogue states. Back in the 2000s, most Americans would stare in bafflement at any policy analyst who dared to bring up the threat of space warfare to them. Of course, given the group-think model that dominates in Washington, few policy experts would waste their time—or risk their careers—worrying about such future threats when everyone was obsessing about preventing another 9/11.

BLACK SWANS AND GREY RHINOS: UNDERSTANDING STRATEGIC SURPRISE

A black swan event is an unforeseen event with extreme and unpredictable outcomes. Since September 11, 2001, the world has been concerned

---

26 David A. Deptula, William A. LaPlante, and Robert Haddick, "Modernizing U.S. Nuclear Command, Control, and Communications," *The Mitchell Institute for Aerospace Studies,* February 2019, pp. 25–33. http://www.mitchellaerospacepower.org/nc3

27 "Report of the Commission to Assess United States National Security Space Management and Organization," U.S. Government Printing Office, 28 March 2001. https://www.govinfo.gov/content/pkg/CHRG-107shrg81578/html/CHRG-107shrg81578.htm

about another black swan event on the magnitude of the terrorist attacks on 9/11. Contrary to popular opinion, though, 9/11 was not a black swan event. It was, in fact, foreseen by several people. Those few who warned about the dangers al-Qaeda posed to the United States, unfortunately, were often marginalized and ignored by an uninventive national security bureaucracy. When viewed through this prism, then, it becomes clear that 9/11 was more akin to a "grey rhino" event. According to Michele Wucker, this is a "highly probable, high impact yet neglected threat . . . Grey rhinos are not random surprises, but occur after a series of warnings and visible evidence."[28] And just like actual rhinos, a grey rhino event can trample a country that ignores the stampede headed its way for too long. Ignoring a grey rhino event tends to be a common pattern throughout America's history, whether referring to the Pearl Harbor attacks, the Cuban missile crisis, or the 9/11 attacks.

The 9/11 attacks, being the worst grey rhino event in American history, should have served as a wake-up call to US leaders about the threat posed by asymmetrical warfare. The response to 9/11 should have engendered among US policymakers more willingness to think outside the box regarding the identification of potential threats. This is especially true when you consider that al-Qaeda was not the only actor that desired to strike the United States in a spectacular manner during the post–Cold War era. In fact, China had first written the book on how to attack the United States unconventionally in order to win a future conflict against the American superpower. That book, *Unrestricted Warfare,* was published in the late 1990s by two senior colonels in China's People's Liberation Army. Beijing referred to this as "informatized warfare."[29] *Unrestricted Warfare* outlined how and why

---

28 Kelsey Munro, "China Cabinet: Black Swans, Grey Rhinos, an Elephant in the Room," *The Interpreter,* 24 January 2019. https://www.lowyinstitute.org/the-interpreter/china-cabinet-black-swans-grey-rhinos-elephant-room

29 "China Military Power: Modernizing a Force to Fight and Win," Defense Intelligence Agency, 2019. https://www.dia.mil/Portals/27/Documents/News/Military%20Power%20Publications/China_Military_Power_FINAL_5MB_20190103.pdf

the Chinese military would use asymmetrical warfare to overpower the conventional military dominance of the United States.[30]

The Chinese paved the way. Not long thereafter, when Vladimir Putin rose to power in Russia, Moscow also began to view US military power as a potential threat. Moscow had disagreed with Washington's decision to attack the Balkans throughout the 1990s. The Russian government had opposed the "double expansion" of the European Union and the North Atlantic Treaty Organization (NATO) into what Moscow viewed as its traditional sphere of influence. What's more, the Russians, having been laid low by the fall of the Soviet Union at the end of the Cold War, were powerless to stop the Western advance. The Russian foreign policy elite began talking about the need to create a multipolar world, wherein there would be many great powers—all balancing against each other—while endeavoring to preserve spheres of influence near their own territories. In Russia's case, this meant ensuring that they were the unchallenged regional hegemon in both eastern Europe and central Asia.[31] Just as the Chinese strategic planners have understood since the late 1990s, Russian strategists today comprehend that they could never challenge the United States in a conventional war—not without the proverbial tables being tipped in their favor. So, Russia has striven to turn those strategic tables to their favor. Either the West would treat them as a fellow great power and respect Russian wishes, or Moscow under the Putinist regime would *make* the Western powers treat Russia with respect.

Since 2013, Russian strategic doctrine has explicitly embraced unconventional military power. Yes, Russia has a robust nuclear force. And it's true that Russia's tank forces still outnumber the Western tank

---

30 "PLA Colonels on 'Unrestricted Warfare': Part I, A November 1999 Report from the U.S. Embassy Beijing," Federation of American Scientists, accessed 5 November 2019. https://fas.org/nuke/guide/china/doctrine/unresw1.htm

31 Jeffrey Mankoff, *Russian Foreign Policy: The Return of Great Power Relations* (New York: Rowman and Littlefield, 2009), 14.

forces that would be arrayed against any potential Russian invasion of Europe.[32] At the same time, though, Russia did not simply want parity with the United States in their "near-abroad." They wanted to dominate the regions closest to their border in the same way the United States dominates the Western Hemisphere. If the West would not allow for this at the diplomatic table, Russia would have little recourse but to create its sphere of influence by force. All of Russia's relevant strategic doctrines made copious references to waging short, sharp conflicts—sort of a modern-day blitzkrieg—into enemy territory to achieve attainable, tangible strategic results.[33]

Shockingly, the generation of American leaders who had come up in the wake of the horrific 9/11 attack ignored the blatant threats that rival countries, like Russia or China, posed to American satellite constellations. Following the 9/11 attacks, the United States vowed to "Never Again" allow for a similar, horrible attack to recur. The country waged a costly global "war on terrorism," predicated in part, on the idea that the United States would be proactive in preventing such threats from rising ever again. The United States, meanwhile, has long pursued a policy of preventing any rival power from coming to dominate the Eurasian landmass. Somehow, though, the Russian and Chinese military threats have been completely missed. That both China and Russia told the world of their revanchist intentions to strike boldly, in unconventional ways for decades—and that the United States did little to prepare to withstand such asymmetrical attacks—will be a pox on all the mansions of America's political elites.

And just as it did with 9/11, the US government has some idea that something big is coming down the pike. Al-Qaeda had been around for

---

32 David A. Shlapak and Michael W. Johnson, "Outnumbered, Outranged, and Outgunned: How Russia Defeats NATO," *War on the Rocks,* 21 April 2016. https://warontherocks.com/2016/04/outnumbered-outranged-and-outgunned-how-russia-defeats-nato/

33 Dara Massicot, "Anticipating a New Russian Military Doctrine in 2020: What It Might Contain and Why It Matters," *War on the Rocks,* 9 September 2019. https://warontherocks.com/2019/09/anticipating-a-new-russian-military-doctrine-in-2020-what-it-might-contain-and-why-it-matters/

at least a decade before 9/11 occurred. In fact, it could trace its roots back to the Mujahideen anti-Soviet resistance in Afghanistan, which US intelligence supported. Al-Qaeda had engaged in what was clearly an escalating global offensive against the United States—even attacking the World Trade Center in 1993 with a truck bomb. Osama bin Laden declared his *fatwa* against the United States, where he outlined in clear terms, for all people to comprehend, his problems with the West. Bin Laden also explicitly stated his intentions to wage unremitting warfare upon the West until his politico-strategic aims were met. Elements of the US intelligence community tracked some of the soon-to-be 9/11 hijackers coming from a terrorist powwow in Malaysia in the summer of 2000 and into California.[34] The CIA had failed to alert the FBI in a timely manner of what those al-Qaeda suspects were doing in the United States. America's unresponsive, post–Cold War bureaucratic system, coupled with an unwillingness to grasp the significance of an unconventional threat, like al-Qaeda, was mostly to blame for the attacks on September 11, 2001. In fact, in August 2001, just as then president George W. Bush was readying to leave Washington, DC, for his home in Crawford, Texas, the intelligence agencies issued a report with the explicit declaration that "Bin Laden [was] determined to strike in [the] US."[35]

Of course, many insist that the necessary questions of "When? Where? How?" were woefully unanswered by the infamous August 2001 memo. These partisans insist that the lack of context in that memo prevented the George W. Bush administration from responding effectively to those warnings. This, as they say in my home state of Florida, is hogwash. September 11 was a grey rhino event that was completely avoidable. Defenders of the Bush administration and the intelligence agencies insist

---

34 John Lehmann, "CIA Bungle: Agents Tracked Hijackers but Told No One," *New York Post,* 3 June 2002. https://nypost.com/2002/06/03/cia-bungle-agents-tracked-hijackers-but-told-no-one/

35 Andrew Glass, "George W. Bush Receives Receives Bin Laden Memo: Aug. 6, 2001," *Politico,* 6 August 2009. https://www.politico.com/story/2009/08/george-w-bush-receives-bin-laden-memo-aug-6–2001–025834

that sufficient warning never came. These people are either lying or they are wrong. To compound matters for anticipating future crises, many of the same people are—and will be for some time—still employed by the US government. America's national security bureaucracy, despite hosting a bevy of conferences on the subject since 9/11, has never properly internalized its failures leading into—and on that—terrible September day. Due to this, America's national security bureaucracy remains incapable of identifying and preventing another 9/11-like grey rhino event.

This same mentality pervades US government officials about the possibility of a sudden and devastating attack against US satellite constellations. So, in the war game scenario presented above, it makes perfect sense that US forces would get their asses handed to them by a Russian or Chinese attack on inadequately defended US satellite constellations. Of course, ultimately, 9/11 was particularly heinous because it targeted not only a military site (the Pentagon), but civilians as well. In fact, most of the victims and targets of the 9/11 attacks were civilians divorced from the US military machine that bin Laden and his fanatical followers loathed.

A sudden attack on US space systems, on the other hand, especially by a state, like Russia, would likely be confined (at least initially) to the military realm. In this way, then, the term, "Space Pearl Harbor" is far more apt than a "space 9/11," since Russia would first target only US military satellites in orbit before they tried to disrupt civilian systems. Only if the US government refused to negotiate with Russia after they conducted Space Pearl Harbor and then invaded parts of Europe would Moscow entertain notions of disrupting civilian life in the United States.

SIMILARITIES BETWEEN RUSSIA IN THE 2020S AND JAPAN IN THE 1940S

The situation between Russia and the United States is far more akin to that of the United States and imperial Japan leading into the Second World War. Of course, the strategic cultures are distinct. What's more, they are separated by geography and time. But the rationale is eerily the same. Ideologically, Russia believes that it has an inherent right to extend its sphere of influence into territories that it has historically dominated. By dominating these European territories, as you've seen, Russia's leaders believe they could create a smaller and more easily defensible border compared to their current territorial boundaries. Japan wanted to accomplish something similar with their Co-Prosperity Sphere in Asia. Moscow has even created the Eurasian Economic Union, which is a thinly veiled attempt at reestablishing the Russian Empire of old using the language of postmodern Western technocrats. Therefore, the similarities between the Japanese situation in 1941 and those of Russia by 2022 should not be ignored. Despite all evidence to the contrary, US policymakers refused to take seriously either Japan's strategic intentions or their capabilities. Few in the US intelligence community truly believed that Japan would strike out against the larger and more distant United States.

Though Russia is larger than the United States, on paper, it is nowhere near as powerful in either the economic or conventional military domains. Plus, many observers believe that the presence of large arsenals of nuclear weapons on both sides would prevent any conflict between the two sides from occurring. These analysts are wrong. The fact is, Moscow and Washington do not share the same opinion on the efficacy of nuclear war. This is evident in the fact that, since the Obama administration's New START Treaty with Russia in 2011, the Russians have not only modernized their nuclear arsenal, but they have also enhanced the abilities of their tactical nuclear weapons arsenal.[36] These smaller nuclear weapons would be used to "soften up" the defenses of any NATO force that stood against a Russian invasion of Europe.[37] The

---

36 Mark B. Schneider, "The Russian Nuclear Weapons Buildup and the Future of the New START Treaty," *Real Clear Defense*, 1 November 2016. https://www.realcleardefense.com/articles/2016/11/02/the_russian_nuclear_weapons_buildup_110294.html

37 Christopher Woody, "Russia Reportedly Warned Mattis It Could Use Nuclear Weapons in Europe, and It Made Him See Moscow as an 'Existential Threat' to the US," *Business Insider*, 14 September 2018. https://www.businessinsider.com/russia-warned-mattis-it-could-use-tactical-nuclear-weapons-baltic-war-2018-9

United States, meanwhile, has fallen woefully behind in the nuclear arms race because most American leaders believe that nuclear warfare is an irrational endeavor and, therefore, it should not be contemplated.[38]

That is why few in Western circles believe Moscow would dare lash out against the United States in the military realm. Yet, that is precisely what Moscow's leadership intends to do, the more untenable that diplomacy with the US-led Western alliance becomes. There exists in the West a strategic schizophrenia: on the one hand, few want to fight Russia. On the other hand, few want to concede anything to Russia at the diplomatic table. In the meantime, the institutions that would be most capable to deter Russia, NATO and the EU, are increasingly weak. NATO has enjoyed some increases in support since the rise of Donald J. Trump to the presidency. Yet, under current conditions, by the time 2022 rolls around, NATO will still be unable to deter any Russian invasion into Europe without concerted backing—and bloodshed—from the United States. The EU, meanwhile, is fragmenting and teetering toward collapse. As this occurs, most Western strategists fail to acknowledge that Moscow will not simply sit idly by while Washington and Brussels decide whether to treat Russia as a friend or foe.

As Roberta Wohlstetter wrote in her history of the Pearl Harbor attack:

> The [US war planners] envisaged a strong possibility [in June 1940] of concerted offensive action by Japan and the Soviet Union in the Far East. On American participation in the war as a belligerent, they argued in no uncertain terms that it would be quite "unreasonable" in light of the "long-range national interests of the United States. Our readiness to meet such [totalitarian] aggression on its own scale is so great, so long as the choice is left to us, we should avoid the contest until we can be adequately prepared."[39]

---

38 Paul Bracken, *The Second Nuclear Age: Strategy, Danger, and the New Power Politics* (New York: Henry Holt, 2012), 216–24

39 Roberta Wohlstetter, *Pearl Harbor: Warning and Decision* (Stanford, CA: Stanford University Press, 1962), 84.

During this period, the War Department was incensed that President Franklin D. Roosevelt insisted on sending munitions to the British in their fight against the Axis powers at the expense of US military readiness and the president's decision to allow for the US Pacific Fleet to be moved from California to the forward base of Pearl Harbor in Hawaii. Wohlstetter continued by pointing out that "the War Department staff believed that such a show of strength . . . might be taken by the Japanese government as a *casus belli*. It would act as a deterrent 'only so long as other manifestations of government policy do not let it appear that the location of the fleet is only a bluff.'"[40] As Wohlstetter assessed, "apparently, to the planners, it was a bluff. America, they believed, was totally unprepared to meet a hostile Japanese reaction [to the Pacific Fleet being deployed to Pearl Harbor]. The president and the State Department, however, were favorably disposed to demonstrations of apparent strength."[41] Today, Russia increasingly believes—correctly, in many cases—that NATO's force is hollow and that the Americans are bluffing about their intentions to militarily defend Europe.

Of course, Japan *did* take the American movements in the Pacific as a sign that US foreign policy toward Japan was becoming increasingly antagonistic.[42] At that point, the United States had enacted onerous economic sanctions against Japan for the illegal invasion of Manchuria and the "Rape of Nanjing" massacre. At that time, China was viewed as an area of strategic interest to the United States. Japanese aggressions there were seen, especially by the State Department, as a direct threat to US national interests in the region.[43] When Nazi-occupied France

---

40 Wohlstetter, 84.

41 Wohlstetter, 85.

42 Ian W. Toll, "A Reluctant Enemy," *New York Times,* 6 December 2011. https://www.nytimes.com/2011/12/07/opinion/a-reluctant-enemy.html

43 "'China Lobby,' Once Powerful Factor in U.S. Politics, Appears Victim of Lack of Interest," *New York Times,* 26 April 1970. https://www.nytimes.com/1970/04/26/archives/china-lobby-once-powerful-factor-in-us-politics-appears-victim-of.html

allowed for Japan to annex its former colonies in Indochina, Japanese power in Asia increased considerably, becoming a greater threat to both the declining British Empire and the distracted United States. In response, President Franklin D. Roosevelt froze "all Japanese assets in America. Britain and the Dutch East Indies followed suit. The result: Japan lost access to three-fourths of its overseas trade and 88 percent of its imported oil. Japan's oil reserves were only sufficient to last three years, and only half that time if it went to war and consumed fuel at a more frenzied pace." Ultimately, Japan was placed into a use-it-or-lose-it mentality. Either they could "back off of [their] occupation of Southeast Asia and hope the oil embargo would be eased—or seize the oil and further antagonize the West, even into war."[44] This, more than anything, was a decision that locked Japan into conflict with the United States.[45] Japan, a small country with few natural resources and grand regional ambitions, needed an ever-increasing array of natural resources to fuel its rise. Oil was the sine qua non for Japan's budding "Co-Prosperity Sphere" across Asia.

By depriving Japan of oil and international trade, Washington did not deter Japan from greater hostile action. Instead, Washington inspired Japan into becoming more aggressive—and reckless. American strategists at the time were more convinced that Japan would attack either the Soviet Union or the British-held Singapore long before Japan could muster the forces needed to strike against far-flung American holdings—particularly those in Hawaii. What few policymakers realized was how committed Japan's ruling war party was to territorial expansion. US policymakers were also insensitive to the fact that with each new embargo and economic sanction (or humiliation), Washington

---

44 History.com Editors, "United States Freezes Japanese Assets," History, 16 November 2009, https://www.history.com/this-day-in-history/united-states-freezes-japanese-assets

45 Sebastien Roblin, "Study This Picture: This Is Why Japan Attacked Pearl Harbor (and Dragged America Into World War II)," *The National Interest,* 2 December 2018. https://nationalinterest.org/blog/buzz/study-picture-why-japan-attacked-pearl-harbor-and-dragged-america-world-war-ii-37712

empowered the radical militarists in Japan, often at the expense of those who strove for more peaceful and stable relations with the United States.[46] The American economic sanctions, coupled with the move of the US Pacific Fleet from its base in California out to Hawaii in 1940, Tokyo assumed that the Americans were readying to strike. Given Tokyo's interpretation of these American moves, the Japanese leadership intended to strike first before the American juggernaut could be fully stirred from its slumber. Japan's target, ultimately, were the US Navy and Army Air Corps facilities on the Hawaiian Islands.

Yes, some civilians were killed during the Pearl Harbor raid. But the Japanese targets were primarily military in nature. And the Japanese targeted the facilities that would provide the logistical support for any sustained American military campaign in the Pacific against Japan. Shortly after their assault on Pearl Harbor, the Japanese quickly turned their ire to the US holdings closer to their territory, such as the Philippines. Japan even managed to land forces on Alaska's outlying Aleutian Islands.[47] The goal was to roll back the *military* reach of the United States from territory that the Japanese wanted to hold, in order to give Japan the time it needed to make minced meat of their local rivals, who were no match for Japan's hardened military. The Russians today (or, in 2022) share similar ambitions and territorial concerns that the Japanese of 1941 held. Russia believes their expansion is warranted to create greater strategic depth in order to better protect the "core" of their country. The core of Russia is the cities very near the present European border, like their capital of Moscow or St. Petersburg.

Given Russia's long history of foreign invasion, such Russian territorial ambitions should not surprise American policymakers. The fact

---

46 "The Failed Attempt to Avert War with Japan, 1941," Association for Diplomatic Studies and Training, 27 November 2013. https://adst.org/2013/11/the-failed-attempts-to-avert-war-with-japan-1941/?fbclid=IwAR3kkbL1PAuPe400nood-eDdepz8EkpbfnWFJgwtbEtnyI08tRuBKC-Iac8

47 Mark Thiessen and Mari Yamaguchi, "75 Years Later, 'Forgotten' WWII Battle on Alaskan Island Haunts Soldiers," *Army Times,* 27 May 2018. https://www.armytimes.com/veterans/2018/05/27/75-years-later-forgotten-wwii-battle-in-alaska-haunts-soldiers/

that most US policymakers appear unable or unwilling to acknowledge Russian territorial concerns should be worrying to all. Just as with the Japanese empire in the run-up to the Second World War, US foreign policy toward Russia is aggravating, not ameliorating, the threat posed to US national interests in a key strategic region. By the 2020s, the policies of the United States will likely be seen as direct threats to Russian national security, and the Russian leadership will decide to strike the Americans with uncompromising vigor. For their part, the Russian leadership has sent clear signals for decades about their concerns over US actions in Europe. Further, Moscow had made plain their intentions to pursue a policy of regional hegemony in Europe, the Middle East, and central Asia. And the US intelligence community knew how Russia would behave in response to consistent US policies of economic sanctions and perceived Western hostility near Russia's borders.

Of course, like the United States at Pearl Harbor, American policymakers will still be caught by surprise when Russia drops the hammer on Europe. American policymakers simply ignored the years of warnings coming from Moscow. Or they refused to acknowledge that Russia would actually strike back, after Washington's continued provocations—just as Washington refused to comprehend that Tokyo would not abide by strong American diplomatic and economic pressure placed on their rising empire. Evoking Wohlstetter again:

> If our intelligence system and all our other channels of information failed to produce an accurate picture of Japanese intentions and capabilities, it was not for want of the relevant materials. Never before have we had so complete an intelligence picture of the enemy. And perhaps never again will we have such a magnificent collection of sources at our disposal.[48]

America's leaders had the most accurate portrayal of Japanese strategic intentions and general capabilities before the Japanese attack

---

48 Wohlstetter, 382.

on Pearl Harbor. Of course, US leaders did not understand that the Japanese would effectively push the limits of their technological capabilities in order to achieve what they hoped would be a knockout blow to the larger, though distracted, Americans. Just as US forces today have been constantly rubbing up against Russian forces, whether it be at sea, in the air, in cyberspace, or on land—notably along Europe's border and within chaotic Syria—American units were having increasing hostile contact with Japanese forces, notably in the southern Pacific well before Japan attacked the United States. Yet, Washington refused to acknowledge that Japan was bristling under the weight of American sanctions. Whatever heinous actions Japanese forces had taken in China (and they were particularly gruesome), as the War Department assessed at the time, it did not serve US national interests to antagonize Japan—which is precisely what Washington was doing by insisting upon its policy of heavy-handed sanctions coupled with intensified military deployments to the Pacific. Washington did this while seeking a diplomatic resolution with Japan. These actions, especially against a militarized foe with a long history of aggression, like Japan, had the opposite of their intended effect.

THE CRISIS OF THE 2020S AND AMERICAN STRATEGIC VULNERABILITY

The only way for the United States to avoid a Space Pearl Harbor, then, is to take President Trump's reforms for the creation of a space force more seriously and support an influx of federal funding for *all* space-related projects. There is a new space race occurring, in which America still retains a lead, but that lead is withering away due to the strategic indifference on the part of several previous American presidential administrations and Congresses. This consistent strategic indifference on the part of American leaders on space policy has deprived the Pentagon, NASA, and a coterie of other agencies charged with America's expansive space operations of vital innovation and leadership at the institutional level. As time progresses, American rivals continue enhancing their space capabilities—and the threat to American space systems—relative to those capabilities the United States possesses. If trends progress as they have, then, strategic disaster for the United States in space is assured.

A few years ago, I asked a friend at the National Geospatial Intelligence Agency (NGA) why the government was so indifferent about defending America's satellite constellations—and what might be done to better defend these critical systems in orbit. He looked away for a moment and then shook his head, saying he did not know. I knew him well enough to know he was uneasy with the discussion. Still, I pressed him for his opinion on the matter. After a few sips of his scotch, my friend glumly replied, "It's going to take getting our asses kicked up there to wake people up." So, in effect, the organizations that were created to anticipate grey rhino events are now fostering the conditions for such catastrophes to happen. And as I told my friend who worked at the NGA: judging from the "reforms" imposed on the national security state following 9/11, there's no guarantee that the government's response to a Space Pearl Harbor will be any more effective than what the Bush administration's nonsolutions to preventing another 9/11 were. After Pearl Harbor, the United States marshalled its energy and went on to defeat the Japanese empire that had attacked the United States in a few bloody but short years. After 9/11, however, America simply created more stifling bureaucracy in response to those attacks—and the global war on terror continues unabated after twenty years.

The president has signed into law the creation of a sixth, independent branch of the United States military dedicated to space. Some in the Pentagon have responded as best they can to heed President Trump's calls. Even among Trump's allies, though, the concept is often met with scorn. Representative Dan Crenshaw (R-TX) told me at an event in early 2019 that he was skeptical of the creation of a space force because he did not think it was necessary to "create more bureaucracy." But reorganizing the US military so that there is a dedicated branch committed only to defending the United States in space is desperately needed. It is the equivalent of God telling Noah to build the ark *before* the rains began.

To prevent a Space Pearl Harbor, then, the Pentagon must ensure it has the ability to protect existing satellites and then replace them with more survivable ones. Sadly, both the Russians and the Chinese have already reorganized their space forces. The militaries of both Russia and China are prepared to fight—and win—the next space war. America's forces are unprepared. We are exposed. *Vulnerable.* Because America is vulnerable in space, US rivals increasingly seek to exploit that vulnerability. Thus, a Space Pearl Harbor is coming. It's simply a question of when and who. As you've seen in this chapter, Russia is the most immediate threat to American space systems because of their current strategic situation.

2

THE CASE FOR SPACE

IN THE 1920S, the first modern, chemical-propelled rocket was launched by Robert Goddard in the United States. Soon, the Soviet Union would begin testing its own rockets. Ultimately, despite their initial investments in chemical-propelled rockets, neither the United States nor the Soviet Union fully embraced the technology until after the Second World War ended in 1945. It was Nazi Germany that came to respect the potency of rockets as a strategic weapon. In time, both the United States and the Soviet Union would regret not having invested more fully in this technology before the outbreak of the Second World War, as it created a massive gap in their defenses.

Noted German scientist Werner von Braun was the father of the Nazi V-2 rocket and similar systems. Yet, the Nazis were never able to fully realize the strategic potential of rockets in time to alter the outcome of the Second World War for them. For his part, Werner von Braun is reported to have agonized that his rockets worked brilliantly, but they were landing on the wrong planet![1] Von Braun, like so many scientists who would succeed him, envisaged using massive chemical-propelled rockets to transport men to the Moon (and beyond). Unfortunately for the scientific community, though, the desire for manned space exploration could never be fully divorced from strategic and military reality. Rather than being two distinct missions, as the Cold War between the United States and Soviet Union would soon prove, space exploration and space defense were fused; one could not happen without the other.

Recognizing that they had come to the rocket game too late, the Americans and Soviets engaged in an epic race to acquire as much data, materiel, and scientists from the recently defeated Nazi Germany—the world's only real rocket power at that point—as they could. This became known as "Operation Paperclip" in the United States. Once Moscow and Washington had amassed the knowledge they had taken from the defeated Nazi Germany, the two sides immediately understood that rocket technology could transform their militaries. What's more, the two competing powers realized that whoever got advanced rocket technology first would have the capability to place personnel and weapons high above Earth—giving that side a potent advantage over the other. Ultimately, rockets would be used to launch nuclear weapons, satellites, and astronauts into space. They were the backbone of both the American and the Soviet strategies for fighting—and winning—the Cold War.

Rockets became doubly more important with the advent of nuclear weapons. Initially dropped from bombers, nuclear weapons after the

---

1 Alejandro De La Garza, "How Historians Are Reckoning with the Former Nazi Who Launched America's Space Program," *Time*, 18 July 2019. https://time.com/5627637/nasa-nazi-von-braun/

Second World War were married to rockets of various ranges. Soon, the United States built an entire arsenal of medium-to-long-range ballistic missiles. By 1949, the Soviet Union had acquired their own nuclear weapon. And from that point, the great nuclear arms race was on. While the Americans had better quality weapons at the start of the Cold War, the Soviets were committed to building higher quantities of such devastating weapons.

Rocket science, though, required a different skill set than the previous age of industrial warfare. Rather than creating more workers to man the factories, the Soviets needed to educate more scientists to build rockets and other newfangled pieces of technology. In the 1950s, Moscow began training as many of their citizens as possible in areas of science and math. The Soviets believed that if they could produce more scientists of better quality than their Western rivals, then they could ultimately produce better technology. This technology would allow the Soviet Union to win the nuclear arms race *and* the space race—all of which would grant Moscow total victory in the Cold War.

With the creation of rockets and the advent of nuclear weapons, the basis for space travel was established. It would only be a matter of time before humanity took to the stars with as much élan as it had traveled the high seas, flown through the air, or crossed large tracts of land. In the famous baseball film *Field of Dreams*, Kevin Costner's character is compelled to build a baseball field to save his community. In what turns out to be a message from God, Costner is told that if he "builds it, they will come." This proved to be true in the film. Similarly, I believe that if humans can get somewhere, they will eventually go there. And once humans go somewhere, they will eventually fight over that place. It is, after all, human nature. Since human nature is flawed but fixed, eventually, a war in space will occur—it's a question of when, not if.

YEAR ZERO

October 4, 1957, is Year Zero in what would become known as the space race. It was on this fateful day that the USSR launched humanity's first satellite, Sputnik. The satellite was intended for military purposes; it demonstrated that the Soviet Union had a new and (at that point) unmatched strategic capability that many Americans had feared would directly threaten the United States. Outright panic about Soviet space capabilities and intentions soon gripped the American people.[2] Then-President Dwight D. Eisenhower had to address the issue—and he needed to have a long-term plan at the ready in a short period of time.

Many military strategists grew concerned that Sputnik was just the start of a new Soviet military push to acquire the all-important strategic high ground of space. The United States had benefited from its dominance in air. But space was physically higher than the blue skies over Earth. American planners began worrying that the Soviets would place weapons systems in orbit or other military equipment with which to threaten the United States. How would the United States respond? Could the United States respond without risking open warfare—a war that the United States might lose, no less? Space, after all, made geographical distances that once favored the United States less important. At that time, the Soviet Union, which ordinarily could not threaten the territorial security of the United States, could launch nuclear missiles into space. Those missiles would, within a few minutes after launch, return to Earth over a target somewhere in the United States and wreak untold amounts of devastation—with little warning and little in the way of defense.

For example, while Sputnik was a basic communications satellite, the Soviets intended to do more with their space program than simply place communications satellites in orbit. Sputnik was merely a proof of concept: Moscow wanted to demonstrate that they could do things that would not only threaten the United States but that the United States could not defend or counter. The Reds had designs to place nuclear weapons platforms in orbit; they even infamously tested

---

2 Eric Mack, "How NASA Was Born 60 years Ago from Panic over a 'Second Moon,'" *C-NET.* 29 July 2018. https://www.cnet.com/news/how-nasa-got-its-start-60-years-ago-sputnik-eisenhower/

nuclear weapons in low-Earth orbit in the 1960s.[3] The Soviets also designed a battle station for Earth orbit in the form of the Almaz space station program. In fact, the Soviet Union's space program produced a variety of concepts that, had it not been for the unstable economic system of the USSR, would have seriously threatened the United States.[4]

THE SPACE RACE WAS A RACE FOR BRAINS

When public outcry over the Sputnik launch reached a crescendo, President Dwight D. Eisenhower addressed the American people with a plan of his own for competing with the Soviets in the newly discovered strategic domain of space. From that point forward, President Eisenhower encouraged American young people to study math and science in order to better compete with the Reds. The US government copiously invested taxpayer money into creating scholarships and public-private partnerships, as well as building a large network that bridged the scientific community, public education, and national security, all in an effort to coordinate and synthesize a strong response to the Soviet advances in rocketry and space technology.

In one famous speech made in November 1957, President Eisenhower decried the lack of American students studying science and math relative to their Soviet rivals. In his speech, Eisenhower argued:

> The Soviet Union now has—in the combined category of scientists and engineers—a greater number than the United States. And it is producing graduates in these fields at a much faster rate. Recent studies of the educational standards of the Soviet Union show that this gain in quantity can no longer be considered offset by lack of quality.

---

3 Callum Hoare, "WW3: Why US Feared Soviet Nuclear Missile Attack from SPACE After Secret Launch," *Express,* 19 May 2019. https://www.express.co.uk/news/world/1128487/ww3-nasa-sputnik-us-soviet-union-nuclear-space-war-eisenhower-khrushchev-spt

4 Boris Egorov, "What Weapons Did the Soviet Union Plan to Use in a Space War?" *Russia Beyond.* 12 April 2018, https://www.rbth.com/science-and-tech/327998-weapons-soviet-union-space

This trend is disturbing. Indeed, according to my scientific advisers, this is for the American people the most critical problem of all. My scientific advisers place this problem above all other immediate tasks of producing missiles, of developing new techniques in the Armed Services. We need scientists in the ten years ahead. They say we need them by thousands more than we are now presently planning to have.[5]

The United States struggled to compete in the opening phases of the space race precisely because it lacked an integrated strategy for creating the next generation of scientists to propel the United States further ahead of the Soviets in the rocket age. Early investments that the Soviets had made into their space program paid off well, placing the Americans back on their proverbial heels. Not only did the Soviet Union deploy the world's first satellite, but they also launched the first human into space: cosmonaut Yuri Gagarin. Beginning with the Eisenhower administration through that of President Lyndon B. Johnson, the United States poured billions of taxpayer dollars and resources into recruiting, educating, and developing the world's best scientific minds for use in its space program.

Yet, it was always time that the United States seemed to be running out of. The United States was playing a game of perennial catch-up with its Soviet rivals in space, where the stakes were unbelievably high and the margin for error was painfully small. This was a strange place for the American people to have found themselves in, just a few short years after having soundly defeated the Axis powers in the Second World War.

Besides, although the Soviets were making their historic strides in space, the size of the Soviet Union's economy was a fraction of the size of the American economy. Unlike the United States, the Soviet Union was still recovering from the devastation that the Second World War had

---

5 Dwight D. Eisenhower, "Radio and Television Address to the American People on 'Our Future Security.'" American Presidency Project. 13 November 1957. http://www.presidency.ucsb.edu/ws/?pid=10950

wrought on their territory. After all, with the exception of Pearl Harbor and the battle of the Aleutian Islands off the Alaska coast in 1943, the physical destruction of the Second World War had not affected the United States the way that it had damaged the rest of the world—notably the Soviet Union. Plus, the Soviet system was driven by a tyrannical, central authority that dictated the daily lives of all who fell under its control. And under the hyper-paranoid rule of Joseph Stalin, the Soviet Union had suffered an immense brain drain, as Stalin either imprisoned or murdered some of the Soviet Union's greatest leaders and thinkers. But after Stalin had died, his successor, Nikita Khrushchev, was nowhere near as paranoid as Stalin had been. Khrushchev was committed to ensuring that the Soviet Union not only caught up with their American rivals, but that the USSR, not the USA, ruled the high ground of space.

From this victory, the Soviets had the all-important first-mover advantage in space. In economics, according to the website *Investopedia*, "a first mover is a service or product that gains advantage by being the first to market. Being first typically enables a company to establish strong brand recognition and customer loyalty before competitors enter the arena."[6] A similar concept can be applied to geopolitics—particularly in the area where geopolitics meets technological innovation. In the case of the space race during the Cold War, the Soviet launch of Sputnik enabled the USSR to be the first to enter into an unclaimed, strategically important domain. It had also granted immense prestige to the Soviet Union, which, until that point, had been viewed as a turgid, backward, desolate place—which, of course, it was. From that moment, the Soviets were able to leverage their initial move into space with a succession of other events that enhanced Soviet power relative to the United States. This, in turn, made the United States and its allies more vulnerable.

Of course, there is also the "first mover *dis*-advantage." As *Investopedia* outlines, "Other businesses can copy and improve upon first movers'

---

6 *Investopedia*, s.v. "first mover," accessed 15 September 2018, https://www.investopedia.com/terms/f/firstmover.asp

products, thereby capturing first movers' share of the market. Also, often in the race to be the first out of the gate, a company may forsake key product features to expedite production. If the market responds unfavorably, later entrants capitalize on first movers' failures to produce a product that aligns with consumer interests." This was what happened with the Soviet space program, although it only became a disadvantage *after* the United States fully committed itself to catching up to the Soviets in space. A series of technical setbacks and political changes in Moscow allowed for the United States to inevitably take the lead in the space race by landing on the Moon first. However, the Soviet space capabilities continued to be enhanced over the years. In fact, after losing the Moon race, the Soviets focused entirely on developing space station capabilities. Because of their focus on space station operations, the Russians cultivated an expertise in building and developing space stations that they retain even today, long after the Soviet Union collapsed.

THE FANTASY OF SPACE AS A WEAPONS-FREE SANCTUARY

Despite the military implications of space exploration, many Americans sought to marginalize the militarization of space in favor of focusing on the civilian aspects of the scientific exploration of space. Even though Washington opted to focus on the peaceful side of space exploration by forming NASA, the strategic need for a military space program in the United States was ever-present—especially in the midst of the Cold War. Strangely, though, most American elites refused to acknowledge that space was, in fact, a place of strategic competition.

For example, the head of RAND Corporation's missile division, James Lipp, famously argued in 1949 that dropping bombs from orbiting satellites or space stations would be impractical as an offensive measure.[7] At that time, Lipp, like many American elites, did not understand that satellites were not the end of human military activity

---

7 James Lipp, "Conference On Methods for Studying the Psychological Effects of Unconventional Weapons," RAND Corporation, pp. 93–97, http://www.dtic.mil/dtic/tr/fulltext/u2/108425.pdf

in space. They were the *start* of human military actions in space. Even President Eisenhower, who understood that space had strategic value in the Cold War, embraced the Lipp notion that space should be viewed as a weapons-free sanctuary—simply because space activities were so expensive and technically difficult at that time. This was why Eisenhower supported the initial creation of the civilian NASA program as opposed to simply letting the Air Force manage all space policy.

Naturally, at the same time that many US leaders and scientists wanted to keep space as a weapons-free sanctuary, the Soviets were increasing their investment in the militarization of space, prompting the Americans to begrudgingly do the same. The Soviets tested a nuclear weapon in space, stoking fears that the Reds were planning to place nukes in orbit. This forced the United States to reciprocate. With each hostile act in space, it stoked an equal—*or greater*—hostile reaction in space from the other side of the Cold War. In the process, each side's space capabilities were enhanced. Over time, whether Washington's policymakers liked it or not, the United States could do more in space than they previously thought possible.

Senator John F. Kennedy (D-MA), like Republican President Eisenhower, was skeptical of the costs and utility of the nascent American space program. But by the time JFK campaigned for the presidency in the election of 1960, even he had become an ardent supporter of an active American space program. As both the Soviets and Americans invested their considerable resources into developing space capabilities, the more space became militarized, the less relevant the sanctuary view of space became.

So, though American elites have tried to prevent the militarization of space over the years, American policymakers could never fully divorce the civilian-run NASA from its military roots. It is in humanity's nature to compete and war with itself. We know this from thousands of years of recorded human history. Attempting to prevent such natural human actions in space was a painfully shortsighted position. The sanctuary view of space has hamstrung the necessary development of space as an economic and strategic asset for decades. This was an especially damaging outlook to take at the same moment that the Soviets were embracing a more robust military capability in space.

The United States never fully realized its military potential in space because many powerful interests within government and academia bought into the utopian view that space was a weapons-free sanctuary. Thus, these powerful, aligned interests not only helped to stymie attempts to militarize space, but they also prevented a timely American weaponization of space. As you will see in this book, these utopians supported treaties and programs that were designed to stunt, rather than expand, America's military space capabilities—even as the Soviets exploited loopholes in those treaties, or outright ignored the agreements for their own ends! Had it not been for the internal failures of the Soviet political and economic system, it is likely that the Cold War in space would have played out far differently—and more dangerously for the West—than it had.[8]

Today, such utopian thinking still pervades the space policy community and has needlessly complicated the ability for the United to defend itself from newer threats, like that of China. For example, when the Soviet Union fell and the Cold War ended, the United States military advocated for what was known as space dominance. American military leaders strove to enhance America's inherent advantages in space. Yet, despite presidential rhetoric, the US government took little initiative in ensuring that the United States remained the most dominant force in space. Because of this weakness, America's rivals today are able to threaten American space systems, such as the intricate satellite networks that orbit the planet, and there is little that the United States can presently do about it.

In fact, the United States is susceptible to a 9/11, or Pearl Harbor–type surprise attack in space even more so than it was during the heady days of the Cold War.

---

8 Dwayne A. Day and Robert G. Kennedy III, "Soviet Star Wars," *Air and Space,* January 2010, https://www.airspacemag.com/space/soviet-star-wars-8758185/

SATELLITES: ALL OF AMERICA'S
STRATEGIC EGGS IN ONE BASKET

Since the utopian sanctuary view of space has been so influential in American space policy planning, the United States increasingly relied on satellites while giving little thought about how to defend them from attack. Something similar transpired with the internet: Americans rely on the internet today. Yet, it is an increasingly unsecured domain. Satellites are also instrumental for things like navigation—think of the Global Positioning System, or GPS. They allow for instantaneous financial transactions to occur. Some satellites provide for key weather monitoring around the world. But the most important feature that satellites provide is bandwidth. In an increasingly signals-dependent global society, one can never have enough bandwidth. The loss of bandwidth would slow down both military and civilian life and bring them to a halt.[9] It is no stretch to argue that without America's satellite constellations in orbit, our modern society would not function today.

Meanwhile, America's human spaceflight capability has been erased. In 2011, the Obama administration presided over the last NASA space shuttle flight. After that final flight, not only did the space shuttle program end, but so too did America's manned spaceflight program. While America still trains astronauts today for space travel, it has no replacement vehicle for the space shuttle. Thus, in order to send astronauts to the International Space Station (ISS), the United States is forced to rent seats on Russian spacecraft. Not only is this a strategic weakness for American rivals to exploit, but it is also hugely embarrassing. Imagine if, at the height of the British Empire, London began dismantling its potent navy to save some money and instead relied on the French navy to protect its interests and transport its citizens to Britain's various colonies. It would have been unfathomable.

---

9 "What Are Satellites Used For?" Union of Concerned Scientists, 15 January 2015, https://www. ucsusa.org/resources/what-are-satellites-used

But that is precisely what the former Obama administration did in 2011 when it ended the space shuttle program. While NASA's space shuttle program was a problematic investment (only because it prevented the development of other spaceships that could have gone farther into the solar system), it cost the taxpayer a meager $1.6 billion per launch, and $209 billion over the course of thirty years![10] Compare this figure to the cost of America's inconclusive Mideast wars, $5.9 trillion, and you see that $209 billion over thirty years is a much better investment than $5.9 trillion over twenty years.[11] The space shuttle program allowed the United States to maintain a strategic capability that few other states possessed: the ability to place its own personnel and equipment into the strategic domain of space at will. Today, after seven years of not having an indigenous manned spaceflight capability, the United States has created a critical strategic gap for itself in space. This is a gap that American rivals are racing to exploit. Meanwhile, despite the Trump administration's renewed focus on space, little has been done in the way of building a replacement spacecraft for the space shuttle. There is much talk coming from Washington, as there always is, and little real action.

Because American policymakers have disincentivized the development of space, there has been little interest in investing limited resources in developing space beyond satellites. This has also had a stultifying effect on scientific research, the fount of economic opportunity today.[12]

---

10 Mike Wall, "NASA's Shuttle Program Cost $209 Billion—Was It Worth It?" Space.com, 5 July 2011, https://www.space.com/12166-space-shuttle-program-cost-promises-209-billion.html

11 Neta C. Crawford, "United States Budgetary Costs of the Post-9/11 Wars Through FY2019: $5.9 Trillion Spent and Obligated," Brown University Watson Institute of International and Public Affairs, 14 November 2018, https://watson.brown.edu/costsofwar/files/cow/imce/papers/2018/Crawford_Costs%20of%20War%20Estimates%20Through%20FY2019.pdf.

12 "Why Are Science and Technology Critical to America's Prosperity in the 21st Century?" in *Rising Above the Storm: Energizing and Employing America for a Brighter Economic Future* (Washington, D.C.: The National Academies Press, 2007), accessed December 30, 2019. https://www.nap.edu/read/11463/chapter/4

After all, there is a direct correlation between the decline in federal research and development funding since the end of the Cold War, and the decline of America's once-dominant position in space. More troublingly, at the precise moment that the United States has abandoned the strategic high ground of space, the Chinese have expanded their investment in developing space as a strategic asset for China's overall push to become the new global hegemon by 2049.

As Ye Peijian, the head of China's moon program, said in 2019:

> The universe is an ocean, the moon is the Diaoyu Islands [better known as the Senkaku Islands in the West], Mars is Huangyan Island. If we don't get there now even though we're capable of doing so, then we will be blamed by our descendants. If others go there, then they will take over, and you won't be able to go even if you want to. This is reason enough [to go to the moon and beyond].[13]

China has invested so heavily into advanced technologies capable of supporting a powerful space program that KPMG's recent quadrennial innovation tech hub survey predicted that Shanghai, not Silicon Valley, would be the world's leading technological innovation hub by 2020.[14] Should the United States lose its innovation advantage to China, it will inevitably lose its dominant position in the high tech-dependent strategic domains of space and cyberspace. If those trends continued unabated, ultimately, China would become the world's superpower. Beijing's moves in the technology sector, international finance, and the world's economic realm are increasingly essential to understanding the nature of China's threat to the US-led world order.

---

13 Brendon Hong, "China's Looming Land Grab in Outer Space," *Daily Beast,* 22 June 2018, https://www.thedailybeast.com/chinas-looming-land-grab-in-outer-space?ref=scroll

14 "U.S. and China Lead Tech Innovation and Disruption, Even as Innovation Spreads Globally: KPMG Report," PR Newswire. 5 March 2017, https://www.prnewswire.com/news-releases/us-and-china-lead-tech-innovation-and-disruption-even-as-innovation-spreads-globally-kpmg-report-300418042.html.

Today, the United States finds itself in a similar position that it found itself on October 4, 1957, when the Soviets launched Sputnik into orbit. The only difference is that most Americans have failed to recognize how dangerous a position the country is in. Much like the Soviets, the Chinese are steadfastly committed to gaining strategic parity with the United States in orbit. And ultimately, China hopes to displace America as the dominant space power on the planet. Whereas the United States responded immediately to the Soviet threat in space, the American response to the Chinese in space has been one of abject ambivalence. As Matthew McConaughey's character in the 2014 film, *Interstellar*, bemoaned, we in the United States "used to look up at the sky and wonder at our place in the stars. Now we just look down and worry about our place in the dirt."

Since 2003, when a Chinese *taikonaut* performed China's first space walk, the rising world power has consistently developed robust space capabilities. Such capabilities are being created explicitly for military purposes. The Chinese are committed to exploiting space as a means of spreading its strategic influence across the Earth. Further, just to prove that the United States is in a second space race, Beijing has developed the capability to deny the United States access to space in the event of a conflict between China and the United States, with what's known as "counterspace" tactics. This is the real nature of the second space race: to degrade and destroy America's space capabilities in order to deny American forces access to the surveillance and communications satellites they require to effectively fight the Chinese military in any conflict, which would undoubtedly be waged near Chinese territory in the Asia-Pacific.

In fact, one can pinpoint Year Zero in the second space race. That would be January 11, 2007. On that fateful day, China launched what was known as an anti-satellite (ASAT) weapon into orbit from the Xichang Satellite Launch Center in the Sichuan Province. The target was a derelict Chinese weather satellite. It was destroyed by a missile. Beijing did not alert other countries that had satellites nearby, violating international legal agreements outlining the safe use of space. Not only was the Chinese action irresponsible, but it has had lasting impacts on spaceflight operations around the world. Because nothing ever stops moving in space, the debris the ASAT weapon test created remains orbiting the Earth. According to experts, it is the largest single debris field ever created by human activity in orbit. It has become a permanent threat to safe space operations.

The Chinese leadership knew that their ASAT test was dangerous. Beijing's leaders also understood that their refusal to alert the world's space powers of their ASAT test raised the risk to the satellites of other countries in orbit. China did not care. With their destructive ASAT test, the Chinese military signaled to the United States that they had the capabilities to threaten any American system in orbit. What's more, Beijing was implying that they were reckless enough to risk damage from any attack cascading and impacting the satellites of other states, in the event of a conflict between China and the West. In essence, Beijing was telling Washington (and the world) that, no matter what, they were aggressively moving into orbit, and they would do so according to their own rules rather than the agreed-upon international rules for space operations. The ASAT test in 2007, though, was but the start of China's expansion into the strategic high ground of space.

Although Washington has responded to China's provocations with ASAT tests of their own, merely responding with in-kind weapons tests will not dissuade the Chinese from trying to conquer space. In fact, the American response often looks weak in the eyes of the Chinese leadership. It encourages Beijing to redouble its attempts to displace the United States in space, as they assume America's leaders will always tie one strategic arm behind their back while China is free to implement their strategies for space dominance without any real hindrance. Today, China has successfully placed and maintained two space stations. On those stations, Tiangong 1 and Tiangong 2, China conducted experiments with radical technology that have direct military applications. Beijing plans on having an even more complex and permanent military space station in orbit by 2020.[15] Meanwhile, China intends to place a laser on their space station.[16] Many legal scholars argue that such a system would violate the Outer Space Treaty of 1967, which prohibits offensive weapons from being placed in orbit. But it is believed that a laser in orbit could help clear dangerous debris that now orbits the Earth after decades of human activity there.[17] Of course, like all space technology, such a system is dual use in nature. While it can theoretically be used to provide greater safety for humans, satellites, and spacecraft in orbit in case of the outbreak of war, such a system can also be used to target critical American satellite constellations, which is precisely what the Chinese intend on using it for.[18] Many analysts, such as Peter Singer, remain skeptical that the Chinese will place a laser powerful enough to disrupt the normal functions of American satellites in orbit. Yet, the Chinese government (as well as others, such as Russia, as you will see in later chapters) routinely lie about their strategic intentions in space. The fact is, once placed in orbit, the United States could not guarantee that the Chinese did, in fact, place a laser in orbit only strong enough to remove dangerous debris.

All of this has occurred at a time when America's space policy was adrift, which is why the Trump administration has (thankfully) publicly acknowledged the threat to American systems in space and shepherded the creation of an independent Space Force. The bureaucracy in Washington, though, continues giving short shrift to the Trump

15 Andrew Jones, "Successful Long March 5 Launch Opens Way for China's Major Space Plans," *Space News,* 27 December 2019, https://spacenews.com/successful-long-march-5-launch-opens-way-for-chinas-major-space-plans/

16 Brandon J. Weichert, "Space Weapons Trump Peace with Russia," *American Spectator,* 18 July 2018, https://spectator.org/space-weapons-trump-peace-with-russia/

17 Quan Wen, et al., "Impacts of Orbital Elements of Space-Based Laser Station on Small Scale Space Debris Removal," *Optik* 154 (February 2018): 83–92.

18 Jeffrey Lin and P. W. Singer, "Is China's Space Laser for Real?" *Popular Science.* 15 February 2018, https://www.popsci.com/china-space-laser#page-2

administration's more expansive space policy agenda.[19] Meanwhile, bureaucratic inertia appears to be slowing down the necessary movement toward making satellites harder to destroy, let alone placing actual weapons systems in orbit that might defend the United States from nuclear-armed rogue states, like Iran or North Korea.[20] Even with the creation of a space force—an essential first step on any path to ensuring America's continued dominance of space—a much longer-term plan for national technology research and development is needed.

With the United States' once-dominant position as the undisputed master of space now in question, even allied countries, like India, are militarizing space. As you will see in later chapters, in 2019, an Indian ASAT test in orbit also created a "mess" in Earth's orbit akin to the one the Chinese had created.[21] Other states, like France and Japan, are also getting more involved in military space operations, clearly assuming that there is an advantage for them to gain in space, as the United States appears to be losing its status as the preeminent space power. The US became the gatekeeper of access to space after the Cold War ended. Over thirty years, though, the gates are open, and the flood waters are rising to dangerous levels for the United States.

It would have been as if President Eisenhower's response to the Soviet Union's launch of Sputnik was to make a few speeches but not follow through with actual policies meant to make the United States more competitive against the Soviets in space. Or worse, for Eisenhower to have made a few speeches calling for more action to

---

19 Brent Ziarnick, "Defense Department Status Quo Is a Weak Argument Against Space Force," *The Hill*. 29 September 2019, https://thehill.com/opinion/national-security/463558-defense-department-status-quo-is-a-weak-argument-against-space.

20 Bill McLoughlin, "Iran vs. US: Tehran Could Hit America with a 'Dirty Bomb Terror Attack' Warns Expert," *Daily Express*, 21 June 2019, https://www.express.co.uk/news/world/1143301/Iran-news-terror-oil-tanker-attack-US-drone-middle-east-latest.

21 Phil Stewart, "U.S. Studying India Anti-Satellite Weapons Test, Warns of Space Debris," Reuters, 27 March 2019, https://www.reuters.com/article/us-india-satellite-usa/u-s-studying-india-anti-satellite-weapons-test-warns-of-space-debris-idUSKCN1R825Z.

respond to the growing Soviet threat in space, only to have his dreaded military-industrial complex opt to ignore or undermine his presidential directives to take greater action.[22] When the Soviets launched Sputnik, the United States responded with a full-throated effort to fund scientific education programs, to encourage public-private cooperative agreements to build better space systems, and to create the institutional capacity for ensuring the United States was not left behind in space. As America did this, it also sped up the launch of its own satellites and personnel into orbit—ultimately launching men to the Moon before the Soviets could get there. The plan was to follow up the Apollo missions with an even greater, more sustained mission set. These plans were never realized, as Washington (wrongly) believed it had other priorities. Even after the collapse of the Soviet Union, when the United States was the unquestioned global hegemon, America's space program lurched directionless from mission to mission, rarely accomplishing anything remotely as impressive—or meaningful—as it had during the heady days of the Cold War.

What few recognize is the fact that, without a sustained effort to maintain and increase America's presence in space, both at the civilian and military level, the capability is extremely hard to get back once policymakers reverse their decision to cancel a program. This is why few space programs manage to be reconstituted once a political decision is made to cancel them. Just think about it logistically: when a program is canceled, all of the support elements are mothballed with it, with few exceptions. This means that not only are engineers lost, but so too are the people and industries that create the support for these precious space programs. When the space shuttle program was nixed in 2011, my home state of Florida—along what's known as the "Space Coast," where Cape Canaveral is located—lost hundreds of thousands of jobs.[23]

---

22 Dwight D. Eisenhower, "Military-Industrial Complex Speech, Dwight D. Eisenhower, 1961," The Avalon Project, https://avalon.law.yale.edu/20th_century/eisenhower001.asp

23 Denise Chow, "Shuttle Workers Face Big Layoffs as NASA Fleet Retires," *Space*, 22 July 2011, https://www.space.com/12391-space-shuttle-program-nasa-workers-layoffs.html

These were not only the NASA engineers and scientists who worked on the space shuttle, but the local service industries, as well as the attendant suppliers who specialized in maintaining the shuttle program. Once these capabilities were gone, bringing them back becomes a costly, time-consuming endeavor. NASA is learning this hard truth today as it struggles to bring forward a new generation of manned spacecraft.[24]

This post–Cold War, bipartisan listlessness has become entrenched throughout the bureaucracy to the point that Washington's response to China's *even more* aggressive actions in space over the last decade have been timid. Not only is the US government's investment into scientific research and development at historic lows, but America's spaceflight capabilities have yet to be revitalized. Go back to the previous paragraph and think about the gutting of the space shuttle program and its support industries. It takes years to build up such capabilities and to craft a reliable logistical supply chain and to maintain the human capital needed to run the program. We will be recouping those losses for years to come, especially as Washington refuses to prioritize America's space program.[25] Whereas China has invested over the last twenty years in the scientific fields needed to create the next generation of scientists and thinkers, the United States has opted instead to do the opposite.[26] We are all set to pay the price for this dithering.

Beijing has determined that space and its supporting high-tech industries are an inextricable component of their goal for global

24 Alex Stuckey, "A Few Good Men," *Houston Chronicle,* 2 May 2019, https://www.houstonchronicle.com/local/space/mission-moon/article/NASA-is-stalled-Apollo-era-legends-13788864.php

25 Loren Grush, "Lawmaker Who Helps Fund NASA Questions the Agency's Need to Get to the Moon by 2024," *The Verge,* 16 October 2019, https://www.theverge.com/2019/10/16/20917171/nasa-artemis-program-moon-jose-serrano-jim-bridenstine-budget-2024

26 David P. Goldman, "The Digital Age Produces Binary Outcomes," *American Affairs* 1, no. 1 (Spring 2017), https://americanaffairsjournal.org/2017/02/digital-age-produces-binary-outcomes/

hegemony by 2049.[27] China has moved boldly toward overtaking the United States in both space and the high-technology sector.[28] Yet, the United States has ignored China's space and technology ambitions and the ambitions of others for too long. Had Eisenhower and his presidential successors not maintained a steadfast commitment to competing with—and defeating—the Soviets in the first space race, the Cold War might have ended differently. Today, the quasi-war with China will end badly for the United States unless American leaders recognize the threat posed by China in space and take active, drastic measures to counteract the Chinese in the strategic high ground of space.

Just as an American, Robert Goddard, created the first chemically propelled rocket, so too has the United States created the technology needed to dominate space. Yet, just as with the rocket before the Cold War, the United States did not understand the importance of space technology—and how easily the United States can be knocked off its dominant place in orbit. Losing space to China or another hostile power, like Russia, would have even more dangerous ramifications for America's dominant position on Earth. The Soviets almost took space in the Cold War. Now, the Chinese appear poised to accomplish the task.

American primacy in space is not a fait accompli. Like everything, it requires will and knowledge to remain dominant. Thus far, America's leadership has expressed neither the will nor the knowledge to ensure its strategic dominance of space. If the situation is not reversed as drastically as Eisenhower reversed America's position in 1957, Americans will be threatened by an array of Chinese weapons, economic opportunities in space will be lost to China, and the world will be forced to make itself comport with the preferences of the Chinese Communist Party.

---

27 Namrata Goswami, "Waking Up to China's Space Dream," *The Diplomat*, 15 October 2018, https://thediplomat.com/2018/10/waking-up-to-chinas-space-dream/

28 Associated Press, "China Lands on the Moon's 'Dark Side' in a Move to Become a Space Power," *Washington Post*, 3 January 2019, https://www.washingtonpost.com/lifestyle/kidspost/china-lands-on-the-moons-dark-side-in-a-move-to-become-a-space-power/2019/01/03/aa9ea21a-0f69-11e9-831f-3aa2c2be4cbd_story.html.

To paraphrase Ye Peijian, if we don't make a greater investment now into space, even though America is capable of doing so, then we will be blamed by our descendants for not having done so. And if others, such as the Chinese, go into space without the United States, then Americans will not be able to go there even if they wanted to. Some readers may argue that it is dangerous to view space and technology in such zero-sum terms, although US rivals, notably China, view space and technology development in such zero-sum ways. And the enemy *always* gets a vote.

3

CHINA ADDS OUR TECHNOLOGICAL

DISTINCTIVENESS TO THEIR OWN

IF YOU HAVEN'T BEEN ABLE TO TELL BY NOW, I'm a total science fiction nerd. (Who else could dedicate themselves to writing a full book on the potential for a real space war?) One of my favorite television series of all time was *Star Trek: The Next Generation*. And in my opinion, one of the best story lines on that television series was the Borg story line. At the end of the show's third season, the crew of the USS *Enterprise-D* encounters a race of cyborgs who fly around the galaxy, target advanced alien races, and assimilate the advanced technology of those various alien races into the Borg's own technology. What's soon discovered is that the Borg also assimilate the alien races themselves, turning them into mere worker drones in their endlessly homogenizing collective. There was no individual Borg, as each individual was part of a shared hive mind that collectively existed together. The Borg were also not innovative. Instead, they were imitative. By absorbing the advanced technology of other races, the Borg enhanced their own technology. Once absorbed, the Borg collective could then innovate from there, repeating the cycle as needed. But the innovation process was always initiated by someone else.

China is like the Borg of *Star Trek*.

This isn't just a Communist Party modus operandi in China; this Chinese M.O. goes back centuries. Whether speaking about the development and expansion of Chinese civilization from a collection of mud huts along the Yellow River basin to the massive country that it is now, China has assimilated foreign groups and ideas into their own culture, and after fusing together, they've pressed onward, stronger and more durable than before. Throughout China's long history, even if a ruling Chinese dynasty lost to an invader, like the Manchus, the Chinese culture had the last laugh. Repeatedly throughout their history, the outside invaders inevitably were subsumed by the larger numbers of Chinese people and culture. Historically, Chinese technological development followed a similar pattern.

After China was colonized by Western empires in the nineteenth century, and cleaved into tiny bits by those powers, the Chinese were desperate to catch up to the Western powers technologically so that they could ultimately oust those forces from their lands. In 1878, the radical Chinese nationalist Zhang Zhidong wrote an essay titled "Quan Xue Pian," which translates to "Exhortation to Study." In this groundbreaking essay, Zhidong elaborated on the concept of *ti-yong*, which meant to "keep China's style of learning to maintain societal essence and adopt Western learning for practical use." This concept of *ti-yong* is not merely an interesting bit of Chinese history. In fact, it is a hallmark of China's strategy for full-spectrum dominance in the twenty-first century.[1] Today, as millions of Chinese students go abroad to Western universities and companies, learn everything they can, and then return to China with that knowledge to teach it to their countrymen, it would appear that Zhang Zhidong's exhortation to study is being taken to its extreme.

For decades, China has targeted the advanced technology and capabilities of other countries—particularly the United States—and sought to pilfer that technology and incorporate it into their own. China did this first with American manufacturing firms. They are presently doing it to American high-tech firms as well. During the Cold War, Communist China was initially aligned with the Soviet Union. From 1949 until the late 1960s, China and the USSR had a close working relationship. Ultimately, that alliance collapsed and allowed for the United States to convert Communist China into a strategic partner against the Soviet Union. One of the reasons for the Sino-Soviet split had to do with the fact that Moscow knew Beijing had been engaged in a sustained campaign of industrial espionage against the Soviet Union. Once the Soviets turned on the Chinese and the United States became China's new global partner, China merely replicated their industrial espionage campaign that they had been waging against the Soviets and applied it to the "free" trade-loving West.

Since the initial openings between Red China and the United States in the 1970s during the Nixon administration, China has experienced a renaissance in its economic growth and technological development. In a short forty years, China evolved from an agrarian backwater, ruled by the totalitarian Mao Zedong and his heinous cult of personality, into an advanced economy seeking to compete with—and eventually—displace the United States as the preeminent global power. It has done this not through the use of bullets and bombs. Instead, it has accomplished this task through entirely unconventional methods. Beijing has

---

1 Brandon J. Weichert, "China's Exhortation to Study (and Conquer)," *American Spectator,* 19 March 2018, https://spectator.org/chinas-exhortation-to-study-and-conquer/

used international trade, legal conventions, perception management in foreign countries, and cyber and industrial espionage to its utmost advantage. China has engaged in a long-running campaign of wooing advanced Western—*usually American*—companies into doing business in China. Once in China, the government in Beijing often requires these foreign companies to share advanced methods and trade secrets with the Chinese government. With these trade secrets, Chinese firms then work to build up an indigenous capability so that they can ultimately remove the need for the Western firms and compete with them globally. Beijing's goal is to enrich China and give Chinese citizens access to high-paying jobs and international prestige, at America's expense.

Until recently, few in the West have tried to prevent China from doing business this way. After all, we are not at war with the Chinese. Clearly, though, the Chinese are at war with the West. Of course, defenders of trade with China insist that the Chinese are merely using the American-built "free" trade system to their advantage. And whenever China is caught stealing Western intellectual property, most Westerners assuage their concerns by claiming that it just proves how China will never be able to best the United States or its Western allies. The Western firms losing their intellectual property to China usually ignore the violations because their companies stand to make so much more money from continued business in China in the near term. Should Western firms complain about the strong-arm practices of China's government, then these foreign firms risk losing access to the vast and lucrative Chinese market. As one CEO told me at an off-the-record briefing that I gave near Capitol Hill in 2018, "If my company doesn't do business with China then my competitor will." This short-term, zero-sum thinking embodies the problem posed to American national security policy experts: their incentives are fundamentally different from those of both their Chinese rivals as well as the interests of American corporate leaders.

Besides, most Westerners think, only an innovative state could dominate the world today. No mere imitator could ever dream of rising to global dominance in such a short time period as the United States has managed to do. Let them keep copying us, these Western skeptics of China's rise say. China will never be able to best us. Westerners are mollified by the false notion that much of what China produces is nowhere near as good as what the West can produce. They have quantity compared to our quality—but consumers will always value quality over quantity.

These panda huggers are wrong.

YES, IMITATORS CAN DEFEAT INNOVATORS

In his 2010 book, *Copycats: How Smart Companies Use Imitation to Gain a Strategic Edge*, Ohio University Fisher School of Business professor Oded Shenkar analyzed whether innovative corporations were truly more profitable than imitator firms. What he found was that the pioneering firms often lost in the long run to their imitative competitors. According to a study conducted from 1948 to 2001, innovators enjoyed only 2.2 percent of what was then the present value of their innovations whereas the rest went to imitator firms.[2] Let that sink in. In other words, 97.8 percent of all profits from new innovations went to the firms that either ripped off the original innovators or merely perfected the products that innovators created. Therefore, regardless of the industry, imitators, not innovators, usually win the day.

You may be wondering, *how is it that imitators tend to be so successful compared to their innovative competitors?* After all, we Americans share a meritocratic view—the *Field of Dreams* concept, if you will—on innovation: if you build it, they will come. If you build it first, though, you will likely benefit more. Yet, reality is different from the expectation. Shenkar explains:

---

2 William D. Nordhaus, "Schumpeterian Profits in the American Economy: Theory and Measurement," National Bureau of Economic Research, April 2004, pp. 34–35, https://www.nber.org/papers/w10433.pdf

With the innovator and pioneer paving the way (and paying for it), the imitator enjoys a free ride. It saves not only on research and development but also on marketing, because customers have already been primed to use the novel product or service. The imitator avoids dead ends, whether a losing bet on a dominant design, such as Sony's Betamax VCR format, or an innovative prescription drug that proves not to work . . . With the benefit of hindsight, imitators capitalize on the shortcomings of early offerings.[3]

These trends apply as much to nation-states as they do corporations, although Shenkar would likely refute these claims. In the case of China, which has mastered fusing together of national and corporate interests, analysts like the economic writer Ted C. Fishman have dubbed the country "China, Inc." In China, Inc., the imitation cycle is taken to all new heights. Harvard professor Michael Porter has spoken favorably of what's known as "innovation clusters." An innovation cluster is exactly as it sounds. It is the concentration of firms from a given industry combined with supporting industries, operating in tandem, all to create a competitive advantage. As Shenkar explains, "Clusters have been commended for their power to support innovation by providing the infrastructure, knowledge, and intellectual exchange that are helpful for the incubation of new ideas. Examples include Silicon Valley, Route 128 in Boston, Cambridge in the United Kingdom, and Herzliya-on-the-Sea in Israel." However, Shenkar argues that "imitation clusters" exist also. These imitation clusters tend to save firms—in this case, China, Inc.—money by lowering costs of research, development, and production that innovators often bear. These clusters also allow information sharing between imitators and the industries that support them (and the governments that support them). Most imitation clusters are organized, especially in China, along industry groupings. Cell

---

3 Oded Shenkar, *Copycats: How Smart Companies Use Imitation to Gain a Strategic Edge* (Boston: Harvard Business School, 2010), 8–9.

phones were mass produced in Shenzhen, which, in turn, is where much of China's work on the revolutionary new 5G (fifth-generation wireless technology) networks is being done, an area that many observers, such as David P. Goldman, believe the Chinese are besting the United States in.[4]

More dangerously, China has announced that its Beidou (meaning "Big Dipper") GPS constellation will be completed by the middle of 2020. Beidou will not only rival the US Air Force-dominated Global Positioning System constellation, but it will also potentially capture a huge portion of the world's telecommunications market share. Already, 70 percent of Chinese smartphones are ready to use Beidou. And that's just the start. Soon, most of Asia will likely rely on Beidou. Plus, thanks to Chinese advancements in the aforementioned 5G technology, Beidou represents "a significant challenge to the US hegemony over the telecommunications infrastructure."[5] Coupling Beidou with China's homegrown 5G networks could "command a lion's share of the market for new telecommunications services." Just remember when the US originally pioneered the telecommunications revolution after the Second World War. It gave US intelligence services and companies considerable leverage over the rest of the world—advantages that Washington still enjoys today. That is slowly being eroded by the supposedly imitative, copycatting Chinese.

As you will see in a later chapter, Russia has also built its own, rudimentary GPS system, GLONASS. That system could potentially marry up to Beidou, forming a potent capability for China's military. Presently, the United States can complicate the plans of other militaries by denying them access to military-grade signals from the US-controlled GPS network. By creating their own alternative to the US system, Beijing is not

---

4 David P. Goldman, "US Ban Won't Derail Huawei's European 5G Rollout," *Asia Times,* 19 May 2019, https://www.asiatimes.com/2019/05/article/us-ban-wont-derail-huaweis-european-5g-rollout/

5 Jonathan Shieber, "China Nears Completion of Its GPS Competitor, Increasing the Potential for Internet Balkanization," *Tech Crunch,* 28 December 2019, https://techcrunch.com/2019/12/28/china-nears-completion-of-its-gps-competitor-increasing-the-potential-for-internet-balkanization/

only giving themselves economic advantage over the United States but strategic leverage as well. Should a military crisis erupt between the two powers, China's military could have access to GPS capabilities—while potentially denying the United States access to similar systems, by destroying or holding hostage the American GPS constellation.

Look at how China has become a dominant player in the development of artificial intelligence. Far from being home to mere "copycats," Taiwan-born technologist Kai-Fu Lee believes that China has adopted a "cutthroat" entrepreneurial environment that could rival anything in Silicon Valley or on Wall Street.[6] And artificial intelligence promises to transform the world—notably the space sector. Right now, most space programs have numerous unmanned space projects underway, from satellites to probes exploring various parts of the solar system. Manned spaceflight, while integral for a nation's prestige, is not presently the most ubiquitous part of a space program. In fact, China's recent Chang'e-4 lunar rover, which made headlines for being the first man-made object to visit the dark side of the Moon was entirely automated. China also plans to send an automated sample return mission, Chang'e-5, to the Moon within a year. It has similar plans for an initial automated mission to Mars. Even as countries and private space companies increasingly invest in manned spaceflight capabilities over the next decade, automated space missions will remain important. Inevitably, as artificial intelligence progresses—particularly with the advent of quantum computing—it is the space sector where the first major advances in applicable artificial intelligence will likely be made.

It's also important to briefly understand what artificial intelligence is. Without going too deep into this budding technology, when most people refer to artificial intelligence, they are, in fact, referring to machine learning—that is, getting machines to learn in much the same way that humans do in order to rapidly improve and adapt from

6 Kai-Fu Lee, *AI Superpowers: China, Silicon Valley, and the New World Order* (New York: Houghton Mifflin, 2019), 25.

experience rather than explicit inputs from programmers.[7] China wants to dominate this emerging tech market as much as it does the other components of the tech industry. And it is well positioned to do so. Beijing has an ample amount of human capital in that cutthroat entrepreneurial environment that will prove more than competitive against rivals in the United States. What's more, China is attracting Western talent and tech companies, like Google, to set up shop in China and conduct artificial intelligence research there.

Until 2019, Google was the world's leading AI research firm. The Chinese firm Baidou claims to have displaced Google (though this is debatable) by creating the world's first machine that can understand the human language.[8] Regardless, Google has maintained a strong relationship with the People's Republic of China. At the same time, Google refused to work with the Pentagon's Project Maven, which was meant to use artificial intelligence to more efficiently analyze large amounts of drone footage. Any deal Google makes with China on artificial intelligence development would allow China's military access to the breakthroughs Google and its Chinese partners enjoy in their joint-AI research projects. Artificial intelligence, particularly the method that is now being favored for machine learning—using neural network approaches to mimic the human brain and human learning styles—would help to create some of the most devastating weapons.[9] What's more, it would give a space program that embraced this technology considerable advantages when attempting to complete a mission in the cosmos.

---

7 "What Is Machine Learning?" *Expert System,* 7 March 2017, https://expertsystem.com/machine-learning-definition/

8 Anthony Cuthbertson, "China's Baidu Dethrones Google to Take AI Language Crown," *Independent,* 27 December 2019, https://www.independent.co.uk/life-style/gadgets-and-tech/news/ai-china-baidu-artificial-intelligence-google-a9261691.html

9 Peter Thiel, "Good for Google, Bad for America," *New York Times,* 1 August 2019, https://www.nytimes.com/2019/08/01/opinion/peter-thiel-google.html

Swarms of space stalkers or other space weapons could be commanded by Google-provided AI systems. Or, Beijing could use AI to enhance their oppressive national surveillance network—which is something that China's state security services are absolutely planning for. China also plans to build an automated lunar outpost over the next few years, in anticipation of the placing of their taikonauts at that lunar outpost by 2024. Artificial intelligence would help greatly. All of this is to say that the purportedly imitative China is as competitive—if not more—in the area of artificial intelligence development as the United States. Evoking Kai-Fu Lee, "If artificial intelligence is the new electricity, Chinese entrepreneurs will be the tycoons and tinkerers who electrify everything from household appliances to homeowners' insurance . . . These will be deployed in their home country and then pushed abroad, potentially taking over most developing markets around the globe."[10] I maintain that it will be in the space sector, though, where this technology is innovated rapidly—and the United States is being left behind. This will not only positively impact China's artificial intelligence sector, but it will also allow for China to dominate the high ground of space quicker.

As more proof of the advantage that being an imitator of the United States' innovator provides China, just look at the budding field of quantum computing. Beginning in 2010, most computer scientists recognized that humanity was pushing the limits of growth in the silicon-based computer chips that our society relies upon. Moore's law, which was crafted by the father of silicon-based computer chips (and the founder of Intel), stated that the number of transistors in a dense integrated circuit doubles roughly every two years. This was an observation and projection of a historical trend rather than a natural or physical law. And given that nothing in this world can grow exponentially forever, it stood to reason that humanity would inevitably reach the end of Moore's law. Today, we are reaching that end. Many computer firms,

---

10 Lee, *AI Superpowers*, 25.

such as HP, have striven to enhance the existing capabilities of silicon chips in order to extend the growth of this technology.[11] But these are merely Band-Aids. Inevitably, for humanity to achieve the robust computing power it seeks, an entirely new form of computing will be needed.

In 2006, Google began investing in a potential replacement for the silicon-based computing that we've relied on since the 1960s: quantum computing.[12] Though development has been slow, Google is believed to have achieved "quantum supremacy" over their rivals, such as IBM, which also invested in their D-Wave quantum computer and is building an entire quantum computing research center in New York.[13] Of course, given Google's penchant for sharing research and technology with China, such as they are doing with artificial intelligence and cloud computing research, there is no guarantee that Google would not seek to entice China's government into opening their coveted billion-person market exclusively to Google products in exchange for helping China's budding quantum computing program along.

Many technologists assert (correctly, I believe) quantum computing will be as disruptive of a technology as the cotton gin or automobile was. Yet, despite some headway being made, most American tech firms have been very conservative in the development of this radical new computing technology. Google has gotten ahead of their corporate competition, but the research is expensive, and they could probably cut down on their R&D time by working with the less scrupulous Chinese. After all, in China, there has been an explosion in research and development of

---

11 Martin Giles, "Intel's New 3D Chip Technology May Help Prolong Moore's Law," *MIT Technology Review,* 12 December 2018, https://www.technologyreview.com/f/612587/intels-new-3d-chip-technology-may-help-prolong-moores-law/

12 Madhumita Murgia and Richard Waters, "Google Claims to Have Reached Quantum Supremacy," *Financial Times,* 20 September 2019, https://www.ft.com/content/b9bb4e54-dbc1-11e9-8f9b-77216ebe1f17.

13 Frederic Lardinois, "IBM Will Soon Launch a 53-Quibit Quantum Computer," *Tech Crunch,* 18 September 2019, https://techcrunch.com/2019/09/18/ibm-will-soon-launch-a-53-qubit-quantum-computer/

quantum computing technology—so much so that in 2020 the Chinese Academy of Sciences plans to open the world's largest quantum computing center in Anhui province.[14]

While Western firms, like Google and IBM, have carefully invested in developing quantum computers, China has gone all-in. Thanks to their initial investment in quantum computing, they managed to invent the world's first version of quantum internet as well as the quantum radar.[15] Quantum technology relies on a concept in physics known as "quantum entanglement." This phenomenon was popularized in the 1930s by none other than Albert Einstein, who described it as "spooky action at a distance."[16] Basically, quantum entanglement states that an atom or an electron or a proton can exist in two places at once. Quantum technology manipulates the quantum entangled atom in order to accomplish advanced computational tasks. Regarding the quantum internet, this new technology uses quantum entanglement to send unhackable information at instantaneous speeds across vast distances.[17] The Chinese launched their Micius satellite into orbit in 2016 and conducted the first live test of their quantum internet a year later, sending an encrypted communication across a record distance.[18] Should this technology become as popular as the Chinese hope for, not

---

14 Gordon G. Chang, "U.S. Must Put a Ban on Google Helping China Develop a Global Digital Dictatorship," *Daily Beast,* 26 March 2019, https://www.thedailybeast.com/google-snubbed-the-pentagonbut-not-the-chinese-military

15 Rafi Letzter, "China's Quantum-Key Network, the Largest Ever, Is Officially Online," *Live Science,* 19 January 2018, https://www.livescience.com/61474-micius-china-quantum-key-intercontinental.html

16 Griffith University, "Quantum 'Spooky Action at a Distance' Becoming Practical," Phys.org, 5 January 2018, https://phys.org/news/2018-01-quantum-spooky-action-distance.html

17 Anil Ananthaswarmy, "The Quantum Internet Is Emerging, One Experiment at a Time," *Scientific American,* 19 June 2019, https://www.scientificamerican.com/article/the-quantum-internet-is-emerging-one-experiment-at-a-time/

18 Michael Irving, "Global Quantum Internet Dawns, Thanks to China's Micius Satellite," *New Atlas,* 23 January 2018, https://newatlas.com/micius-quantum-internet-encryption/53102/

only would world communications be revolutionized, but America's intelligence community would have a difficult time learning their rivals' secrets because of the virtually unhackable nature of quantum internet transmissions.[19] What's more, since China created the infrastructure for this new form of communication, like the United States with the internet, it is likely that China will be able to corner the new global market on quantum internet communications—giving China's security and intelligence services great strategic advantages over the United States.

A similar story can be told for China's development of quantum radar. Presently, the United States has significant advantages over the rest of the world in the development of stealthy, fifth-generation fighter planes, such as the F-35 Joint Strike Fighter. Over the course of twenty years, the Department of Defense has spent $1.5 trillion on building what many believe is essentially a flying computer chip.[20] It is a fighter plane that purportedly can do everything. Its proponents argue that the F-35 took so long to develop and deploy because the Pentagon bundled as much advanced technology into its airframe as it could to keep the F-35 ahead of its competitors for years to come. Unfortunately, few in the DOD anticipated that another country—especially imitative China—would garner the capability to see through the protective stealth of the F-35.[21] It's unlikely that China would have been able to achieve such a feat, though, had it not been for China's aforementioned sustained investment into quantum technology research. With their new quantum radar, China's military can detect any aircraft operating

---

19 Josh Chin, "China Makes Leap Toward 'Unhackable' Quantum Network," *Wall Street Journal*, 15 June 2017, https://www.wsj.com/articles/scientists-make-leap-toward-quantum-network-1497549676

20 "F-35 Jet Fighters to Take Integrated Avionics to a Whole New Level," *Military & Aerospace Electronics*, 1 May 2003, https://www.militaryaerospace.com/computers/article/16709227/f35-jet-fighters-to-take-integrated-avionics-to-a-whole-new-level

21 Jamie Seidel, "China's Claim It Has 'Quantum' Radar May Leave $17 Billion F-35 Naked," *News. com.au*, 27 February 2017, https://www.news.com.au/technology/innovation/inventions/chinas-claim-it-has-quantum-radar-may-leave-17-billion-f35-naked/news-story/207ac01ff3107d21a9f36e54b6f0fbab

in the stratosphere approaching what they view as their territory before that aircraft can threaten Chinese forces.[22]

As Sébastien Roblin of the *National Interest* explained in 2018:

> A quantum radar functions by using a crystal to split a photon into two entangled photons . . . Then the radar beams one half of the entangled pair outwards, and monitors the corresponding effects on their entangled partners. If the beamed particles bump into, say, a stealth fighter, the effect on the beamed photon would be visible on the un-beamed partner photon as well. Then the photons which register a "ping" are sorted out from the unaffected photons to form a sort of radar image.[23]

Because of this cutting-edge, completely revolutionary new technology, existing countermeasures meant to defend against modern radar systems are believed to be ineffective. What's more, "the entangled photons would not be detectable—giving detected targets little warning—and would not be traceable back to their source. This means that, unlike a traditional air defense radar, which exposes itself to detection whenever it actively emits radio waves to search for aircraft, a quantum radar does not broadcast its presence and make itself vulnerable to anti-radiation missiles."[24] Of course, many in the West are skeptical of the Chinese claims. After all, not only is there a large gap between developing such a technology and mass producing it, but stealth fighters were never truly undetectable. In fact, existing technology can, at times, be used

---

22 Jeffrey Lin and P. W. Singer, "China's Latest Quantum Radar Could Help Detect Stealth Planes, Missiles," *Popular Science,* 11 July 2018, https://www.popsci.com/china-quantum-radar-detects-stealth-planes-missiles

23 Sébastien Roblin, "Quantum Radars Could Unstealth the F-22, F-35 and J-20 (or Not)," *National Interest,* 10 May 2018, https://nationalinterest.org/blog/the-buzz/quantum-radars-could-unstealth-the-f-22-f-35-j-20-or-not-25772

24 Roblin.

to detect US stealth planes.[25] The difference is that the Chinese now have a working prototype they can experiment with. As they experiment with this new technology, China's military can perfect it and, eventually, mass produce reliable versions of the quantum radar. And while US stealth technology was never truly "invisible," quantum radar ensures that stealth planes will be rendered naked.

None of this would have been possible without China's initial investment into quantum computing technology, which the West had already been pioneering as early as 2006. Herein lies another aspect about imitators: if you give them enough time to imitate, inevitably, they start to innovate. As we've seen with both the quantum radar and quantum internet, while still in their nascent forms, imitator China has now started to out-innovate the West in technologies that have severe national security implications. Should this trend continue, in another decade, it is likely that China will have stripped the United States of its technological and, possibly, economic advantage and be on par with their American foes . . . all while moving ahead of the West.

REMEMBER SLATER THE TRAITOR

Another matter that skeptics of China's rise take issue with is the entire idea that an imitator could rise to challenge the United States—and potentially defeat the innovative United States. In this case, it is important to remember the true history of the United States' technological development during the Industrial Revolution. For the United States behaved in its early years exactly as China does today. Just as China steals US technology and assimilates it into their own, catapulting themselves up the development ladder, all in an effort to successfully challenge—and inevitably beat—the United States, America pilfered manufacturing practices from the British Empire.

---

25 Dario Leone, "That Time Serbia Shot Down An American F-117 Stealth Fighter," *National Interest,* 1 August 2019, https://nationalinterest.org/blog/buzz/time-serbia-shot-down-american-f-117-stealth-fighter-70546

In 1793, a short seventeen years after the United States declared its independence from the British Empire, a young Englishman named Samuel Slater emigrated from England to Rhode Island. Once he arrived, Slater built a series of textile factories based on the advanced proprietary English model that he had learned as a young man working in those factories back home. Today, Slater is known as the "father of the American factory system."[26] His textile mills were ultimately the inspiration for Eli Whitney's great invention, the cotton gin, that helped to spur the Industrial Revolution. In England, however, he is known as "Slater the Traitor" because he brought Britain's advanced industrial secrets to their upstart American rivals.[27]

Ultimately, Slater's contribution to American industry would help lay the groundwork for the day when the United States would displace the British Empire as the world's dominant power. As you can see, when Slater first opened his textile factories along America's east coast, the British Empire still dominated the world's industry—and continued to dominate the world for decades to come. To the British, their American cousins were just copycats. So long as Britain continued to innovate, they would always outpace the United States. But so long as the United States kept imitating the British Empire, the Americans were able to keep pace. As the United States enhanced its industrial might and continued growing as a nation, inevitably, the United States leapfrogged the British, and the game, as they say, was up for Britain.

A similar pattern is at play today regarding China.

All of this affects the new arms race in space. Space development, whether it is in the civilian or military sectors, feeds off of—and into—the development of other technologies that most people take for granted. To get a better sense of this fact, download the NASA magazine *Spinoff*, which has been in publication since the 1970s. *Spinoff* was created to

---

26 "Samuel Slater," PBS.com, https://www.pbs.org/wgbh/theymadeamerica/whomade/slater_hi.html

27 Kat Eschner, "How Industrial Espionage Started America's Cotton Revolution," *Smithsonian Magazine,* 20 December 2017, https://www.smithsonianmag.com/smart-news/how-industrial-espionage-started-americas-cotton-revolution-180967608/

illustrate to the public how important NASA's mission is for the continued advancement of our society. Each edition of *Spinoff* highlights a different set of technologies that have been made possible thanks to NASA investment and involvement. Without the space program, things like cloud computing, invisible braces for your teeth, the turbofan engine, cochlear implants, blood pumps, the wireless headset, and a variety of other consumer goods that most take for granted today would not have come into being (or, at least, would have taken far longer to be invented than they ultimately did).[28]

Just as the massive investment in the Manhattan Project during the Second World War led to great increases in the America's technological capabilities, and as NASA's Apollo program led to thousands of various spin-off technologies that helped to influence the explosion of the computing revolution begun in the 1970s, a sustained space development program can today lead to a new wave of extraordinary technological expansion in the United States. Right now, China is the one leading the world in terms of having a comprehensive national strategy for space and high-tech development. There has been a concomitant expansion in Chinese technological growth across the board as a result. In fact, China has become so good at technological development that in 2017, KPMG's Global Tech Innovation Hub Survey predicted that Shanghai would be the world's leading tech innovation hub by 2020. The survey in question is conducted by KPMG every four years, and it consists of KPMG asking eight hundred of the world's leading tech executives to anticipate where they believe the top global tech innovation hubs will be in four years' time. China and the United States took the lion's share of the top ten. However, the fact that Shanghai topped the list should be disconcerting.[29]

The imitators are now becoming the innovators.

---

28 Christopher McFadden, "23 Great NASA Spin-off Technologies," *Interesting Engineering,* 1 October 2018, https://interestingengineering.com/23-great-nasa-spin-off-technologies

29 "Revenue Growth Is the Top Metric to Measure the Success of Innovation," KPMG, 26 April 2018, https://home.kpmg/sk/en/home/media/press-releases/2018/04/revenue-growth-is-top-metric-to-measure-innovation-success.html

4

CHINA HAS THE GUTS TO TRY

WHEREAS WE IN THE WEST have become skeptical of new things, China has that frontier-style, leap-without-looking mentality that most Americans used to possess. Today, there is a strict separation between the public and private sectors in the United States. Similarly, for the last few decades, the federal research and design budget has been precipitously declining in the United States. Thus, the sort of brash technological innovation that took place in decades past is no longer happening as it once did.

Today, we have the Defense Advanced Research Projects Agency (DARPA) and a smattering of other federally funded research arms, but nothing like in the glory days of Bell Labs.[1] Back then, the United States government could make the initial investment in a given area of research. The fact that the US government was making the initial investment in the development of infrastructure for new technologies meant that venture capital in the private sector could be encouraged to invest. After all, with federal money backstopping a project, private-sector investment in risky technology development would be a safer bet than if the private sector was doing all of the lifting. Following the initial investments from the government secured an initial wave of investment from venture capital firms, the real innovation would occur. Soon, other industries would get in on the action as the original technology became profitable, and that was basically how the modern technology sector in the United States got to the level where it is today.

Far too often in the United States today, the men and women in both government and the private sector have forgotten the crucial marriage of initial state funding and private sector innovation, which redounded to strategic advantages for the United States. Many erroneously assume that some variation of laissez-faire capitalism—libertarianism—resulted in the technological revolution the United States has experienced for the last several decades. These assumptions are wrong. And such assumptions color the vision of leaders in both the public and private sectors to such a point that no real progress is made in the development of new technologies that might lead to a new wave of technological innovation.

Writing in *American Affairs Journal*, Robyn Klingler-Vidra argued:

> The state has been a central, even necessary character in the American tech industry from the beginning. The government's role in financing the technology underlying today's Silicon Valley giants (notably Apple and Google) was detailed in Mariana Mazzucato's book *The*

---

1 Walter Isaacson, "How America Risks Losing Its Innovation Edge," *Time,* 3 January 2019, https://time.com/longform/america-innovation/

Entrepreneurial Spirit (2013). For Mazzucato, the growth of these firms is not a product of unfettered capitalism, but rather a case of "When John Maynard Keynes met Adam Smith": the government initiates what the private sector cannot, and then market forces can flourish. This necessary role of the state in opening the door to the free market corresponds with Karl Polanyi's view in *The Great Transformation* (1944)—the idea of a liberal market utopia is a fallacy, as "even laissez faire is planned." The most sophisticated neoliberals themselves have been conscious, at least privately, of the primary role of the state in organizing and facilitating markets; but popular neoliberalism has either drifted toward libertarian views or emphasized only neoliberalism's deregulatory elements.[2]

This marriage between the public and private sectors to hasten lasting technological development is something that the Chinese have perfected. Of course, China today operates on a state capitalist model, which is fundamentally different from the Western version of capitalism, although China's state-owned enterprise model is more reliant on their government than anything in America's past (and that could be their undoing in the long run).[3] China has, however, managed to catch up to the United States *because* of this state capitalist model.

China's government covers the risks by owning many SOEs involved in the development of risky or experimental technology, meaning Chinese firms can afford to exist in the red far longer than any Western private company can. Chinese corporate leaders can follow a long-running plan for R&D and wait for the world to demand the new technology.[4] The assumption is that while some may never take

2 Robyn Klingler-Vidra, "Building the Venture Capital State," *American Affairs Journal* 2, no. 3 (Fall 2018), https://americanaffairsjournal.org/2018/08/building-the-venture-capital-state/

3 Ian Bremmer, *The End of the Free Market: Who Wins the War Between States and Corporations?* (New York: Portfolio, 2010), 43–84.

4 "China Issues S&T Development Guidelines," GOV.cn, 9 February 2006, http://www.gov.cn/english/2006–02/09/content_183426.htm

off, many will. And when those innovations do catch on, China will be able to get ahead of their competition. This is how China has remained competitive against the West in technology. It is true that China's current leader, Xi Jinping, has encouraged the termination of many state-owned "zombie enterprises."[5] Still, even "private" Chinese firms operate to both earn profits and enhance Chinese strategic interests. Specifically, this fusion of big Chinese corporations with China's big government is how Beijing plans on besting the United States in both the ongoing technology race as well as the strategic high ground of space.

WHOEVER CREATES A CHEAPER WAY TO TRAVEL IN SPACE WINS

Take, for example, the development of the experimental electromagnetic drive (EmDrive) by British inventor Robert Shawyer in 2001. No one in the West believed the technology was feasible. Essentially, the EmDrive is nicknamed the "impossible drive" because it produces thrust from nothing—thereby violating Newtonian physics.[6] The Chinese acquired plans for the engine and slowly developed it over the last several years. While the West ignored technology and lampooned its creator, China studied it.

In 2016, Dr. Chen Yue, the director of commercial satellite technology for the China Academy of Space Technology (CAST), a Chinese Communist Party organ, claimed to have a working version of it.[7] If the EmDrive works as advertised, it could transport a manned spacecraft

---

5 Cissy Zhou and Finbarr Bermingham, "China Steps Up Efforts to Close Failed Zombie Companies by 2020, but Faces Harsh Economic Reality," *South China Morning Post,* 7 February 2019, https://www.scmp.com/economy/china-economy/article/2185186/china-steps-efforts-close-failed-zombie-companies-2020-faces

6 "Roger Shawyer Explaining the Basic Science Behind #EmDrive," YouTube video, 14:35, Nick Breeze, 31 August 2016, https://www.youtube.com/watch?v=wBtk6xWDrwY

7 Dom Galeon, "China Claims They Have Actually Created an EM Drive," *Futurism,* 20 September 2017, https://futurism.com/china-claims-they-have-actually-created-an-em-drive

from the Earth to neighboring Mars in just seventy days.[8] China is most certainly interested in beating the United States to Mars, although it is likely that the Chinese envisage using this technology a bit closer to home. According to studies, if the EmDrive works as it is designed, it would cut fuel costs for existing satellites and spacecraft by as much as half.[9]

Now, imagine if you will, placing such a system on a new satellite. The ability to maneuver in orbit would be greatly improved compared to existing satellites, and it would be harder to track. There is some concern that the Chinese may have already tested this experimental drive on their Tiangong-2 space station in 2016.[10]

Beyond what many naysayers believe to be the fanciful obsession with EmDrive, China has been investing copiously in another alternative space propulsion system, known as the Hall-thrust engine, colloquially known as an ion drive. Presently, most spacecraft rely on chemical-propelled engines. These are expensive and somewhat unwieldy. The ion drive, like the proposed EmDrive, is designed to cut down on the onerous costs associated with conventional spacecraft engines. The Hall Thrust Ion Engine uses electricity collected by solar panels for fuel. This solar-power-produced fuel is then used to create an electromagnetic field to create thrust in space. According to China's Ministry of Science and Technology, which oversaw the 502 Research Institute, a subsidiary of the China Aerospace Science and Technology Corp (a state-owned enterprise that reports to the Ministry of Science

---

8 Horia Ungureanu, "EmDrive Engine Can Theoretically Bring Us to Mars in 70 Days: What's Next After Peer Review?" *Tech Times,* 3 September 2016, https://www.techtimes.com/articles/176203/20160903/emdrive-engine-can-theoretically-bring-us-to-mars-in-70-days-whats-next-after-peer-review.htm

9 David Hambling, "Why DARPA Is Betting a Million Bucks on an 'Impossible' Space Drive," *Popular Mechanics,* 2 November 2018, https://www.popularmechanics.com/space/rockets/a24219132/darpa-emdrive/

10 Brian Wang, "China Says Tests of Propellentless EMDrive on Tiangong-2 Space Station Were Successful," *Next Big Future,* 21 December 2016, https://www.nextbigfuture.com/2016/12/china-says-tests-of-propellentless.html

and Technology—there's that pesky "China, Inc." again!) the new ion thruster will "outperform all of the ion thrusters used on satellites or spacecraft that are currently in use."[11] While this may sound like nothing more than Chinese imperial bluster, there is some reason to believe this claim—and even more reason for American space policy analysts to be concerned.

The mechanics of the Hall thruster are relatively, if you'll pardon the expression, straightforward: the electromagnetic field produced by solar power basically changes propellant materials like xenon into fast-moving ions, particles with either a positive or negative charge. These ionized particles leave the craft, creating enough force in the process to move a spacecraft or satellite in the other direction. Herein lies the ion thruster's superiority over conventional chemical-propelled thrust such as rockets: the ion thruster does not require high temperatures to produce the propellant in the way that most chemical-produced thrust demands. This then makes the ion thruster much lighter than its chemically propelled alternatives, which, therefore, makes space travel cheaper and, therefore, more feasible. As a corollary to this, the ion thruster can be as much as ten times more fuel efficient than conventional rocket engines.

The problem with older versions of the Hall accelerator ion thrusters was that "by no means all the atoms of gas passing through the electron chamber [were] converted into ions." Due to this, then, most ion thrusters, while powerful and efficient, still did not live up to their full potential. Rather than accept this reality, Chinese scientists "proposed the shape of the nozzles from which the gas enters the engine working chamber, so that the gas jet will not move perpendicular to the direction of thrust, as in conventional engines, but is twisted into a spiral."[12]

---

11 Stephen Chen, "Electric Vehicles in Deep Space: China Hails Its New Ion Thruster for Rockets as the World's Best," *South China Morning Post,* 13 January 2016, https://www.scmp.com/tech/science-research/article/1900601/electric-vehicles-deep-space-china-hails-its-new-ion-thruster

12 Anastasia Sinitskaya, "Chinese Engine Tested in Orbit," *Asgardia: The Space Nation,* 10 September 2019, https://asgardia.space/en/news/Chinese-Ion-Engine-Tested-in-Orbit

China took existing technology that both the Americans and Russians possess and enhanced the hell out of it—when few others were even considering such modifications. Now, Beijing's space forces have a strategic advantage over the Americans.

The ion thruster is usually used for communications satellites in high-altitude orbits performing long-term missions. The fastest recorded ion thruster can accelerate in space to thirty kilometers (eighteen miles) per second. The new Chinese variant of the ion thruster is believed to beat that record by a *minimum of 30 percent*! It is true that many other advanced countries were perfecting similar technology, but Beijing got there first in 2016, thus winning yet another component of the ongoing second space race. The Chinese tested a working version of the ion thruster in orbit aboard a small satellite in September 2019.[13] The United States is still working on its version. Don't worry; China will just wait around for the Americans to catch up.

An important aspect of the ion thruster is that it would allow satellites to move continuously in orbit. Under present conditions, according to one Pentagon official, a Chinese spy satellite utilizing this engine would make it difficult for the Pentagon, using its current detection methods, to keep an eye on it. Or, a space stalker might be able to approach one of America's sensitive military satellites undetected and then either damage it or intercept a data stream meant for that sensitive American satellite. The possibilities with this engine are nearly endless.

These issues boil down to a simple problem, the likes of which have proven to be a major hindrance to every country or company seeking to have a greater presence in space. The cost of fuel and the limits of conventional chemical-propellant-based engines, such as rockets, make operating in space a far costlier endeavor than either private sector investors or taxpayers are comfortable with. Thus, space has been the purview of the governments of a few wealthy countries and billionaires, like Elon

---

13 DM Chan, "Pocket-Sized Chinese Thruster Has Big Future," *Asia Times,* 11 September 2019, https://www.asiatimes.com/2019/09/article/pocket-sized-chinese-thruster-has-big-future/

Musk or Jeff Bezos. The engines that China has been investing in are all meant to reduce the cost of transportation in space—which would be a major factor in facilitating the expansion of human development of space as both a strategic domain as well as a domain for economic development. Suddenly, more dangerous and harder-to-detect satellites not only become a possibility, but so too does the possibility of an economically viable lunar or asteroid mining operation, since transportation costs and time would become more economically feasible.

Just remember: if a country or company can reduce the cost of fuel and increase the distance spacecraft can travel, then that company or country will have achieved a significant victory in the second space race. Presently, China appears poised to be the leader of such technology. Thankfully, though, as you will see in later chapters, the United States under the Trump administration is starting to respond to this serious threat.

5

HOW CHINA PLANS TO WIN THE

RESOURCE WAR IN SPACE

THE CHINESE COMMUNIST PARTY'S CLAIM to political legitimacy today is not ideological. Before China opened itself to the West and allowed for market capitalism to take hold there, the CCP maintained its political legitimacy as the one true power in China because it was ideologically Communist. Once they abandoned Communism in favor of state capitalism, however, the CCP's legitimacy rested on the claim that they presided over a continuous period of incredible economic growth. Should this no longer appear true to the Chinese people, then, the CCP's days as the dominant political power in China would be over. A key component of the Chinese economic growth model is having access to large amounts of cheap energy.

Presently, 60 percent of all of China's vital imports travel through maritime trading routes. Specifically, much of China's oil is imported and travels to China via these crowded oceanic trading routes. Not only are these essential trading routes crowded with international shipping, but they also form what are known as "critical chokepoints."[1] These are parts of the maritime trading routes where geography basically squeezes trade into a relatively small area, such as in the Strait of Malacca.[2] Due to the geographical realities of these tight chokepoints in international maritime trade, the United States Navy could, at any time, implement what's known as a strategy of "offshore control" by blockading these critical chokepoints.[3] By engaging in offshore control, the US Navy, with its global reach, would be strangling China's economy from afar. China lacks a global navy. Therefore, defeating the US Navy's offshore control strategy would be difficult. While China is rapidly developing its own navy to counter the US Navy, developing a force capable of challenging a competent navy, like America's, will take time.

Thus, China has been seeking to diversify its energy portfolio so that they do not have to rely as heavily on nonrenewable energy sources being imported over these distant vulnerable maritime trading routes. Nuclear fusion is a key element to their diversification program. The establishment of mining colonies on the Moon and beyond could provide China with the fuel source their nuclear fusion reactors need to power their civilization's incredible rise to global dominance. So, too, would space-based solar energy. Any one of these technologies could not only help China defeat the threat of America's offshore control capabilities, but

---

1 "Maritime Chokepoints: The Backbone of International Trade," *Ship-Technology*, 3 October 2017, https://www.ship-technology.com/features/featuremaritime-chokepoints-the-backbone-of-international-trade-5939317/

2 China Power Team, "How Much Trade Transits the South China Sea?" *China Power*, 27 October 2017, https://chinapower.csis.org/much-trade-transits-south-china-sea/

3 T. X. Hammes, "Offshore Control Is the Answer," *Proceedings*, December 2012, https://www.usni.org/magazines/proceedings/2012/december/offshore-control-answer

it might also swing the balance of power in China's favor, as they, not the United States, would pioneer the development of a new form of affordable, innovative, renewable alternative energy source.

The Chinese made history in 2019 by placing an unmanned rover on the dark side of the Moon, with two attendant satellites in Lagrangian Point-2 (an orbit between Earth and the Moon).[4] Much like their aforementioned ASAT test in 2007, the Chang'e-4 mission to the dark side of the Moon was a proof of concept for the Chinese: Beijing wanted to show the world that they, too, could get to the Moon. They also wanted to make history by landing on the previously unexplored dark side of the lunar surface. Further, the Chinese brought along with them the tools needed to grow cotton on the Moon and to test the lunar surface.[5] China wants to test the lunar surface to prove whether it is abundant in natural resources, notably an isotope known as helium-3.[6] If the Moon has an abundant supply, the Chinese want to establish a lunar mining colony and essentially strip-mine the Moon. There are, of course, other natural resources on the Moon that could be mined and provide China with great wealth, but ^{3}He is important because of its implications for energy production.

Presently, China is one of only a handful of countries spearheading research into nuclear fusion.[7] In the West, people have spoken

---

4 Brandon J. Weichert, "Trump's Most Unhelpful Moon Tweet," *Space News,* 17 June 2019, https://spacenews.com/trumps-most-unhelpful-moon-tweet/

5 Gwyn D'Mello, "China Has Grown Cotton on the Far Side of the Moon, and It's the First Country in the World to Do It," *India Times,* 16 January 2019, https://www.indiatimes.com/technology/science-and-future/china-s-lunar-lander-has-successfully-managed-to-grow-cotton-on-the-far-side-of-the-moon-360483.html

6 Vikram Mansharamani, "China Wants to Mine the Moon for 'Space Gold,'" PBS, 31 March 2016, https://www.pbs.org/newshour/economy/china-wants-to-mine-the-moon-for-space-gold

7 David Stanway, "China Targets Nuclear Fusion Power Generation by 2040," Reuters, 12 April 2019, https://www.reuters.com/article/us-china-nuclearpower-fusion-idUSKCN1RO0NB.

dreamily about nuclear fusion but never managed to realize it.[8] This is likely due to a lack of sufficient investment and leadership at the national policy level, although, in the last decade, private-sector investment into nuclear fusion research has increased to historic highs—and continues growing.[9] Still, China is ahead of the curve.[10] If nuclear fusion is viable, then it will revolutionize the energy sector, as it will provide abundant, clean, and sustainable energy for humanity and allow for us to wean ourselves off of purported "dirty," nonrenewable fossil fuels, such as oil and natural gas.

Current methods for nuclear fusion rely on deuterium-tritium reactions to spark nuclear fusion.[11] Yet, the deuterium-tritium reaction is too difficult to rely on, and the chance of radioactive leak is high. Helium-3 is believed to be a better alternative to the deuterium-tritium mix.[12] The only problem is that there are scant amounts of He-3 available on Earth, meaning that the cost for acquiring such a rare commodity is too high for it to be useful in providing cheap energy for all. But on the Moon, where there is no protective atmosphere, the isotope is believed to exist in abundance—having been emitted by the sun over the course of billions of years and absorbed by the lunar rocks during that time. Whoever manages

---

8 Jonathan Tirone, "Nuclear Fusion," *Washington Post,* 20 June 2019, https://www.washingtonpost.com/business/energy/nuclear-fusion/2019/06/20/c6bd5682-938d-11e9-956a-88c291ab5c38_story.html

9 Akshat Rathi, "In Search of Clean Energy, Investments in Nuclear-Fusion Startups ARE Heating Up," *Quartz,* 26 September 2018. https://qz.com/1402282/in-search-of-clean-energy-investments-in-nuclear-fusion-startups-are-heating-up/

10 David Anderson and Shira Polan, "China Made an Artificial Star That's Six Times as Hot as the Sun, and It Could Be the Future of Energy," *Business Insider,* 19 December 2018. https://www.businessinsider.in/china-made-an-artificial-star-thats-6-times-hotter-than-the-sun-and-it-could-be-the-future-of-energy/articleshow/67164428.cms

11 Rodrigo Antunes, "Tritium: A Challenging Fuel for Fusion," *Euro-Fusion,* 8 November 2017, https://www.euro-fusion.org/news/2017-3/tritium-a-challenging-fuel-for-fusion/

12 Christopher Barnatt, "Helium-3 Power Generation," *Explaining the Future,* 30 January 2016. https://www.explainingthefuture.com/helium3.html

to claim the Moon, establish mining colonies, and get the mined ^{3}He down to Earth to be used in nuclear fusion reactors, will not only have revolutionized energy production but will also enjoy immense strategic—and economic—advantages over their rivals on Earth.

Even without the Helium-3, though, the Moon is a veritable bonanza of minable natural resources, all of which would confer great economic and strategic benefits to the country that achieved dominance on the Moon. What's more, many of these natural resources—such as ice—would be needed to support permanent human settlements on the Moon. For the colonization of the Moon (or any celestial body in our solar system) to work, it needs to be self-sufficient. Therefore, harvesting the natural resources on the Moon for sustaining human existence would be the first step any country would need to take toward not only colonizing the Moon, but also coming to dominate it. China has bold plans to do just this. In fact, Beijing has altered its plans for a manned lunar colony at the Moon's southern pole, according to NASA scientist Dr. Jim Rice. Initially, China wanted to establish a manned colony on the Moon's south pole around 2030. After the Trump administration announced its plans to create a space force as well as to reinvigorate NASA's manned spaceflight program, intending to return its astronauts back to the Moon by 2024, China announced it would colonize the Moon by 2024. Since the Moon's southern pole receives nearly continuous sunlight, the possibility to harvest large amounts of solar energy to power human settlements throughout the Moon is high. Much like Bedouin tribesmen staking out the limited waterholes in the Arabian desert, threatening to kill anyone who dares trespass over such valuable real estate, China likely wants to get its taikonauts to the Moon's south pole, first. Once there, China will build a large solar energy harvesting operation, using similar illegal territory-grabbing methods that China has perfected in the South China Sea. This will be done to deprive their American rivals of access to the cheapest, most abundant source of energy that could best support long-term, sustainable human habitats on that forbidding world.

So, while many American leaders would insist upon establishing colonies on the Moon, irrespective of China's monopolization of vital lunar real estate, Beijing will have denied the United States access to strategic regions of the Moon. Of course, it is possible that the United States and its allies would simply develop alternative energy sources to power their manned lunar colonies, yet the problems would redound to cost. After all, as I've articulated, the most prohibitive factor in greater space development is the cost of such a vast undertaking. To properly develop space, one must allocate large amounts of resources, talent, money, and most important, time. As you've seen throughout this work, the United States rarely plans for long-term missions in space anymore.

It's just too expensive and requires too much concentration to do anything long-term these days. The Chinese, on the other hand, plan for the long term and are willing to not only take risks to gain strategic dominance in space, but also to lose money and potentially lives in these risky undertakings in the short term, in order to enjoy long-term economic and geopolitical advantages. Once China's lunar colony is established in regions that have long-term exposure to sunlight, they will have positioned themselves to be the dominant force on the Moon— which would have negative knock-on effects (at least for Washington) for the international political and economic system on Earth.

Further, in the realm of alternative energy production, China has also been assiduously investing in the creation of space-based solar power. When most people hear about solar power, they think of the reflective panels that directly absorb the sun's energy during daytime. This technology is, in fact, precisely what most believers in man-made global warming insist will help prevent Earth's continuous warming. Along with wind power, global warming activists believe solar power is a renewable source of energy, and is therefore less harmful to the planet's natural environment.[13] Of course, what these global warming activists

---

13 Ellie Anzilotti, "The Green New Deal Is a Chance to Make Clean Energy Accessible to All," *Fast Company*, 27 June 2019. https://www.fastcompany.com/90366185/green-new-deal-100-percent-clean-energy-will-help-economy

rarely admit is that under current conditions, solar power is not a viable alternative to fossil fuels because the energy this technology produces is both insufficient (as well as more expensive) than the conventional forms of nonrenewable energy, such as oil, coal, or natural gas.[14] The main problem with current Earth-based solar energy is a technical problem known as "intermittency." In essence, this form of solar power collection is insufficient to meet Earth's energy needs because there is no region on Earth that experiences zero cloud cover and twenty-four-hour sunshine. Plus, the power distribution from these Earth-based collectors is inefficient when compared to space-based solar power. Current solar power designs, therefore, raise the cost of energy on ordinary consumers, which, over time, will have severely constraining impacts on the global economy, likely increasing poverty globally.[15]

It is believed that space-based solar power (SBSP) would eliminate the environmental complications entirely. The country that embraces SBSP first will enjoy a significant strategic advantage over its foes, as a March 2007 report by the National Security Space Office suggested.[16] By placing what's known as a solar collector (a satellite) in geosynchronous orbit, where the satellite will have continuous exposure to sunlight, the collector can absorb enormous amounts of solar energy and transmit it down to collectors located on Earth's surface in the form of microwave radiation. The concentrated and continuous stream of solar energy transmitted to ground-based collectors, however, would not suffer intermittency at all. What's more, its power distribution

---

14 Michael Shellenberger, "The Reason Renewables Can't Power Modern Civilization Is Because They Were Never Meant To," *Forbes,* 6 May 2019. https://www.forbes.com/sites/michaelshellenberger/2019/05/06/the-reason-renewables-cant-power-modern-civilization-is-because-they-were-never-meant-to/#738de505ea2b

15 Brandon J. Weichert, "AOC Isn't Even a Good Environmentalist," *American Spectator,* 18 March 2019. https://spectator.org/aoc-isnt-even-a-good-environmentalist/

16 Joseph D. Rouge, "Space-Based Solar Power as an Opportunity for Strategic Security," *Report to the Director, National Security Space Office Interim Assessment,* 10 October 2007. https://space.nss.org/space-based-solar-power-as-an-opportunity-for-strategic-security/

method could be beamed "almost anywhere across the globe, except for the poles," instantly, according to former NASA and CalTech Jet Propulsion Laboratory engineer John Mankins.

The idea of space-based solar power was first posited by the famous science fiction writer Isaac Asimov as early as 1941. It looked as though it might become reality in 1968, when an American aerospace engineer, Peter Glaser, proposed the first realistic space-based solar power system for NASA. Like so much with NASA during the 1970s, just as the program was starting to bear some fruit, it was canceled and its funding redirected away—likely to something unimportant for the long-term survival of the country. Since that time, John Mankins has "devised multiple concepts to extend the use of solar in space, among them a solar-powered interplanetary transport vehicle and a space-based power system," according to Eric Rosenbaum and Donovan Russo of CNBC.

In fact, Mankins's work received significant attention in the first year of the first term of the George W. Bush administration—so much so that it looked as though NASA was going to build a working prototype based on Mankins's designs. In typical fashion, unfortunately, the initiative was ultimately nixed just as it was gaining traction.[17] The short-term political concerns of the George W. Bush era took precedence. Plus, former vice president Dick Cheney was fiercely skeptical of any attempt to diversify America's energy portfolio away from non-renewable fossil fuels.[18] Since Cheney was given control over the Bush administration's energy policy, the push to build real SBSP capabilities ended prematurely, along with all other aspects of the George W. Bush's original promise from the 2000 presidential campaign to support the

---

17 Eric Rosenbaum and Donovan Russo, "China Plans a Solar Power Play in Space That NASA Abandoned Decades Ago," *CNBC,* 17 March 2019. https://www.cnbc.com/2019/03/15/china-plans-a-solar-power-play-in-space-that-nasa-abandoned-long-ago.html

18 Peter Baker, *Days of Fire: Bush and Cheney In the White House* (New York: Doubleday, 2013), 95–97.

development of new energy technology once in office.[19]

As the United States government allows yet another innovation to slip through its fingers, the Chinese are coming from behind, rapidly investing in space-based solar power. In fact, according to Xie Gengxin, the deputy head of the Chongqing Collaborative Innovation Research Institute for Civil-Military Integration in southwest China, a thirty-three-acre space-based solar power testing facility is being built in Chongqing's Bishan district with an initial investment of 100 million yuan (or, $15 million).[20] This facility will allow for China to "develop transmission technologies while studying the effect of microwaves beamed back to Earth on living organisms," according to Xie. Should this technology work, then, China will have revolutionized the energy industry and could rapidly go from one of the world's worst polluters to the leader in a new international technology upon which the rest of world would come to rely. It could also remove a strategic lever the US Navy has over China's energy sector.

Look at how much clout Saudi Arabia, the Russian Federation, and even the United States today enjoy in world affairs because of their for-tuitous positions atop vast oil and natural gas resources that the rest of the world requires to survive. Imagine if China were to place working solar collectors in geosynchronous orbit to prove that their technology works—just as China has done with its quantum internet satellite, Micius. Once in orbit, China's solar energy collectors would form the bedrock of a new global energy grid. Beijing could then convince the Western countries most worried about man-made global warming (such as Germany) to invest significantly into China's new energy infrastructure. Germany, the true economic power behind the European Union,

---

19 Paul Krugman, "Enemies of the Sun," *New York Times,* 5 October 2015. https://www.nytimes.com/2015/10/05/opinion/paul-krugman-enemies-of-the-sun.html

20 Scott Snowden, "China Plans to Build the World's First Solar Power Station in Space," *Forbes,* 5 March 2019, https://www.forbes.com/sites/scottsnowden/2019/03/05/china-plans-to-build-the-worlds-first-solar-power-station-in-space/#53c361d45c94

has already refused to follow Washington's lead on the matter of 5G internet, when Chinese telecommunications firms, such as Huawei and ZTE, signed deals to supply 5G cellular networks to Germany in 2019.[21] Ultimately, German chancellor Angela Merkel and her political allies have prevented the deals from going through—but that is all they have done. While both Washington and Brussels have insisted that Chinese firms be banned from providing critical services to Europe—which would give China immense leverage over Europe—all Berlin's leaders could do was decide to kick the proverbial can down the road another year.[22] The fact is, it is economically shortsighted for Germany to not use China's 5G: Beijing has a superiority in this critical area, namely because they enjoy the all-important first-mover advantage. Both US and European firms are tragically behind the Chinese in this essential area.[23] Inevitably, many European governments will choose to do business with Beijing, as the political consensus between Europe and the United States naturally deteriorates.[24] The influences of China (and Russia) are on the ascendance in the continent, meaning that America's power in Europe will be further reduced over time. All of this serves as a telling example of what will happen should the Chinese scale up either their quantum internet or space-based solar power capabilities, as I suspect they ultimately will.

---

21 Janka Ortel, "Germany Chooses China over the West," *Foreign Policy*, 21 October 2019. https://foreignpolicy.com/2019/10/21/germany-merkel-chooses-china-over-united-states-eu-huawei/

22 Andreas Rinke and Holger Hansen, "With or Without Huawei? German Coalition Delays Decision on 5G," Reuters, 17 December 2019. https://www.reuters.com/article/us-germany-china-huawei/with-or-without-huawei-german-coalition-delays-decision-on-5g-rollout-idUSKBN1YL22Z

23 Grady McGrergor, "China Is Launching Its 5G Network Ahead of Schedule and on a Spectrum the U.S. Can't Yet Match," *Fortune*, 31 October 2019. https://fortune.com/2019/10/31/china-5g-rollout-spectrum/

24 Noah Barkin, "The U.S. Is Losing Europe in Its Battle with China," *Financial Times*, 4 June 2019. https://www.theatlantic.com/international/archive/2019/06/united-states-needs-europe-against-china/590887/

An innovation like SBSP could potentially rejigger the entire world order away from the United States and toward China—which is precisely what Beijing envisions. We've already seen how Russia has weaponized its natural gas vis-à-vis Europe, and how Moscow has benefited both economically and strategically from that arrangement. Renowned scientist Thane Gustafson has assessed that Europe imports roughly 40 percent of its natural gas from Europe, with that number only growing. Gustafson further postulates that this trend will continue, and Europe's dependence on Russian natural gas will be a "bridge" to a more sustainable energy future. In that future, renewable fuels would provide Europe with most of its power. Such a future is possible. But suppose that the Russian "bridge" of natural gas leads Europe into a world of Chinese-dominated space-based solar power.[25] What could Beijing do to the world order if China had a monopoly on the space-based solar power that most of the planet relied upon in the near future? What kind of prestige would Beijing's totalitarian government suddenly enjoy if Westerners came to believe that China, not the United States or Europe, was pioneering solutions to the supposedly apocalyptic threat of man-made global warming?

This all has grave strategic consequences for the United States and the world order it purports to lead, especially as Beijing moves increasingly toward implementing its goal of cleaving the Indo-Pacific away from the United States. China's leaders plan to do this through unconventional methods. As Sun Tzu once said, "You can fight a war a long time or you can make your nation strong. You cannot do both." China has focused its efforts on building out its economic and technological advantages to make its country strong. China is becoming so strong and competitive that any war against them will be costly. Therefore, China has strategic leverage to influence the world system as it wants.

---

25 Andrew Moravcsik, "Power of Connection: Why the Russia—Europe Gas Trade Is Strangely Untouched by Politics," *Nature,* 2 December 2019. https://www.nature.com/articles/d41586–019–03694-y

Over time, this capability will only increase. Thus far, Beijing's plot to override American influence in what they perceive as their part of the world is working. The development of advanced technology, as well as the creation of a space warfare capability, are essential elements in China realizing their lofty goals. For years, China has studied the rise of the United States; they have processed how America became the dominant power in the world. What's more, Beijing assessed that it was not military power per se that was responsible for America's rise. Instead, it was economics, cultural certitude, and the confluence of industry and technological development that fed into the rise of America's military and global dominance.

China has sought to replicate this strength by effectively absorbing American capabilities into their own. This has worked well for Beijing. In the coming years, one can expect to see how China will use American technology and ideas to defeat the United States in unorthodox ways—such as in the strategic high ground of space.

6

THE RUSSIAN BEAR BLASTS OFF

TO SPACE (AGAIN)

THE STORY OF THE RUSSIAN SPACE PROGRAM picks up where the old Soviet space program left off. During the Cold War, the space race was a military competition above all else. And just because the Soviets were defeated by the United States in the Moon race, this did not end the Soviet quest for dominance in space. It simply meant that the Soviets needed to perfect new space capabilities to challenge the Americans.

If we were to take our *Star Trek* analogy beyond the previous chapters on China, you would see that if China represents the Borg from *Star Trek: The Next Generation,* then the Russians are most certainly akin to the Klingons—a hardened race of warriors who live to challenge other great powers for territorial conquest and glory.[1] Russia, like the Klingons, takes a "good enough" approach to technology development and military strategy. The Russians are so similar to the warrior race of Klingons that, according to Peter Hopkirk, "long after [Peter the Great's] death in 1725 a strange and persistent story began to circulate through Europe about Peter's last will and testament. From his death bed, it was said he had secretly commanded his heirs and successors to pursue what he believed to be Russia's historical destiny—the domination of the world."[2] While this tale is apocryphal, it does seem to encapsulate the Russian spirit of the last several hundred years in terms of its foreign policy, whether it be under the tsars of the Russian Empire, the leaders of the Soviet Union, or Vladimir Putin's *siloviki*.[3]

Is Russian technology good enough to keep pace with (and, as they hope, defeat) America? Can Russia's military remain ahead of its regional rivals and continue to challenge the United States? On some level, yes. Just like the Klingons, the Russians will employ brute force and guile, depending on the situation, in order to push their limits with a rival like the United States. Russian behavior is meant to extract maximum concessions from their foes while appearing to be superior, when the Russians are, in fact, weak.

As Moty Cristal, an expert on negotiation, argues:

> Russians perceive negotiation as a "power game," as a "сила" (force). They will typically present a very tough position at the beginning of a negotiation, and they will offer tough responses to their counterparts even at the final stages of negotiations.

---

1 Roy Wenzl, "Star Trek and Vietnam: When Klingons Were Stand-Ins for the Soviets," History, 10 September 2018. https://www.history.com/news/star-trek-series-episodes-vietnam-war

2 Peter Hopkirk, *The Great Game: The Struggle for Empire in Central Asia* (New York: Kodansha America, 1994), 20.

3 Karina Orlova, "The Siloviki Coup in Russia," *The American-Interest,* 21 September 2016. https://www.the-american-interest.com/2016/09/21/the-siloviki-coup-in-russia/

Although negotiation theorists speak about the overall opportunity, and finding "win-win" outcomes that can benefit both sides, Russians find it difficult to adapt to this negotiation approach. Indeed, the word "victory" itself in the Russian language means that the other side loses or leaves the game. The Russian negotiation mentality is a very strong approach, and a rather inflexible one, which to some extent ignores emotional and psychological considerations often discussed in negotiation theory.[4]

Take this a step further and apply this to overall Russian foreign and military policies. The Russians don't take no for an answer. They don't stop until they get what they want. And what Russia wants, as noted in the opening chapter of this work, is to preserve their territory by expanding their borders to geographical locations in Europe that are easier for Russia's military to defend. Yet, Moscow understands that attacking the United States military without first gaining a strategic advantage over the Americans would be foolish. Given this, the Russians are a very tough competitor. They will keep up until they either get what they want or can no longer fight—as ultimately happened at the end of the Cold War. The Soviet Union could not go on. Just look at Russia's space program, and you will see this mentality at play.

Even as the American space program was ending its mission to the Moon, the Soviets were focusing on developing a host of new space technologies to defeat their American foes. Moscow had envisioned attacking the United States from space if the Cold War ever turned hot. Toward that end, the USSR invested heavily in space-based weapons. As space policy analyst James Clay Moltz assesses, from 1958 to 1962, the Americans and Soviets "together tested nine nuclear weapons in space." In so doing, these tests gave the world the first demonstration of an electromagnetic pulse (EMP). EMP is an energy surge emitted from

---

4 "Culture and Negotiations: Russian Style," *Skolkovo School of Business,* accessed December 31, 2019. https://embahs.skolkovo.ru/en/emba-hs/blog/culture-and-negotiations-the-russian-style/

a nuclear weapon when it is detonated. The EMP wave travels ahead of the destructive nuclear blast and effectively deactivates all advanced, silicon-based technology (of course, even during this time, computers were being developed to operate off silicon-based technology) *without* causing physical damage. Ultimately, the United States and the Soviet Union would develop EMP weapons that emitted the potent EMP surge without needing to detonate a nuclear warhead that would kill and physically destroy everything around the blast zone.

The nine nuclear tests in low-Earth orbit were so destructive that they "eventually disabled seven satellites [in orbit], including US military satellites, the first US civilian communications satellite, and British and Soviet spacecraft." These tests prompted the Partial Test Ban Treaty (PTBT), which was signed by the two superpowers in 1963.[5] While the Outer Space Treaty of 1967 banned the placement of nuclear weapons (let alone warfare in general) in space—in an effort to remove even the possibility that a nuclear attack could occur from space—the Soviets had taken to heart the concept that space was a strategic high ground.[6] From the 1960s until the end of the Cold War, Moscow worked hard to build military capabilities that would deprive the Americans of access to Earth's orbit as well as provide the Soviet Union with the advantages that typically come from gaining control over the strategic high ground.

THE SOVIETS SAW SPACE AS A BATTLEFIELD

The Reds quickly sought to gain strategic primacy in orbit around the Earth. During the 1970s and beyond, both superpowers had come to depend on satellites for many functions. Notably, the United States had spearheaded the use of satellites as a strategic asset during wartime. In the Vietnam War, the Pentagon started using satellites to assist with

---

5 James Clay Moltz, *Crowded Orbits: Conflict and Cooperation in Space* (New York: Columbia University Press, 2014), 28–29.

6 "Outer Space Treaty of 1967," National Aeronautics and Space Administration, accessed 22 June 2019. https://history.nasa.gov/1967treaty.html

their ongoing military operations. Moscow took note, and had, in fact, been devising their own version of this concept known as the "military-technical revolution."[7] Of course, the United States would ultimately use satellites for nearly all of its military and civilian functions by the 1990s.[8] In the 1970s, though, the Soviets were planning to debilitate what was then the nascent American reliance on satellites as a means of weakening the American military threat to the Soviet Union. To achieve this military goal, the Soviets, much like the Chinese today, began investing heavily in anti-satellite weapons.[9]

During this period, the Soviets designed their first co-orbital satellite.[10] These are the satellites known today as "space stalkers," which are meant to tailgate behind other satellites and then attack those sensitive satellites. As noted in the previous section on China's space capabilities, co-orbital satellites tend to be smaller and more maneuverable than conventional satellites. In peacetime, co-orbital satellites are often deployed to tag along with other satellites and, if necessary, remotely repair them.[11] As demonstrated in the opening chapter, these same rudimentary co-orbital satellites can be fashioned into an offensive weapon that will physically attack—and destroy—an unsuspecting communications, surveillance, or early missile warning satellite in orbit.

---

7 Albert A. Nofi, "Recent Trends in Thinking About Warfare," CNA, September 2006, P. 13. https://www.cna.org/CNA_files/PDF/D0014875.A1.pdf

8 Dwayne A. Day, "And the Sky Full of Stars: American Signals Intelligence Satellites and the Vietnam War," *Space Review,* 12 February 2018. http://www.thespacereview.com/article/3430/1

9 Anatoly Zak, "The Hidden History of the Soviet Satellite Killer," *Popular Mechanics,* 1 November 2013. https://www.popularmechanics.com/space/satellites/a9620/the-hidden-history-of-the-soviet-satellite-killer-16108970/

10 Bart Hendrickx, "Russia's Secret Satellite Builder," *Space Review,* 6 May 2019. http://www.thespacereview.com/article/3709/1

11 Tim Fernholz, "The US Is Worried That a Russian Satellite Is Really a Weapon," *QZ,* 16 August 2018. https://qz.com/1361037/the-us-is-worried-that-a-russian-satellite-is-really-a-weapon/

Co-orbital satellites can also be used to nuzzle in beside an American communications or intelligence satellite and passively intercept any signals coming in or out of that satellite, making signals intelligence collection a much easier endeavor. For example, since 2014, a Russian satellite named "LUCH" has "sidled up to a number of other satellites . . . raising concerns that it could be intercepting data."[12] As far back as the 1960s, the Soviet Union was developing such capabilities. Today, such capabilities have only become more advanced and an even greater threat to the United States.

Meanwhile, the Soviets started investing in the development of laser systems. Writing in the *National Interest*, Sebastien Roblin stated that

the Soviet Union began experimenting with lasers in the fifties and sixties. Its first laser weapons, emerging in the seventies, were fixed ground-based systems with the suitably science-fiction names Terra-3 and Omega. Terra-3 encompassed two different devices, installed at the Sary Shagan testing ground in Kazakhstan: a visible ruby laser and an invisible carbon-dioxide laser. Initially conceived in the 1960s to swat down ballistic missiles in the terminal descending phase, following the 1972 treaty banning antiballistic-missile systems, Terra-3 was reoriented towards damage orbiting satellites, though with little success due to inaccurate tracking systems.

[The Omega lasers] were intended to hit aircraft and missiles in the atmosphere. Omega-1 and -2 proved more successful [than the Terra-3 lasers] at striking distant targets, but the system still lacked sufficient hitting power and power generation. Nonetheless, the Omega laser is thought to be the basis for Russia's current development of ground-based laser air defense.[13]

---

12 Nathan Strout, "Russian Satellite Creeps up to Intelsat Satellite—Again," *C4ISRNet,* 3 September 2019. https://www.c4isrnet.com/battlefield-tech/2019/09/03/russian-satellite-creeps-up-to-intelsat-satellite-again/

13 Sebastien Roblin, "Russia's Cold War Super Weapon (Put Lasers on Everything It Can)," *National Interest,* 16 July 2017. https://nationalinterest.org/blog/the-buzz/russias-cold-war-super-weapon-put-lasers-everything-it-can-21553

Of course, there were many setbacks. The Soviet space program, in particular, during this period had a retinue of failures that would have been painfully embarrassing for the regime in Moscow.[14] Some of these failures remained classified until the 1990s, after the USSR collapsed. Still, the Soviet leadership was committed to the prospect of developing some form of space dominance over the Americans, whether getting men to the Moon first or placing giant laser systems in orbit.[15] As Moscow invested in lasers as a potential strategic weapon in space, they also began to merge their budding space station program with space-based "defensive" weapons.

The Almaz space station program was a Soviet military project.[16] Produced under the guise of the Soviet "civilian" space station program, Salyut, this was the Soviet attempt to place military personnel in low-earth orbit for extended durations. Of the seven Salyut stations that were launched in orbit, three of them were, in fact, Almaz battle stations. One of the three Almaz stations was equipped with a large cannon. The weapon was test-fired. The shells fired from the cannon burned up in the atmosphere, preventing any independent analysis. The Soviets even designed the Almaz station that tested the space cannon to reorient itself to withstand the powerful recoil (in space, once momentum is generated, an object cannot stop on its own) to remain in its proper orbit once the weapon was fired. Whether the weapon was a success remains unknown. What is known is that if the Soviet Almaz military space station program had not become mired in political intrigue and budget cuts, the next Soviet military station would have been outfitted with a laser.

14 Richard D. Lyons, "Failures Mark Russian Space Program," *New York Times*, 26 September 1973. https://www.nytimes.com/1973/09/26/archives/failures-mark-russian-space-program-mechanical-failure.html

15 John Wenz, "5 Big, Bold Soviet Space Missions That Never Were," *Popular Mechanics*, 16 June 2015. https://www.popularmechanics.com/space/a16037/failed-soviet-space-missions/

16 Louis de Gouyon Matignon, "The Soviet Almaz Military Space Station," *Space Legal Issues*, 8 May 2019. https://www.spacelegalissues.com/the-soviet-almaz-military-space-station/

Despite the fact that the Soviets ultimately terminated the Almaz military space station program, many of its components had already been built. Thus, Moscow simply repurposed the various parts of the secretive program for use in other military programs. These programs became instrumental in expanding Soviet early missile warning and covert surveillance programs.[17]

THE SOVIETS PREPARE FOR A BLIZZARD IN SPACE

During this time, the Soviets built their *Buran* (Russian for "Blizzard") space shuttle. At first, Western observers believed the *Buran* shuttle to be nothing more than a bad knock-off of the American space shuttle. And while the *Buran* did an unmanned orbital test that was successful, the Soviet Union collapsed before the spacecraft could be developed further or even deployed with a crew.[18] Yet, the *Buran* shuttle was not merely a rip-off of the American version, as Western analysts assumed (one could hardly blame Western analysts for this assumption, since the *Buran* shuttle looked identical in every way to the American space shuttle). Instead, however, the *Buran* was a highly capable spacecraft. In fact, many experts now believe that the Soviets "created the most advanced and versatile space-launch vehicle humanity had ever known."[19] For example, Soviet space historian Bart Hendrickx told CNN that "in the end, the Russians developed a system that was more capable, more versatile and safer than the space shuttle, but by the time it was ready to fly, the Cold War was coming to an end and the project had lost much of its political support."

---

17 Anatoly Zak, "The Almaz Military Space Station Program," *Russian Space Web*. http://www.russianspaceweb.com/almaz.html

18 Jacopo Prisco, "Two Abandoned Soviet Space Shuttles Left in the Kazakh Desert," *CNN*, 21 December 2017. https://www.cnn.com/style/article/baikonur-buran-soviet-space-shuttle/index.html

19 Anatoly Zak, "Did the Soviets Actually Build a Better Space Shuttle?" *Popular Mechanics*. 19 November 2013. https://www.popularmechanics.com/space/rockets/a9763/did-the-soviets-actually-build-a-better-space-shuttle-16176311/

By this time, the Soviet-American dance had become the defining aspect of both countries' space programs. Even in conflict, the Soviet and American space programs were inextricably bound to each other. In the 1970s, despite the fact that the Americans had made it to the Moon before the Soviets, the Soviet space program was on the verge of getting a manned expedition to the Moon. The various pieces necessary for such a launch had been developed by the time that Moscow had terminated their manned lunar program. The reason the Soviets did this was due to the fact that Soviet intelligence had learned of Washington's desire to terminate its Apollo Moon program in favor of the space shuttle program. The American space shuttle program, while being a civilian-led endeavor, had obvious military applications—namely, that such a vehicle could place sensitive equipment in Earth's orbit, such as advanced satellites or be used to transport pieces for an American space station into orbit. Essentially, the space shuttle was a "space truck" meant to haul things (potentially like weapons) into orbit.[20]

The United States Air Force assigned General James A. Abrahamson to manage part of the space shuttle program during the Reagan administration. His entire mission was to figure out the military applications for the space shuttle. He, along with the legendary Hungarian-born physicist Edward Teller, some policy wonks at the Heritage Foundation, and Danny Graham of the Advanced Research Projects Agency (ARPA) feverishly developed concepts for applying space shuttle operations to the national security space policy arena. The Soviets, ever mindful of what their American rivals were up to, strove to replicate—and leapfrog—the Americans in the creation of military applications for their own space shuttle.[21] And unlike the American space shuttle, which was managed by the civilian-led NASA, the *Buran* space shuttle program was a military program.

---

20 Quora Contributor, "Why Didn't the Space Shuttle Program Continue?" *Forbes,* 7 June 2018. https://www.forbes.com/sites/quora/2018/06/07/why-didnt-the-space-shuttle-program-continue/#2034801690ae

21 Nigel Hey, *The Star Wars Enigma: Behind the Scenes of the Cold War Race for Missile Defense* (Washington, D.C.: Potomac Books, 2006), 7–8.

Since losing the Moon race, the Soviets had temporarily lost their confidence and began looking to the Americans to see what their Western rivals were doing. Yet, the Soviets did defeat the Americans again, this time, in the aforementioned space station race with their successful Salyut and Almaz space stations. Although, the Americans had their space shuttle program up and running many years before the Soviets got the *Buran* program online. Despite this, though, the Soviet military envisioned using the *Buran* to deploy weapons into space. This became a priority mission for Moscow, as President Ronald Reagan pushed for the creation of a space-based missile defense system—colloquially known as "Star Wars"—meant to render the Soviet nuclear threat irrelevant.[22]

Moscow reasoned that they, too, could keep pace with Reagan's designs for a space-based missile defense shield by using their more advanced *Buran* space shuttle to deploy space-based lasers that would theoretically destroy any incoming American nuclear missiles that might be launched in the event of the outbreak of war between the Soviets and the United States. The USSR was so intent on deploying space weapons from the *Buran* space shuttle that they even designed the Energia-M heavy-lift rockets with greater lift capacity than the rockets that launched America's space shuttle.[23] In other words, the Soviets planned to place more equipment in orbit to out-deploy their American foes in the great race for strategic missile defense in orbit.

Had the Soviet Union managed to pull itself from the brink of its ultimate destruction, it is more than likely that the race to get weapons into space would have become the next phase of the great space race that defined the Cold War. And given the *Buran* space shuttle's capabilities, it is quite likely that the Soviets would have at the very least managed to keep pace with the American "Star Wars" program. Thankfully, the

---

22 Edwin J. Feulner, "The Strategic Defense Initiative at 34," Heritage Foundation, 22 March 2017. https://www.heritage.org/missile-defense/commentary/the-strategic-defense-initiative-34

23 Anatoly Zak, "Energia-M: The Swan Song of the Soviet Space Program," *Russian Space Web*, http://www.russianspaceweb.com/energia_m.html

Cold War ended not with a nuclear bang but with a political whimper, and the Soviet Union collapsed under the weight of its own internal contradictions. The *Buran* space shuttle and its powerful Energia-M rocket were left to rust in a hangar at the Russian-controlled Baikonur Cosmodrome in the desert of Kazakhstan, never to be returned to service after the *Buran*'s initial automated test flight in 1988. Yet, the basic building blocks for an advanced space program remained in place well after the Cold War ended and the collapse of the Soviet juggernaut.

7

RUSSIA BECOMES THE "WALMART"

OF ROCKETS

THE INITIAL DECADE FOLLOWING THE END OF THE COLD WAR was a time of momentous political and economic changes for Russia. This was doubly true for the Russian space program. Although the Soviet space program was one of two of the world's most advanced space programs at the time of its collapse, the new post-Cold War government of the Russian Federation was in shambles. Former Soviet elites attempted coups against the new government.

Old Soviet institutions were suddenly privatized, and virtually overnight the phenomenon of Russian oligarchs came into being. Meanwhile, the old Soviet space program survived the fall of Communism—but was it was poorly funded. Space programs are among the most expensive programs a national government could fund. Moscow was bankrupt in the 1990s, and its nascent capitalist economy was too unpredictable to provide consistent levels of funding. Its proto democracy, under the command of the alcoholic Boris Yeltsin, was also a mess—meaning that there was no common mission for its ailing space program other than to avoid being totally canceled.

Following the Cold War, Washington policymakers worried about the possibility of loose nukes and other weapons that the Soviets had built to wage—and win—a hot war with the United States falling into the wrong hands. The potential risk of advanced ballistic missile technology being illegally proliferated to rogue states was very high. Thus, Washington proposed a series of initiatives aimed at preventing the dissemination of Russian missile technology from the newly formed, chaotic Russian Federation to rogue states, such as North Korea, or even stateless terrorist organizations, such as al-Qaeda. Entirely new linkages were formed between the West and Russia's flailing space program that would have been inconceivable even during the height of Cold War-era détente. Notably, according to space policy analyst James Clay Moltz, old Soviet space and missile enterprises, such as Khrunichev and Energomash, rapidly moved away from their initial focus on military hardware and into providing Western companies with commercial products. Namely, these storied Cold War-era Russian space firms provided essential pieces of technology for the Pentagon. The former Soviet conglomerates-turned-private enterprises ultimately partnered with American defense contractors, like Boeing and Lockheed Martin, to supply the United States with reliable, advanced, and relatively cheap space equipment.[1]

As Moltz describes the US-Russian post-Cold War relationship on military space policy:

---

1 James Clay Moltz, "Russia and China: Strategic Choices in Space," *Space and Defense Policy* (New York: Routledge, 2009), pp. 272–73.

Efforts by the US government in the Department of Defense Cooperative Threat Reduction Program helped keep Russian missile specialists from emigrating to countries of proliferation concern, sustained the bulk of the Russian space infrastructure, and allowed Russia to continue as a leading participant in international human spaceflight, space science, and space-launch services.[2]

Essentially, the combination of US government funding and the sale of Russian space assets to Western investors created the modern Russian space program. Slowly, Russian space policymakers recognized the inherent economic value for Moscow in the private space sector. Plus, the sustaining of Russia's space program meant that Moscow could remain strategically involved in space development. Over time, American largesse, coupled with Washington's obsessive quest to keep Russia as an enemy from the late 1990s throughout the 2000s, would lead to this advanced Russian space program, along with Russia in general, becoming a strategic liability for the safety and security of the United States.[3]

Yet, the 1990s and early 2000s were a time of great progress both for the private space sector, which relied heavily on Russian space technology, as well as the flowering of Russia's indigenous space program. For example, Energomash is a legacy Soviet rocket company that also produced the Energia rockets for the old Soviet *Buran* space shuttle program. The Russian government owns 80 percent of this firm. Energomash, partnered with a tiny Florida firm called RD Amross, supplies the United States military with the RD-180 and RD-181 rocket engines.[4] Energomash's rocket designs are so simple, powerful, and affordable that the Pentagon uses them to launch sensitive military

---

2 Ibid.

3 Peter Conradi, *Who Lost Russia? How the World Entered a New Cold War* (London: Oneworld Publications, Ltd., 2018) pp. 99–102.

4 Brian Grow, Stephen Grey, and Roman Amin, "In Murky Pentagon Deal with Russia, Big Profit for a Tiny Florida Firm," Reuters, 18 November 2014. https://www.reuters.com/investigates/special-report/comrade-capitalism-rocket-men/

satellites into orbit even when US-Russian relations are at lowest point since the Cold War ended.[5] In fact, Elon Musk, whose SpaceX company is attempting to create a better, cheaper, American-built rocket engine to launch US military satellites in orbit, has described the Russian engines as "brilliant."[6] Meanwhile, a host of private space companies have purchased Russian space equipment, and these Russian machines have become the nucleus of many businesses involved in space.[7]

Most importantly, the opening up of Russia's space program to Western practices effectively allowed for the once-turgid Soviet mindset of central planning to be replaced by the wildcatting Western capitalist mindset. Projects were undertaken by these Russian firms, in conjunction with Western money, which ultimately redounded to the benefit of Russia's overall space sector. By 2002, when the Russian economy started becoming solvent, Russia was poised to continue its steady expansion in space that was prematurely ended when the Soviet Union fell. What's more, Moscow had essentially made much of the West's space program and budding private space sector companies dependent— in one way or another—on Russian technology. This not only created a highly lucrative market for Russia's economy, but it also created critical strategic leverage for Moscow over the West.

Just as Russian-owned-and-operated gas pipelines into Europe became a cudgel for Moscow to clobber the West with during the Georgian and Ukrainian crises in 2008 and 2014, the fact that so many aspects of America's space industry relied upon Russian technology

---

5 Matthew Bodner, "Energomash Raises Alarm Over U.S. Ban on Russian Rocket Engines," *Space News*, 10 July 2018. https://spacenews.com/energomash-raises-alarm-over-u-s-ban-on-russian-rocket-engines/

6 Matthew Bodner, "Can SpaceX and Blue Origin Best a Decades-Old Russian Rocket Engine Design?" *Technology Review*, 26 June 2019. https://www.technologyreview.com/s/613744/spacex-blue-origin-russian-rd180-rocket-engine-design/

7 Loren Thompson, "U.S. Growing Dependent on Russia for Satellite Propulsion Systems," *Forbes*, 14 September 2018. https://www.forbes.com/sites/lorenthompson/2018/09/14/u-s-satellite-makers-turn-to-foreign-sources-for-in-space-propulsion-despite-buy-american-push/#65a60ad51590

became critical leverage for Moscow to lord over Washington during crisis points between the two powers over the last decade.[8] Yet, the essential shifting of the Russian mindset from the Soviet central planning to the free market on matters of space was key. Basically, as Moltz once argued, Russia became the "Walmart" of rocket engines—cheap, effective, and powerful systems became available to many countries and companies around the world, and few could afford to pass up purchasing this technology. Few today can match or outdo the Russian-built systems, though many are trying.

Moscow has been worked into the global supply chain for space. This has occurred for more than twenty years. This is a smaller scale of how the Chinese economy has become essential for maintaining the overall global supply chain. While it is possible for the United States to diversify its reliance on Russian space technology, the fact remains that the United States government quite literally paid and helped to build out the present Russian space capabilities. Because of this, the Russian threat and competitive edge to the United States is severe. It is for this fact that, rather than ranking China as the second-most powerful space program in the world, most analysts insist that Russia remains the second-most advanced and powerful program.

The retention of much of the Soviet capabilities, coupled with the embrace of Western business practices and the decades-long enmeshing of Russia's space program with the West, has ensured that the Russians remain competitive. Despite the fact that Moscow and Washington became tethered to each other in post-Cold War era, and irrespective of the fact that the Russians benefited so much from a close relationship with the West, Moscow still sought to double-dip. Yes, they resisted the urge to sell off everything to the highest bidder, although, Moscow's leadership insisted on proliferating many things to less-than-savory countries during the post-Cold War era.

---

8 Fiona Hill, "Energy Empire: Oil, Gas and Russia's Revival," *The Foreign Policy Centre,* September 2004. https://www.brookings.edu/wp-content/uploads/2016/06/20040930.pdf

8

RUSSIAN EMP AND SATELLITE

JAMMING THREATS

MANY REPORTS IN THE 1990S identified both North Korea and Iran as key distribution points for advanced Russian military technology.[1] It has been learned that the Russians gave the North Koreans their plans for EMP weapons.[2] Because of this, former CIA director James Woolsey

---

1 John A. Lauder, "Russian Proliferation to Iran's Weapons of Mass Destruction and Missile Programs," *Iran Watch*, 5 October 2000. https://www.iranwatch.org/library/government/united-states/executive-branch/central-intelligence-agency/russian-proliferation-irans-weapons-mass-destruction-and-missile

2 Elizabeth Shim, "North Korea's 'Electronic Bomb,' Technology of Russian Origin, Experts Say," UPI, 14 September 2017. https://www.upi.com/Top_News/World-News/2017/09/14/North-Koreas-electronic-bomb-technology-of-Russian-origin-experts-say/6401505408979/

has speculated the possibility that the North Koreans have not only developed their conventional nuclear weapons capability, but also their EMP capabilities (which will be discussed at greater length in a later chapter).[3] What's more, Russia has already proliferated many different weapons and military technology to the Iranians and North Koreans as well as previously to the Saddam Hussein regime.[4]

For example, the Russians were viscerally opposed to the Anglo-American invasion of Iraq in 2003.[5] Due to this fact, Moscow attempted to proliferate specific technologies to American enemies that could be used to disrupt and weaken the overwhelming American military advantage—particularly in the strategic high ground of space. Russian technology firms, such as Aviaconversiya, Ltd., offloaded as much anti-satellite capabilities as possible to potential targets of the US military during the 1990s.[6] This was particularly true in the case of satellite jamming technology. The Iraqis did not have this capability in 1991. By 2003, when Americans invaded Iraq to topple Saddam Hussein, though, the Iraqis had been sold at least six satellite jammers by Aviaconversiya. Not only did the United States enmesh itself with Russia's post–Cold War space industry, but the US Army apparently awarded sensitive contracts to Aviaconversiya in the years leading up to the Iraq War of 2003. In fact, it is likely that Aviaconversiya used the information they gleaned from working with the US Army on sensitive GPS-related issues

---

3 James Woolsey and Vincent Pry, "How North Korea Could Kill 90 Percent of Americans," *The Hill*, 29 March 2017. https://thehill.com/blogs/pundits-blog/defense/326094-how-north-korea-could-kill-up-to-90-percent-of-americans-at-any

4 Associated Press, "North Korea's New Missile Has Russian Fingerprints 'All over' It," *NBC News*, 10 May 2019, https://www.nbcnews.com/news/world/north-korea-s-new-missile-has-russian-fingerprints-all-over-n1004151

5 Thomas Ambrosio, "The Russo-American Dispute over the Invasion of Iraq: International Status and the Role of Positional Goods." *Europe-Asia Studies* 57, no. 8 (2005): 1189–210. http://www.jstor.org/stable/30043987.

6 Frank Vizard, "Safeguarding GPS," *Scientific American*, 14 April 2003. https://www.scientificamerican.com/article/safeguarding-gps/

to build the jammers they ultimately sold to Saddam Hussein's forces![7]

Desert Storm was dubbed as America's "First Space War," largely because of the strategic advantages that satellites provided the United States and its coalition partners in combat against Saddam Hussein's military. Not only did satellite communications facilitate the lightning attack against Saddam's forces that had occupied Kuwait, but America's nascent Global Positioning System (GPS) significantly helped the Coalition forces in their bid to defeat Iraq's military. Or, as then chairman of the Joint Chiefs of Staff former US Army general Colin Powell described the war plan against Iraq's military, "First, we're going to cut it off, and then we're going to kill it [while ripping up the Iraqi air force in its entirety]."[8]

In 2003, the Iraqis attempted to use six Russian-built satellite jammers to disrupt the encrypted signals of America's Global Positioning System in order to neutralize the ability of the US military to conduct surgical strikes against Iraqi targets. Saddam had obviously learned from the epic thrashing he received from the US military in 1991. While the US military maintained that the Russian-built jammers in Iraq were not a serious threat to American forces, the fact that the US military moved so quickly to destroy these six jammers at the very beginning of the conflict indicates that they were likely more of concern than the Pentagon was willing to admit. By targeting America's vaunted GPS constellation, Saddam's forces were hoping to throw American missiles, warplanes, and any other units operating in Iraq's vast desert terrain severely off course, thereby reducing their threat to Saddam's military.

Thanks to redundancies built into America's cruise missiles, such as the Inertial Navigating System (INS) which is a less advanced, though

---

7 Bob Brewin, "U.S. Army Awarded Contracts to Russian GPS Jammer Vendor," *Computer World,* 27 March 2003. https://www.computerworld.com/article/2581651/u-s—army-awarded-contracts-to-russian-gps-jammer-vendor.html

8 Dan Balz and Rick Atkinson, "Powell Vows to Isolate Iraqi Army and 'Kill It,'" *Washington Post,* 24 January 1991. https://www.washingtonpost.com/wp-srv/inatl/longterm/fogofwar/archive/post012391_3.htm

still helpful backup navigation system, as well as the aforementioned Pentagon decision to physically destroy Iraq's small cluster of jammers, the US military never had to deal with being deprived of its GPS functionality in the Iraq War. What few acknowledge, though, is that the Pentagon had anticipated some degree of interference from jammers, which is why the Global Hawk—or, Predator—drones were first deployed in the early 2000s, the same drones that today the military uses for targeted assassinations of suspected terrorists as well as for surveillance missions. The Pentagon envisaged using drones as signal amplifiers in a modern contested battle space, wherein satellite jammers disrupted the normal functioning of America's advanced weapons. Drones would orbit high above a battlefield and would be used to intercept the encrypted military signals coming from the GPS constellation in space and then amplify those signals, sending them down to US forces operating in the other terrestrial strategic domains in order to overcome any disruption.

In 2003, though, the jamming technology was still in its infancy.

Today, this is another matter. Russia has spent nearly twenty years perfecting the technology after Moscow recognized how reliant on satellites the US military has become. That the United States had propped up the Russian technology and space sector during the tumultuous 1990s did not dissuade Russian leaders from continuing to seek new and innovative ways to weaken America's military, conceivably forcing Washington to accommodate Moscow's intentions rather than risk a war in which American technology was useless. Moscow's critique of post–Cold War US foreign policy has been that Washington consistently relies on its hard military power to enforce what Moscow and Beijing and several other countries—as well as France and Germany—view as an unacceptable unipolar world order in which the United States is the global hegemon.[9] America's commitment to upholding its global hegemony makes Russia,

---

9 Elihugh M. Abner, "Putin's Multipolar World and What It Means for U.S. Strategy," Association of the United States Army, 10 July 2017. https://www.ausa.org/publications/putins-multipolar-world

already suffering from an inferiority complex after having lost the Cold War, feel more humiliated than it had already.

Another area that jammers could disrupt the operation of the US military is in the arena of signals intelligence. In the months leading into Desert Storm, according to the Pulitzer Prize–winning journalist Fred Kaplan, the US military had discovered that "Saddam had run fiber-optic cable all of the way from Baghdad down to Basra and, after [Saddam's invasion of Kuwait], into Kuwait City. American intel officers contacted the Western firms that had installed the cable and learned from them the locations of the switching systems." The American forces then bombed those positions. Once those switches were hit, according to Kaplan, Saddam was forced to switch over to a backup network "built on microwave signals." Kaplan then asserts that the National Security Agency (NSA) was prepared for this move and had "positioned a new top-secret satellite directly over Iraq, one of three spy-in-the-sky systems . . . This one sported a receiver that scooped up microwave signals."[10] The proliferation and presence of several mostly Russian-built satellite jammers and anti-satellite weapons in countries such as Iran and North Korea would mean that an operation similar to the one the NSA engaged in during the start of Desert Storm would be an onerous undertaking. It is likely the NSA operation in question saved countless lives by gathering the valuable signals intelligence on Iraqi war plans.

In fact, while Moscow has maintained that they will not take any destabilizing action in the world order, few in the Kremlin acknowledge—or even believe—that supplying advanced weapons, such as jammers or EMP technology, to rogue states is disruptive.[11] What's more, the Russians insist that they will take no aggressive action in space unless prompted by what they argue are aggressive American

---

10 Fred Kaplan, *Dark Territory: The Secret History of Cyberwar* (New York: Simon & Schuster, 2016), 22.

11 Stephen Blank, "Russia's Proliferation Pathways," *Strategic Insights,* December 2008. https://pdfs. semanticscholar.org/61f1/8a380a86513fd2de36c52f8e8b1635b0a38c.pdf

activities in space. The reality is that Moscow's proliferation of advanced technology—especially satellite jamming technology—pose a clear and present danger to the United States in the strategic domain of space.

Nor was Saddam Russia's only customer for satellite jammers. Further, the United States has not been the exclusive victim of satellite jamming attacks. In 2011, for example, reports circulated that Iran was using Russian-built jammers to prevent the signals from a BBC satellite from reaching Iranian viewers.[12] The reason that the Iranian regime did this was to prevent what they view as anti-regime propaganda from being received by their people. In 2009, the cities of Iran erupted in protests over the blatantly rigged Iranian presidential election. While the Iranian government managed to survive the mass demonstrations, Tehran was always uneasy about the dissemination of Western culture—especially media broadcasts—to their people. Of course, Western governments routinely use their awesome media power to broadcast anti-regime programming to the people of oppressed countries, in the hopes that such media broadcasts would encourage resistance and ultimately change to the regime. Yet, the fact remains that Moscow sold Iran satellite jamming technology—which Iran dutifully used to disrupt the natural operations of a foreign satellite passing by overhead.

The Iranian attack with Russian jammers on a British news satellite could have just as easily been employed against American military satellites during a conflict. In fact, there is some evidence that Russia has deployed ground-based satellite jammers in Syria—and has used these jammers against American forces that were fighting in Syria. These jammers have negatively impacted the ability for US forces fighting on the ground to defend themselves because they cut off US forces from communications, radar, and other tools that give American warfighters

---

12 Paul Sonne and Farnaz Fassihi, "In Skies over Iran, a Battle for Control of Satellite TV," *Wall Street Journal,* 27 December 2011. https://www.wsj.com/articles/SB10001424052970203501304577088380199787036

unprecedented abilities to defend themselves in combat.[13] The one upside of Russia's intervention in Syria has been that US military and intelligence officials have been able to see the latest jamming technology that Russia has proliferated to their allies around the world. Yet, many understandably worry that the continued escalation of hostilities between the United States and various Russian partners will lead to a major international crisis between Washington and Moscow. What's more, as the US military develops countermeasures against Russian jamming, Moscow is also able to counteract those countermeasures, and on and on the dance goes. (Dueling overrides are such headaches, aren't they?)

Already, US forces were compelled to kill upwards of two hundred Russian mercenaries belonging to the Russian government-affiliated Wagner Group in Syria in 2018.[14] Here's how unconventional warfare expert Sean McFate describes the pitched battle between Russian mercenaries and US commandos in 2018:

> The opening salvo of artillery was so intense that the American commandos dived into foxholes for protection. After the barrage, a column of tanks advanced on their positions, shooting their 125-millimeter turret guns. The commandos fired back, but it was not enough to stop the tanks.
>
> A team of about thirty Delta Force soldiers and rangers from the Joint Special Operations Command—America's most elite task force—were pinned down at a Conoco gas plant in eastern Syria. Back at headquarters, roughly twenty miles away, a team of Green Berets and a platoon of Marines stared at their computer screens, watching the drone feeds of the battle. They were on a secret mission to defend

---

13 Lara Seligman, "Russian Jamming Poses a Growing Threat to U.S. Troops in Syria," *Foreign Policy*, 30 July 2018. https://foreignpolicy.com/2018/07/30/russian-jamming-poses-a-growing-threat-to-u-s-troops-in-syria/

14 Stepan Kravchenko, Henry Meyer, and Margaret Talev, "U.S. Strikes Killed Scores of Russian Fighters in Syria, Sources Say," *Bloomberg*, 13 February 2018. https://www.bloomberg.com/news/articles/2018–02–13/u-s-strikes-said-to-kill-scores-of-russian-fighters-in-syria

the Conoco facility, alongside Kurdish and Arab forces. No one expected an enemy armored assault.

Attacking them were five hundred mercenaries, hired by Russia, who possessed artillery, armored personnel carriers, and T-72 tanks. These were not the cartoonish rabble depicted by Hollywood and Western pundits. This was the Wagner Group, a private military company based in Russia, whose employees, as with those at many high-end mercenaries, were organized and lethal.

The American commandos radioed for help. Warplanes arrived in waves, including Reaper drones. F-22 stealth fighter jets, F-15E strike fighters, B-52 bombers, AC-130 gunships, and AH-64 Apache helicopters. Scores of strikes pummeled the mercenaries, but they did not waver.

Four hours later, the mercenaries finally retreated. Four hours. No Americans were killed, and the US military touted this as a big win. But it wasn't. It took America's most elite troops and advanced aircraft four hours to defeat five hundred mercenaries. What happens when they have to face one thousand? Five thousand? More?[15]

Had it not been for copious numbers of US surgical strikes from the air and from distant artillery, the American forces there would have been overrun and killed by the Russian mercenaries. Think back to the last two major conflicts the United States has fought in: Afghanistan and Iraq in 2003. Smaller US military units were able to defeat a larger foe, thanks to their heavy reliance on airpower and over-the-horizon artillery. These capabilities become irrelevant if Russian satellite jammers are able to debilitate the effective operations of the critical satellites that link US forces on the ground, at sea, or in the air, with their various commands. Imagine a similar instance to what transpired in Syria between Russian and American forces happening in the not-too-distant

---

15 Sean McFate, *The New Rules of War: Victory in the Age of Durable Disorder* (New York: William Morrow, 2019), 132–33.

future. Yet, this time, the Russians effectively employ satellite jammers. Now, that smaller force of Americans will basically be left to fend for themselves against a much larger and more ferocious Russian or Russian-backed adversary. The same tactics Russia employed against American ground forces in eastern Syria will likely be used against NATO forces in Europe—only on a much larger scale.

Space, viewed from this perspective, becomes a decisive asset in modern combat. The country like Russia that manages to deprive the United States of space will have a disproportionate advantage in combat. Given this, I believe that the United States will suffer its first major military defeat in generations because of a weakness exploited by an American rival in space. It is not only in Syria or the Middle East where US forces are increasingly fighting against Russian or Russian-supported elements. According to US Army General Stephen Townsend of US Africa Command (AFRICOM), the Wagner Group is acting as a "proxy force" for the Russian Federation all across Africa—particularly in places like the Central African Republic, Niger, and Nigeria.[16] The tactics and strategies devised to combat US forces in places like Syria will likely be employed and augmented by Russian forces in Africa, as the new race for greater influence on that continent increases between the major world powers—namely, the United States, China, and Russia). One can expect to see more, not fewer, attempts by Moscow to proliferate and use anti-satellite technology against the United States.

---

16 Kyle Rempfer, "Why This US General Says Russian Wagner Mercenaries in Africa 'Concern Me Greatly,'" *Military Times,* 4 April 2019. https://www.militarytimes.com/news/your-military/2019/04/04/why-this-us-general-says-russian-wagner-mercenaries-in-africa-concern-me-greatly/

9

RUSSIA BUILDS ITS OWN GPS

RUSSIA TODAY IS VERY DIFFERENT from that chaotic place that it was in the 1990s. The rise of Vladimir Putin and his supporters (known in Russia as the *siloviki*) to power in 1999 completely reoriented Russian politics away from the slow-but-promising direction of democracy and back to the traditional Russian form of government: autocracy.[1] In many respects, the United States propped up and helped to create the current Russian military threat in space through its unquestioning level

---

1 Brandon J. Weichert, "Russia Is Not Going to Change," *New English Review,* September 2018. https://www.newenglishreview.org/custpage.cfm?frm=189386&sec_id=189386

of support for the Russian space program in the uncertain 1990s, as well as its merging of key Russian technology conglomerates not only with the wider world, but specifically with the United States military.[2] More dangerously, US foreign policy has exacerbated existing tensions in the Russo-American relationship since the end of the Cold War, which has negatively redounded to the ongoing military competition between the United States and the Russian Federation.

Since the end of the Cold War, Moscow has long opposed the expansion of NATO and the European Union into countries that the Kremlin has historically viewed as having been part of the Russian sphere of influence. Despite the era of good feelings that generally existed between the United States and the Russian Federation, the two sides frequently disagreed on matters of foreign policy. Over time, the disagreements crowded out the good feelings until the relationship soured. This book is not about US-Russian relations. Yet, it bears noting that Russian military advancements—particularly in their space program and related policy areas—became greater and more direct threats to the United States the longer that the political situation between the two powers deteriorated to their present low.

US-Russian relations have slid from a Cold War to a "cold peace."[3] The two sides not only disagreed over NATO and EU expansion, but also over US military intervention in the Balkans and, ultimately, the US invasion of Iraq in 2003. By 2007, when Russia's economy had enjoyed a resurgence, thanks to the historically high price of fossil fuels on the energy market (Russia is a major producer of fossil fuels), Moscow had the ability to not only begin modernizing its ailing force but to begin implementing its own expansionist foreign policy in parts of the world

---

2 Eric Berger, "Russia Embassy Trolls US Launch Industry After New Rocket Engine Sale," *Ars Technica,* 1 August 2018. https://arstechnica.com/science/2018/08/russian-embassy-trolls-us-launch-industry-after-new-rocket-engine-sale/

3 Andrew Marshall, "Russia Warns NATO of a 'Cold Peace,'" *Independent,* 6 December 2014. https://www.independent.co.uk/news/russia-warns-nato-of-a-cold-peace-1386966.html

that it viewed as belonging in its sphere of interest, like Georgia in 2008 and Ukraine in 2014. It was in 2007 that an angry Russian President Vladimir Putin sojourned to Munich, Germany, and gave what is probably the most infamous speech of his political career, wherein he basically outlined how Russia would wage war on the West in order to secure what he believed was Russia's natural defense perimeter by reclaiming influence over most former Soviet states.[4]

Russia further undermined American national interests in the Middle East by militarily intervening to protect Bashar al-Assad of Syria, their much-maligned client, from what was certain to be yet another bloody overthrow of a Mideast dictator. During this fight, as you read earlier, Russia moved several of its space weapons—such as satellite jammers—into the fight. Russia also began expanding its own surgical strike capabilities, which Russia had never been able to fully develop in the confusing aftermath of the Cold War.

As Reuters analyzed in 2015, as the Russian intervention in Syria was escalating:

> In [Russia's] air campaign over Georgia [in 2008] Russia used, primarily Su-24 and Su-25 attack aircraft. Those types have been in operation since the Soviet period. They fired only unguided munitions. This was mainly because the global positioning system, which can be used to guide munitions to their target, was being jammed, while the rival Russian version, called GLONASS, was not then ready for use. For Syria, GLONASS is up and running, and the air force has been using Su-34 attack aircraft, in operation since 2014, equipped to launch KAB guided bombs. Russia has also deployed Kalibr ship-launched cruise missiles fired from the Caspian Sea.[5]

---

4 Thom Shanker and Mark Landler, "Putin Says U.S. Is Undermining Global Stability," *New York Times,* 11 February 2007. https://www.nytimes.com/2007/02/11/world/europe/11munich. html?mtrref=www.google.com&gwh=2A9EBFA12D122970579919D1E784175E&gwt=pay&assetTy pe=REGIWALL

5 Margarita Antidze and Jack Stubbs, "Before Syria, Russia Struggled to Land Air Strikes on Target," Reuters, 26 October 2015, https://www.reuters.com/article/us-mideast-crisis-syria-russia-bombing/before-syria-russia-struggled-to-land-air-strikes-on-target-idUSKCN0SK1WF20151026

Of course, the Russians likely use their guided munitions packages sparingly because such weapons are costlier than the unguided weapons that Russia's air force has used over the years.[6] Still, Syria was a place for Moscow to test its burgeoning alternatives to what was once the uncontested American strategic domain of space. Russia's GLONASS satellite constellation, while nowhere near as advanced or reliable as the American GPS, is simply "good enough" for Russian forces to use in conflict.[7] In the 2008 Russian invasion of Georgia, the Americans were able to jam GPS signals going to the invading Russian forces. This is one reason why the Russian invasion of Georgia was such a brutal affair: even if Russia had a predilection for using guided munitions to cut down on civilian casualties in combat, they could not have used those weapons. In 2015, though, the presence of a good enough Russian alternative to the American-controlled GPS constellation gave Russia some significant strategic advantages.

Like much of Russia's present space capabilities, the GLONASS system has its origins in the Cold War, specifically in the 1970s. When the Cold War ended, funding and political support for the GLONASS system, as for much of Russia's space program, dried up. But everything changed for the fledgling and long-abandoned GLONASS program when Vladimir Putin became leader of Russia. Recognizing the strategic implications of Russia possessing a global positioning satellite network of its own, divorced from the hegemony that the United States enjoyed over the technology, meant that Moscow could move that much closer toward achieving its strategic goals of creating a truly multipolar world, wherein Washington was but one of several power centers along with Moscow. Russia's leaders fully understood

---

6 James Rosen, "Why Is Russia Using 'Dumb Bombs' in Syria?" *Miami Herald*, 6 October 2015. https://www.miamiherald.com/news/nation-world/world/article37968804.html

7 Joseph Trevithick, "Russia Says It Used Precision Guided Munitions in Strikes on Syrian Rebel Drone Makers," *The Drive*, 5 September 2018. https://www.thedrive.com/the-war-zone/23376/russia-says-it-used-precision-guided-munitions-in-strikes-on-syrian-rebel-drone-makers

the strategic vulnerability they faced of having to rely on the United States' GPS system—as became evident during the 2008 Georgia War when, as noted previously, the Russians could not rely on GPS to assist their forces in fighting the US-backed Georgia.

In 2015, the GLONASS system was completed according to the designs of the Russian military. Whereas many Americans often praise their country's history of technical supremacy, GLONASS can compete with GPS, particularly in the Northern Hemisphere. In the GLONASS-GPS debate, the latter will remain dominant for years to come in the commercial navigation sector. But in the military domain, GLONASS will offer Russia a "good enough" alternative to the US-controlled GPS constellation. In fact, according to some reports, it can even offer greater accuracy than its GPS rival at higher altitudes.[8] Not only does the presence of GLONASS offer complications to Russia's longtime American foe, but it also provides Russia yet another opportunity to compete economically against the Americans by offering the world an alternate system that is nearly comparable to the American one. And given the fact that several states—some of them being US allies—dislike having to rely on the United States for such things as GPS, Russia's system just might prove to be the boon that Putin hopes it will be. The present GLONASS constellation of twenty-four satellites is the nucleus of what could be a larger network, particularly if Europe's version of GPS, Galileo, or the Chinese variant, Beidou, were meshed into the GLONASS system, as has been discussed over the last few years by interested parties.

As the historian and foreign policy analyst Mark Episkopos wrote about GLONASS, "More than just a technical alternative, [GLONASS] is for Moscow a strategic alternative to and a counterweight against GPS. In this context, it becomes apparent that the establishment of GLONASS

8 Anurag Bisht, "What Is GLONASS and How Is It Different from GPS?" *Beebom,* 25 August 2016. https://beebom.com/what-is-glonass-and-how-it-is-different-from-gps/

stations is a key part of Russia's influence-building project."[9] Episkopos was referencing how Moscow was using GLONASS to support its controversial operations in the Arctic.[10] Yet, GLONASS is part of a wider Russian design to expand its influence across a variety of areas, notably in space. What's more, the presence of GLONASS increases Moscow's attempts to jam US satellites in conflict zones, such as Syria or Europe. After all, having a reliable alternative available to the American system means that Moscow can safely jam the GPS network without having to risk entirely depriving Russian forces of GPS as well.

In both Syria and Europe, having the GLONASS system available has led to Russian forces actively jamming US forces and their allies. Notably, during a 2019 NATO war game led by Norway, the Russians are believed to have actively jammed Norwegian forces. The Norwegian government has assessed that their critical communications and navigation systems and networks have endured "electronic harassment" from nearby Russian forces operating out of the Kola Peninsula, which is home to Russia's Northern Fleet, as well as their elite Arctic military forces from 2017 through 2019. Not only has Norway encountered electronic jamming of their satellite signals, but so too has nonaligned Finland. Both Norway and Finland are located in the northern frontier of Europe and have a long history of weathering Russian aggression. In fact, since the Cold War, Finland has endured what's become known in international relations as "Finlandization," wherein its independence was maintained but its foreign policy was essentially managed by Moscow.

Today, Finland is independent, but its proximity to Russia and its historical quasi-subjugation by them has prompted Finland's leadership to maintain a neutral policy regarding NATO and Russia. Yet, recently, Finland's government has expressed a willingness to deepen its ties with

---

9 Mark Episkopos, "Russia Has 1 'Weapon' NATO Can't Easily Defeat," *The National Interest,* 2 May 2019. https://nationalinterest.org/blog/buzz/russia-has-1-weapon-nato-cant-easily-defeat-55387

10 Brandon J. Weichert, "America Should Not Abandon the Arctic," *American Greatness,* 17 November 2019. https://amgreatness.com/2019/11/17/america-should-not-abandon-the-arctic/

NATO. These moves have invited the Russian electronic harassment of Finland's forces, just as Norway has been made to endure. In one instance, Russia deployed a squadron of Su-24 warplanes and performed a "mock strike" on a Norway intelligence-controlled radar system in the Arctic.[11] As you've seen, the presence of Russia's GLONASS network in orbit means that Russian forces seeking to conduct a lightning strike on nearby NATO forces, such as those operating in Europe's High North, would allow for Russian forces to block NATO GPS signals without much risk to their own wartime navigation.

Of course, GLONASS has some weaknesses. Its coverage area is nowhere near as wide as that of the GPS, which has existed in some form for nearly thirty years. The world relies on GPS. Under the present configuration, America's GPS network sends out two kinds of signals: one is a widely used unencrypted GPS signal for civilian use. The other is an encrypted signal for military use. The fact that these GPS satellites transmit both signals, though, as the Norwegian government has argued in their official protests over Russian jamming, means that civilian systems could be negatively impacted by Russian jamming as much as military systems are. So, everything from your personal car's GPS-navigation system to large airliners that rely heavily on satellite navigation could be thrown off-course. Shipments of natural gas or oil coming across the Atlantic Ocean from the United States meant to ameliorate Europe's increasing dependence on Russian-produced fossil fuels could also be thrown dangerously off course by a Russian satellite jamming attack. All of these factors could potentially weaken America's influence over Europe. Russian disruption of energy flows from North America to Europe could also raise energy costs on consumers globally, which would only benefit a major fossil fuel producer like Russia.

---

11 Gerard O'Dwyer, "Norway Accuses Russia of Jamming Its Military Systems," *Defense News*, 8 March 2019. https://www.defensenews.com/global/europe/2019/03/08/norway-alleges-signals-jamming-of-its-military-systems-by-russia/

Russian actions in space could impact all of the other, "lower," terrestrial domains. The Russians, like the Chinese, understand that space is a zone of competition. Like Beijing, Moscow has assessed that the United States' overwhelming military might relies heavily on its space systems. America's bipartisan leadership has consistently refused to enhance the defenses of their critical satellite constellations, providing Russia and China with a tempting target.

The fact that the United States has remained utterly intransigent on issues related to NATO and EU expansion has led most Russian leaders to assume that the United States is waging a silent war on Russia meant to encircle, strangle, and ultimately dismember the Russian Federation.[12] When the Soviet Union collapsed, Russia's border with Europe—West Germany—shrank. As this occurred, a succession of Russian leaders insisted to their American counterparts that Washington slow the pace of NATO and EU expansion, fearing for Russia's territorial integrity. Most American leaders scoffed at the fears of Russia's leadership.

Of course, no one in either the United States or Europe desired to invade post-Soviet Russia. Most Russian leaders, however, were raised with the knowledge of how Russia has always been under threat of invasion, notably by the Mongols, then by Napoleon, and then by Hitler. Russian leaders did not take the self-serving assurances of either American or European leaders at face value. For Russian leaders, such as the friendlier Boris Yeltsin or the harsher Vladimir Putin, Russia's post–Cold War weakness was too inviting for the stronger and hegemonic West to simply ignore.[13]

---

12 Brandon J. Weichert, "The True Ambitions of Russian Foreign Policy Today," The Institute of World Politics, 8 August 2017. https://www.iwp.edu/events/the-true-ambitions-of-russian-foreign-policy-today/

13 Benn Steil, "Russia's Clash with the West Is About Geography, Not Ideology," *Foreign Policy*, 12 February 2018. https://foreignpolicy.com/2018/02/12/russias-clash-with-the-west-is-about-geography-not-ideology/

In the wake of Russia's invasion of Ukraine and the annexation of Crimea, a land with a Russian-speaking majority population, the United States poured on onerous economic sanctions. Then, Western intelligence agencies accused Russia of a wide range of covert attacks, ranging from economic espionage to cyberattacks to trolling the Americans during their presidential election in 2016. As this occurred, Moscow found itself increasingly isolated from the West, both by its own actions and by the draconian diplomatic and economic moves of Western leaders—not only those of former president Barack Obama but also of President Donald Trump. Rather than accede to Western sanctions and shaming tactics, the Russians behaved much as the old Japanese empire did under the pressure of similar US sanctions in the 1930s. In the run-up to the Second World War, America had enacted onerous sanctions and an oil embargo upon the resource-starved but expansionistic Japanese empire for their invasion of China. Rather than give in to American sanctions, Tokyo chose greater levels of resistance that eventuated in the dastardly surprise attack on Pearl Harbor on December 7, 1941, pushing the United States into the Second World War. It is foolish for Washington to believe that the sanctions are doing anything other than hardening Russia's resolve to fundamentally upend the post–Cold War, American-led international system.

10

THE BEAR HUGS THE DRAGON

THINGS ARE SO BAD NOW between Moscow and Washington that Russia has run into the waiting arms of neighboring China, a country with whom Russia shares a long and complex history. American actions against Russia have precipitated the birth of a strange Sino-Russian entente that is hardly natural and most unwanted. After all, these two behemoths would form the very alliance on Eurasia that most Western strategists have spent a century trying to prevent. Out of this newfound alliance has come greater levels of cooperation on everything ranging from direct military interaction to shared nuclear energy development projects to joint Russian and Chinese ventures in space.[1] The aforementioned GLONASS system has been joined by China's own budding GPS constellation, known as Beidou.[2] Already, Moscow and Beijing have announced their intentions to band together their GPS networks and form a counterweight to America's dominance in GPS—thereby bringing to space the new relationship that they are building on Earth with all of the horrendous complications for US national security and grand strategy.[3]

In fact, as was noted in the previous chapters on China, the one stumbling block to China's manned lunar program and their space station program had been the failure of the *Long March V* heavy-lift rocket.[4] For two years, this indigenously produced rocket has had a series of technical failures. Due to this, many of China's ambitious space plans were on hold until the technology could be perfected. In August 2019, the head of the Russian space agency, Roscosmos, offered to sell China their heavy-lift rocket. As you've seen throughout this work, Russian rocket engines are cheap, efficient, and reliable.[5] Thanks to American largesse during the 1990s, Russia's space capabilities not only survived the fall of the Soviet Union, but those capabilities were also perfected. During the reign of

1 "Rosatom to Build Four Nuclear Power Plants in China," *Power Engineering International*, 11 June 2018. https://www.powerengineeringint.com/2018/06/11/rosatom-to-build-four-nuclear-power-plants-in-china/

2 Pratik Jakhar, "How China's GPS 'Rival' Beidou Is Plotting to Go Global," BBC, 20 September 2018. https://www.bbc.com/news/technology-45471959

3 Dana A. Goward, "Russia, China Alliance on Navigation Satellites Threatens GPS," *National Defense*, 23 August 2019. https://www.nationaldefensemagazine.org/articles/2019/8/23/viewpoint-russia-china-alliance-on-navigation-satellites-threatens-gps

4 Dave Mosher, "Video Shows a Top-Secret Chinese Space Mission Failing in Mid-Flight—China's Second Rocket Loss of the Year," *Business Insider*, 23 May 2019. https://www.businessinsider.com/china-long-march-rocket-launch-failure-weibo-video-pictures-2019-5

5 Liu Zhen, "Russia Offers Rocket Engine Tech as China's Long March 5 Struggles to Get off the Ground," *South China Morning Post*, 28 August 2019. https://www.scmp.com/news/china/military/article/3024766/russia-offers-rocket-engine-tech-chinas-long-march-5-struggles

Nikita Khrushchev, the Soviets vowed to produce rockets "like sausages."[6] Russia's tight relationship with the West during the 1990s and early 2000s allowed Russia to accomplish this task, becoming the equivalent of a "Walmart" for space technology.[7] Ultimately, in December 2019, the Chinese successfully launched their *Long March V* rocket, thereby paving the way for Beijing to place their modular space station, the *Tianhe-1,* into orbit beginning in 2020, to send a sample return mission to the Moon, to ultimately place a lunar colony on the Moon's south pole by 2024, and to send missions to Mars as well.

Yet, the budding Sino-Russian relationship in space does not end with China's successful launch of an indigenously produced heavy-lift rocket. Relations with the West have soured so badly that both the United States and Europe are looking at alternatives to Russia's famous rocket engines. To counteract this trend, Moscow continues looking east. In 2018, for example, tired of America's refusal to cooperate on matters of space policy, Russia and China entered into a space exploration deal.[8] Of course, the deal is amorphous, covering a wide range of technological developments under the imprimatur of space. Yes, these two great powers can pool their resources together for genuine missions of space exploration. But so too can they combine their resources together for military space operations under the guise of civilian "space exploration." Space programs are inherently dual use in nature: what can be used in peacetime for exploration and scientific discovery can also be used in wartime to wreak havoc on the space systems of a rival.[9]

---

6 Richard Ned Lebrow, "Was Khrushchev Bluffing in Cuba?" *Bulletin of the Atomic Scientists* 44, no. 3 (April 1988): 41–42. https://www.tandfonline.com/doi/abs/10.1080/00963402.1988.11456136

7 Brandon J. Weichert, "The Case for an American Space Corps," *American Greatness,* 3 December 2017. https://amgreatness.com/2017/12/03/the-case-for-an-american-space-corps/

8 Anatoly Zak, "Russia's Space Agency Might Break Up with the U.S. to Get with China," *Popular Mechanics,* 7 March 2018. https://www.popularmechanics.com/space/moon-mars/a19159719/roscosmos-china-collaboration/

9 "Russia, China Sign Space Exploration Deal," *Moscow Times,* 8 June 2018. https://www.themoscowtimes.com/2018/06/08/Russia-China-Sign-Space-Exploration-Deal-a61736

And China is only too happy to welcome Russia into its geostrategic orbit.[10] It is China that will benefit most from their newfound relationship with Russia—a relationship only made possible by decades of American strategic ignorance.[11] The agreement between Roscosmos and China's National Space Agency (CNSA) builds upon agreements made between the two space powers going back to 2014. These agreements promise cooperation on a long list of essential strategic issues related to space, ranging from "cooperation in space and in the market for space navigation" as well as manned lunar exploration.

The fact is, the Russian economy is in free-fall since the global price of oil has plummeted following highs reached in 2014.[12] On top of that, Russia's GDP is the equivalent of Italy's.[13] As the West refuses to deal with Russia, Moscow is forced to pivot to their east in all aspects. Given that China, despite its own economic woes, is still viewed by many as the next great economic juggernaut, Moscow is now seriously considering some form of integration into China's growing economic and geopolitical sphere. Space is but one exemplar of the growing Sino-Russian entente cordiale. And once Russia turns fully toward China, it will be very difficult for Moscow to stop it and return to their preferred level of cooperation with the West. By the time Moscow realizes that Beijing intends to absorb their ailing Russian neighbor into China's new world order, Russia will have already become far too enmeshed in that order. Like the Borg, once China assimilates a nation, removing oneself from

---

10 Artyom Lukin, "Putin's Silk Road Gamble," *Washington Post,* 8 February 2018. https://www.washingtonpost.com/news/theworldpost/wp/2018/02/08/putin-china/

11 Brandon J. Weichert, "How to Create Your Own Enemies," *American Spectator,* 15 June 2019. https://spectator.org/how-to-create-your-own-enemies/

12 Ben Aris and Ivan Tkachev, "Long Read: 20 Years of Russia's Economy Under Putin in Numbers," *Moscow Times,* 19 August 2019. https://www.themoscowtimes.com/2019/08/19/long-read-russias-economy-under-putin-in-numbers-a66924

13 Lauren Carroll, "Graham: Russia 'Has an Economy the Size of Italy,'" *Politifact,* 27 July 2014. https://www.politifact.com/truth-o-meter/statements/2014/jul/27/lindsey-graham/graham-russia-has-economy-size-italy/

Beijing's collective is nearly impossible—and very costly.

Due to Russia's shrinking economy, Russia needs rich partners to conduct complex space operations and to enhance their high-tech sector, which Putin very badly wants to do.[14] Politics has made for strange bedfellows in the case of Moscow and Beijing. Russian leaders are aware that China yearns to push Russia out of the Far East.[15] Yet, these Chinese ambitions are longer term and unworkable at present—especially as the United States presents these two powers with a common foe. As difficult as it is to grasp, the Russians would prefer to cooperate with the United States, especially in space. The present political conditions prevent this from being possible. So, rather than abandon their space dreams, Moscow is willing to share their relatively advanced capabilities with the wealthier, equally ambitious—and anti-American—Chinese. Russia has the technology that China needs to conduct advanced space operations, and China has the wealth to fund it. To many in both Moscow and Beijing, this is a logical pairing.[16] Yet, both Russia and China understandably continue regarding each other with suspicion.[17] And Russia in particular continues hoping for a thawing of relations with the West, since they realize that further enmeshing themselves in the Chinese regional order will merely make their country subordinate to Beijing's increasingly expansionist whims.[18]

---

14 Ashlee Vance, "This City in Siberia Could Become the Next Silicon Valley," Bloomberg, YouTube, 1 December 2016. https://www.youtube.com/watch?v=66MJEzkU8HE

15 Ivan Tselichtchev, "Chinese in the Russian Far East: A Geopolitical Time Bomb?" *South China Morning Post.* https://www.scmp.com/week-asia/geopolitics/article/2100228/chinese-russian-far-east-geopolitical-time-bomb

16 Andrew Jones, "China, Russia to Cooperate on Lunar Orbiter, Landing Missions," *Space News,* 19 September 2019. https://spacenews.com/china-russia-to-cooperate-on-lunar-orbiter-landing-missions/

17 Nyshka Chandran, "'Serious' Rivalry Still Drives China-Russia Relations Despite Improving Ties," *CNBC,* 14 September 2014. https://www.cnbc.com/2018/09/14/china-russia-ties—more-rivalry-than-allaince.html

18 Holly Ellyatt, "Are Russia and China the Best of Friends Now? It's Complicated, Analysts Say," *CNBC,* 27 September 2019, https://www.cnbc.com/2019/09/27/russia-and-chinas-relationship—how-deep-does-it-go.html

The Russian and Chinese cooperation in space is still in its infancy. It can be curbed. Dmitry Rogozin, the head of Roscosmos, has made clear his preference to remain a partner of the United States in space. Although, Rogozin and the Putin regime have also made plain their intentions to be in a position of leadership—and a true partner—in any future NASA mission, such as the proposed lunar gateway.[19] Failure on the part of NASA to reward Russia with a significant partnership role in any new mission would likely mean that Moscow would deepen their ties with China's space and high-tech sector.[20] That is why Washington should take a much more conciliatory tack with Moscow on certain issues, especially regarding space. Far from pushing Russia away, the United States should embrace Russia, thereby negating any need for Moscow to flock to China's orbit in space.[21] The great tragedy in US-Russian relations is that the current tensions need not exist: both countries have many great opportunities that can only be exploited together—particularly in space. Presently, though, relations appear doomed, and the prospect of a US-Russian war is high.

The longer the US-Russian relationship remains at its current low point, and the more that the Sino-Russian alliance strengthens, the more danger both Russia and China will pose to the United States. For example, Beijing and Moscow have been coordinating on a, frankly, bizarre and devilish little project to heat the Earth's atmosphere. The ionosphere is the "layer of the Earth's atmosphere that is ionized by solar and cosmic radiation. It lies 75–1,000 km (46–621 miles) above the Earth . . . The ionosphere has major importance to us because, among other functions, it influences radio propagation to distant places on

---

19 Jeff Foust, "Russian Official Sounds Skeptical Note About Gateway," *Space News,* 2 October 2018. https://spacenews.com/russian-official-sounds-skeptical-note-about-gateway/

20 Doug Messier, "China Ready to Cooperate on Orbital Space Station," *Parabolic Arc,* 15 January 2019. http://www.parabolicarc.com/2019/01/15/china-ready-cooperate-russia-orbital-space-station/

21 Brandon J. Weichert, "Russia and America Must Build a Lunar Space Station," *Space News,* 4 December 2017. https://spacenews.com/op-ed-russia-and-america-must-build-a-lunar-space-station/

the Earth, and between satellites and Earth. For the very low frequency (VLF) waves that the space weather monitors track, the ionosphere and the ground produce a 'waveguide' through which radio signals can bounce and make their way around the curved Earth."[22]

The Russian-Chinese experiment to heat the Earth's ionosphere was conducted a total of five times in 2018 over Europe. The first experiment was conducted in June 2018, and it caused physical disturbances to a region of the ionosphere that was as large as Great Britain. This joint Russian and Chinese program disturbed an area over the small Russian town of Vasilsursk in eastern Europe. According to reports, the town "experienced an electric spike with 10 times more negatively charged subatomic particles than surrounding regions." During another Russian and Chinese test on the ionosphere, the temperature "in high altitude increased more than [212 degrees Fahrenheit] because of the particle flux."[23]

Much like the rest of Russia's current whizbang technology, the basis for the project can be found in the Soviet Union of old: the atmosphere above Vasilsursk was heated by a facility known as Sura, which was built by the Soviet Union's military in the Cold War. When the Sura facility was first constructed in 1981, according to Stephen Chen of the *South China Morning Post*, Soviet scientists were able to "manipulate the sky as an instrument for military operations, such as submarine communication." Essentially, much like satellite jamming or space stalkers, the goal of the joint Russian and Chinese project in the ionosphere is to disrupt satellite communications between a rival force such as the US military or NATO.

---

22 "Tracking Solar Flares," Solar Center at Stanford University, accessed January 1, 2020. http://solar-center.stanford.edu/SID/activities/ionosphere.html

23 Stephen Chen, "China and Russia Band Together on Controversial Heating Experiments to Modify the Atmosphere," *South China Morning Post*, 17 December 2018. https://www.scmp.com/news/china/science/article/2178214/china-and-russia-band-together-controversial-heating-experiments

Yet, there are many in the scientific community who are "nonplussed" by what they view as the fear scenarios presented above. Christopher Scott, a University of Reading space and atmospheric physics professor, is not concerned about the Russian tests. In fact, various scientists in the West appear to be missing the point. While it is true that the Russian and Chinese tests used an insufficient amount of energy to have a lasting impact, this was not the ultimate goal of the Russian and Chinese test.[24] The intention of the test was to prove whether a facility, like Sura in Vasilsursk, could disrupt the functionality of transmissions to and from satellites, such as GPS, in space over a specific geographic area.

You've already seen how both Russia and China view GPS as an American strategic advantage worth nullifying should a crisis ever erupt between their countries and Washington. Clearly, both Beijing and Moscow are satisfied enough with the conclusions of those involved with the project who argued that "the detection of plasma disturbance . . . provides evidence for likely success of future related experiments." The Sura facility remains in operation today, and experimentation continues. As Professor Guo Lixin, a Chinese scientist who was not involved with the program but who is a "leading scientist on ionosphere manipulation technology in China," told Stephen Chen of the *South China Morning Post*, "the joint experimentation was extremely unusual" since the "technology involved is too sensitive."[25] You see, despite their flirtation with each other, as stated previously, both Russia and China eye each other with suspicion, which is understandable after centuries of hostility between the two neighbors. They are, however, willing to work together against the US.

Western scientists pooh-poohing the Russian and Chinese tests as being insufficient are missing the wider strategic implications of the Sino-Russian experiments. The Western scientists—naysayers—care only

---

24 Sarah Sloat, "Russia and China Teaming Up to Heat the Atmosphere Is 'Not That Exciting,'" *Inverse,* 23 December 2018, https://www.inverse.com/article/51972-ionosphere-russia-china-experiments

25 Chen, "China and Russia Band Together."

about the technical details *as they understand them.* America cannot allow the naysayers to get US and European policymakers to essentially whistle past the grave on such an important issue. China and Russia are both stringently opposed to US foreign policies in their regions. Both countries have identified America's satellite constellations as the Achilles' heel of the US military. Thus, both China and Russia have developed strategies and robust counterspace capabilities to deny the United States access to space in war. What's more, China and Russia have both experimented with various forms of satellite jamming techniques. Their ionosphere-heating experiments could be just one more method both countries may use on US forces the militaries of either China and/or Russia face.

There's also a bit of mirror imaging going on in the minds of most Western scientists who are belittling the joint Russian and Chinese project. Meers Oppenheimer is a prominent Boston University scientist specializing in the ionosphere. Oppenheimer denigrates the Sino-Russian ionosphere tests. The High-Frequency Active Auroral Research Program (HAARP) has existed in Alaska for many years. And the US military has conducted experiments there on the Earth's ionosphere similar to what the Russians and Chinese did over Vasilsursk in 2018. Yet, Oppenheimer argues that because the United States has been unable to achieve success in its own experimentation with Earth's ionosphere, the Russians and Chinese will fail as well. Oppenheimer believes that the Russians and Chinese are, in fact, using their ionosphere-heating project in the same way that American scientists are currently using HAARP: to allow scientists in both Russia and China to better "understand what's going on in the upper atmosphere" and how "the ionosphere can disrupt GPS communication, and how space currents flow through it." The bias of Western experts, the naysayers, is revealed in those words. In essence, if the United States and its Western partners cannot use the ionosphere the way that Russian and Chinese strategists want to, then no one will be able to. This is an ignorant assumption being made by very well-educated people in the West. It could be your kids in uniform paying the price for such uninformed Western presumptions.

Oppenheimer is right about one thing: the Russians and Chinese *are* trying to learn how "the ionosphere can disrupt GPS communication." He is wrong about the "why." Whereas American and European scientists have consigned their efforts to the esoteric realm of technical pursuits, leaving the prospect of GPS disruption in the hands of Earth's fickle sun, Chinese and Russian experts believe in human agency. They want to heat the ionosphere in order to know how to aggravate this distant region of Earth's atmosphere to the point that US forces are denied access to their GPS navigation satellite systems. The breakdown between Moscow and Washington has prompted Russian leaders to enter a wartime footing against the West. Beijing is following a similar course regarding its relationship with Washington. Both Russia and China are partnering to prevent themselves from being subordinated to what they view as an unacceptable global-techno American imperium.

On their own, China and Russia are frightful competitors for the United States—particularly China. Russia has been a nuisance for some time. The idea, though, that both Russia and China could team up and singularly focus on tearing down America's hard-won global dominance is even scarier. Despite Russia's recent flirtations with China, the West remains a more attractive partner. But the longer that Russia and China dance with each other, the more likely that these two powers will simply become wedded to each other. The deeper their ties become, the less attractive the United States and the West will be as viable alternatives. And once the Sino-Russian partnership is solidified, unlike the breakdown between the Soviet Union and China in the Cold War, it is quite unlikely that their threat to the United States—particularly in the strategic high ground of space—will be ameliorated as easily as they were in the Cold War.

The United States must first ensure that its space systems can survive the sort of attacks that Russia and China are both planning to conduct on American satellite constellations. Then, Washington must work hard to reduce Russia's paranoia about American strategic intentions while attempting to seduce Russia back into the Western fold. The path forward with Russia will be hard, but a peace with Moscow is possible—and far more preferable than the alternatives. One thing is clear, though: Russia's space ambitions and their ability to threaten the United States in—and from—orbit is only intensifying.

11

ROCKET MAN GOES NUCLEAR

"ROCKET MAN," thundered US President Donald J. Trump, while standing before a rapt audience of foreign diplomats in the green granite amphitheater of the United Nations General Assembly, "is on a suicide mission for himself and for his regime. The United States is ready, willing, and able [to totally destroy North Korea]."[1] The American leader was castigating the rogue regime of North Korea for its reckless pursuit

---

1 Donald J. Trump, "Trump to UN: 'Rocket Man' on a Suicide Mission," CNN, YouTube, 1:57–2:07. https://www.youtube.com/watch?v=8mstcJDHaGE

of an illicit nuclear weapons and ballistic missile program.[2] Further, Trump was letting the world know that, one way or the other, North Korea would be denuclearized.

President Trump had been engaged in a long-running war of words—and, most important, tweets—with Kim Jong-un. During this time, the North Korean nuclear weapons program was advancing at an incredible rate.[3] The regime was launching ballistic missiles into the air over the Sea of Japan with alarming constancy.[4] With each test, the North Koreans learned more about how to develop and launch increasingly complex—and longer-range—nuclear missiles.[5] As the threat intensified over the course of late 2016 and early 2017, President Trump refused to back down, and, in typical Trump fashion, created the hilarious nickname of "Rocket Man" with which to publicly mock the North Korean leader.[6] Pyongyang's tireless pursuit of nuclear weapons also allowed the totalitarian Kim regime to explore other weapons of mass destruction, ones that would directly threaten the United States' vital satellite constellations in orbit, such as electromagnetic pulse (EMP) weapons.

Ultimately, through a chance diplomatic maneuver in 2018, President Donald Trump was able to get Kim Jong-un to meet him

---

2 Jeff Daniels and Mike Calia, "Trump: North Korea's 'Reckless Pursuit' of Nuclear Weapons Could Soon Threaten the US," *CNBC,* 30 January 2018. https://www.cnbc.com/2018/01/30/trump-.html

3 Sarah Kessler, "Obama Warned Trump That North Korea Would Be His Biggest Problem," *Quartz,* 4 March 2017. https://qz.com/924760/donald-trumps-biggest-problem-will-be-north-korea-obama-warned/

4 Josh Smith, "How North Korea's Latest ICBM Test Stacks Up," Reuters, 28 November 2017, https://www.reuters.com/article/us-northkorea-missiles-technology-factbo/how-north-koreas-latest-icbm-test-stacks-up-idUSKBN1DT0IF

5 Tetsuro Kosaka, "As North Korea's Missile Arsenal Grows, So Does Its Nuclear Threat," *Nikkei Asian Review,* 19 October 2019. https://asia.nikkei.com/Spotlight/Comment/As-North-Korea-s-missile-arsenal-grows-so-does-its-nuclear-threat

6 Greg Hadley, "'Rocket Man': Trump Uses Belittling New Nickname for Kim Jong Un in UN Speech," *McClatchy DC,* 19 September 2017. https://www.mcclatchydc.com/latest-news/article174107001.html

for a one-on-one summit in Singapore.[7] From that point, the two sides ratcheted down tensions, though they still remain high. In late December 2019, in fact, North Korea threatened the Trump administration yet again with hostilities unless the White House continued with what were then the stalled negotiations between the United States and North Korea. By New Year's Day in 2020, Kim Jong-un had vowed to restart his country's nuclear weapons tests while working to introduce "a new strategic weapon."[8] So, the prospect of any long-term peace deal existing between the United States and North Korea remains low, as Washington will not desist from its demand that North Korea dismantle its nuclear weapons program. Pyongyang, meanwhile, maintains that its possession of a nuclear weapons program is the only thing standing between itself and a US-led regime change military operation.[9] Whether the long-term prospects of peace between North Korea and the United States come to fruition or not remains unclear. As the president might say, "We'll see." Yet, so long as the Kim regime continues to exist, it will still retain the capabilities that North Korea has built up for decades—and its political mood might not remain as sanguine about the prospects of peace as it currently appears.

GOING ROGUE

The scourge of rogue states seeking nuclear weapons has been with the United States for decades. Following the Cold War, the fear of nuclear weapons proliferation was high. A host of malign actors, ranging from Saddam Hussein's Iraq to Iran to North Korea, were all seeking nuclear

7 Brandon J. Weichert, "Musings in Singapore," *The Weichert Report,* 11 June 2018. https://theweichertreport.com/2018/06/11/musings-in-singapore/

8 Robert Gearty, "Kim Jong Un Threatens to Renew Testing Nuclear Weapons, Long-Range Missiles," *Fox News,* 1 January 2020. https://www.foxnews.com/world/kim-jong-un-testing-nuclear-weapons-long-range-missiles

9 Uri Friedman, "The Word That Derailed the Trump-Kim Summit," *Atlantic,* 24 May 2018. https://www.theatlantic.com/international/archive/2018/05/libya trump-kim/561158/

arms. In the case of North Korea, there had been interest from the Kim regime going back to the early 1990s to acquire a nuclear weapons program. After all, North Korea already had a horrific chemical and biological weapons program.[10] Yet, that program did little to elicit the kind of visceral reaction from the West that the prospect of a nuclear-armed North Korea did. And in the case of rogue regimes, much of what they do is generate a fearful response from their American foes. This was particularly true of the starving and isolated North Korea, which routinely uses the threat of regional war to get foreign aid from the West, in order to prop up its failing regime.

Previous American presidents have oscillated between totally ignoring the North Korean threat and giving in to the threat to make the problem go away. Rarely does the latter work. In fact, ignoring North Korea doesn't work. It's a perpetual problem. And as has been proven in recent years, Pyongyang's nuclear weapons threat has only grown, regardless of whatever tactic past US leaders have employed against North Korea. Since 2006, North Korea has had rudimentary nuclear weapons capabilities.[11] Pyongyang then achieved the vaunted ability to miniaturize their nuclear warheads in 2013, according to an assessment by Jeffrey Lewis of the James Martin Center for Nonproliferation Studies.[12] With miniaturization technology, the North Koreans could place powerful nuclear warheads atop an intercontinental ballistic missile that could, theoretically, reach the shores of the United States or Europe.

What's more, it is likely that Pyongyang received technical assistance not only from the rogue Pakistani nuclear scientist A. Q. Khan but also

10 John M. Donnelly, "The Other North Korean Threat: Chemical and Biological Weapons," *Roll Call*, 12 June 2018. https://www.rollcall.com/news/politics/the-other-north-korean-threat-chemical-and-biological-weapons

11 Staff, "North Korea Nuclear Tests: What Did They Achieve," BBC, 3 September 2017. https://www.bbc.com/news/world-asia-17823706

12 Peter Grier, "Are North Korea's Nuclear Weapons Small Enough to Fit a Ballistic Missile?," *Christian Science Monitor*, 8 April 2013. https://www.csmonitor.com/USA/Politics/DC-Decoder/2013/0408/Are-North-Korea-s-nuclear-weapons-small-enough-to-fit-a-ballistic-missile

from China and, quite possibly, Russia. Thanks to these efforts, North Korea did not follow the same pattern of nuclear weapons development that either the United States or the Soviet Union had followed during the Cold War. North Korea was able to then develop their nuclear weapons program according to their own pace, thanks to the assistance they received from Khan's network. The assistance Pyongyang received from China, Russia, and Iran further allowed Pyongyang's scientists to try new things—all in an effort to expedite the multistep process to create a viable nuclear weapons capability.[13]

From 2013 until 2019, the North Korean program was focused on building an ICBM capable of reaching America. Meanwhile, Pyongyang continued mass producing nuclear weapons of varying yields, testing them in underground bunkers while working to share that technology through the international black market to fellow rogue states, like Iran.[14] Yet, their lack of adequate ballistic missile technology kept hampering the North Korean progress on nuclear weapons.[15]

Under the guise of North Korea's space agency (yes, a country that cannot feed its own people and has no electricity except for what their elite receive has a space agency), Pyongyang began testing various ballistic missiles. The North Koreans argued that their space program was separate from their illicit nuclear weapons program. Unlike the nuclear program, North Korea insisted, the space program was intended for civilian use. Of course, as you've seen throughout this book, all space programs are *dual use in nature.* The North Koreans know this as well. Their claims are especially perplexing considering that in 2006 the

13 David Albright, "North Korean Miniaturization," *38 North,* 13 February 2013. https://www.38north.org/2013/02/albright021313/

14 Bruce E. Bechtol Jr., "North Korea's Illegal Weapons Trade," *Foreign Affairs,* 6 June 2018. https://www.foreignaffairs.com/articles/north-korea/2018–06–06/north-koreas-illegal-weapons-trade

15 Alex Lockie, "If North Korea Wants a Reliable ICBM, It's Going to Have to Fire One Towards Somebody," *Business Insider,* 3 August 2017. https://www.businessinsider.com/north-korea-icbm-reliable-overfly-2017–8

director of the scientific research department of North Korea's National Aerospace Development administration, Hyon Kwang-il, stated that "even though the US and its allies try to block our space development, our aerospace scientists will conquer space."[16]

So much for the whole "we come in peace" thing.[17]

"North Korea has an active space program that is closely related to its missile program, which has made significant progress in recent years," opines the Center for Strategic and International Studies' (CSIS) 2018 assessment on North Korea's space threat. CSIS offers the caveat typical of all think tank assessments: "Still, many experts doubt that the few satellites launched by North Korea perform all of the functions that the North Korean government claims."[18] The writer of the piece should add, "we hope." Frankly, North Korea has managed to defy all of the assumptions that experts have made about their technical capabilities.[19] It would not surprise me, as an observer of these matters, if the experts were inadvertently downplaying North Korea's threat. This is especially true in the unconventional arena. And what is more unconventional than a backwards state, like North Korea, possessing a space program that could threaten the United States?

Let's trace the North Korean space agency's recent successes. In 2012, on what would have been the one-hundredth birthday of Kim il-Sung, the founder of the Kim dynasty, his grandson and current North Korean leader Kim Jong-un, oversaw the successful launch of a North

---

16 Debalina Ghoshal, "North Korea's Toxic Space Program," Gatestone Institute, 22 October 2018. https://www.gatestoneinstitute.org/13157/north-korea-space-program

17 Eric Talmadge, "Possible Peace Declaration Looms Large over Trump-Kim Summit," *AP*, 20 February 2019. https://www.apnews.com/75876690900c45d2a77213b0d020235a

18 "Space Threat 2018: North Korea Assessment," *Center for Strategic and International Studies: Aerospace*, 12 April 2018. https://aerospace.csis.org/space-threat-2018-north-korea/

19 David E. Sanger and William J. Broad, "How U.S. Intelligence Agencies Underestimated North Korea," *New York Times*, 6 January 2018. https://www.nytimes.com/2018/01/06/world/asia/north-korea-nuclear-missile-intelligence.html

Korean satellite into Earth's orbit.[20] Pyongyang claimed the launch was part of their civilian space program and that it had nothing to do with the country's illicit nuclear weapons program. Of course, most people understood that this was false. After all, the same kind of rocket used to deploy a satellite into orbit can also launch a nuclear weapon at the United States. The successful satellite launch meant that whatever lessons Pyongyang learned would be taken and applied to their budding ICBM program.

With each successful satellite launch, there was an accompanying (or preceding) successful nuclear weapons test underground or another salvo of missiles fired. Each time, Western observers feared for the worst. The North Korean tests were becoming increasingly regular and very powerful. Things were so tense in 2017 and the beginning of 2018 that an emergency management worker in Hawaii accidentally sent out a text message to all Hawaiians urging them to take cover, as there was a North Korean missile inbound. Of course, the kind of missile capable of reaching Hawaii from North Korea is an ICBM. And the type of nuclear warhead that sits atop an ICBM is powerful enough to wipe out nearly everything in Hawaii. Mass panic ensued.[21] There was one infamous cell phone video of a desperate father cramming his daughter into a manhole on a Hawaiian street, believing that she might be able to survive a nuclear blast underground.[22]

Few people at the time of the alert questioned its authenticity because for nearly a year North Korea had been launching missiles at such consistent intervals that most people—particularly Hawaiians—were

---

20 Jethro Mullen and Paul Armstrong, "North Korea Carries Out Controversial Rocket Launch," *CNN,* 12 December 2012. https://www.cnn.com/2012/12/11/world/asia/north-korea-rocket-launch/index.html

21 Zachary Cohen, "Missile Threat Alert for Hawaii a False Alarm; Officials Blame Employee Who Pushed 'Wrong Button,' *CNN,* 14 January 2018. https://www.cnn.com/2018/01/13/politics/hawaii-missile-threat-false-alarm/index.html

22 Staff, "Man Helps Young Girl into Manhole in Hawaii," *CNN,* 0:40, 14 January 2018. https://www.cnn.com/videos/us/2018/01/14/hawaii-residents-false-missile-alert-girl-manhole-sot.cnn

accustomed to the threat and were aware that, inevitably, the North Koreans would launch a nuke their way. Everyone fundamentally understood that once North Korea possessed reliable ICBMs, their threat to the United States would be real. North Korea's space program plays heavily into the development of such weapons.

NORTH KOREA LOVES EMP

According to the Commission to Assess the Threat to the United States from Electromagnetic Pulse Attack (or, the EMP Commission), a properly assembled EMP weapon detonated above the United States could wreak nationwide blackout conditions upon the country. By detonating what's technically known as a "High-Altitude Electromagnetic Pulse" (HEMP) weapon in low-Earth orbit above the United States, a country, like North Korea, could quite seriously send the United States back to a nineteenth-century level of technological development, seemingly overnight.[23] At present, the United States infrastructure is not well suited to responding to the sudden deprivation of a national blackout. As the EMP Commission assessed:

> A nationwide blackout of the electric power grid and grid-dependent critical infrastructures—communications, transportation, sanitation, food and water supply—could plausibly last a year or longer. Many of the systems designed to provide renewable, stand-alone power in case of an emergency, such as generators, uninterruptable power supplies (UPS), and renewable energy grid components, are also vulnerable to EMP attack.
>
> A long-term outage owing to EMP could disable most critical supply chains, leaving the US population living in conditions similar to centuries past, prior to the advent of electric power. In the 1880s, the US population was less than 60 million, and those people had

---

23 Ariel Cohen, "Trump Moves to Protect America from Electromagnetic Pulse Attack," *Forbes,* 5 April 2019. https://www.forbes.com/sites/arielcohen/2019/04/05/whitehouse-prepares-to-face-emp-threat/#22c88976e7e2

many skills and assets necessary for survival without today's infra-structure. *An extended blackout today could result in the death of a large fraction of the American people through the effects of societal collapse, disease, and starvation* [emphasis added].[24]

The North Korean space threat *affects us all.*

In January 2016, North Korea placed its second satellite in a very low orbit around the Earth. What's more, this satellite launch occurred in the aftermath of a powerful North Korean nuclear weapons test. Then there is the issue of the satellites themselves. Just what was North Korea putting into orbit? Pyongyang unconvincingly assured the world that they were merely deploying weather satellites in orbit, although the North Korean satellites in question are not in an optimal orbit for weather satellites. On the other hand, the North Korean satellites *are* in a perfect orbit for an EMP bomb. Also, the North Korean satellites orbit the Earth in a south-to-north trajectory. This is a strange orbit for weather satellites, as they are usually orbited around Earth along an east-west axis to maintain a sun-synchronous orbit. The south-north angle, though, allows for the North Korean satellites—if they were loaded with EMP weapons—to evade America's early warning radars and national missile defenses.[25]

Meanwhile, despite what skeptics say, the fact that the North Korean satellites in question are "tumbling" in space *does not* mean that they are "broken." Instead, if the North Korean satellite is an EMP weapon, it would not need to be overly sophisticated or need to be oriented in the way that weather satellites must be oriented to function properly. All

24 "Assessing the Threat from Electromagnetic Pulse (EMP): Executive Report," *Commission to Assess the Threat to the United States from Electromagnetic Pulse (EMP) Attack,* 9 April 2018. http://www.firstempcommission.org/uploads/1/1/9/5/119571849/executive_report_on_assessing_the_threat_from_emp_-_final_april2018.pdf

25 Daniel Ashman, "North Korean Satellites May Hold Nuclear Bombs," *Epoch Times,* 28 February 2019. https://www.theepochtimes.com/north-korean-satellites-may-hold-nuclear-bombs_2816818.html

that a North Korean EMP satellite would require is a simple antenna to receive a basic command from the ground, such as "detonate." What's more, the very shape of the satellites in question—rectangle shaped, solar panels on all sides, simple antennas on the ends—are optimal for placing egg-shaped nuclear bombs in the rectangular satellite. As NASA scientist Jonathan Nowacki told the *Epoch Times,* "That's not a weather satellite. It has no large sensors." Nowacki also observed that the satellite name in Korean means "Bright Star" when translated in English. *And a thermonuclear reaction most closely resembles a small star.*[26]

The nuclear weapons program has offered Pyongyang's leadership another opportunity to develop more weapons with which to threaten and frighten the world. Despite the fearful responses from both the Soviet Union and the United States after their EMP tests in low-Earth orbit in the 1960s, North Korea appears to be encouraged by the destructive effects of EMP on high tech.[27] North Korea's leadership likely understands that their number-one foe is the United States. And without America's whizbang technological advantages, the US military would not be the threat to North Korea that it currently is.

In fact, Pyongyang is so interested in EMP and the fear it inspires in Western governments that they have expanded their investment into the technology.[28] It was even reported in 2011 that the North Koreans officially built a working EMP bomb and tested it twenty-five miles underground.[29] What's more, in 2004, two Russian generals told the EMP Commission that the design for the Soviet-era EMP weapon

---

26 Ashman.

27 Paul Bedard, "North Korea Reveals Nuclear EMP Attack Plans," *Washington Examiner,* 24 November 2018. https://www.washingtonexaminer.com/washington-secrets/north-korea-reveals-nuclear-emp-attack-plans

28 William R. Graham, "North Korea Nuclear EMP Attack: An Existential Threat," *38 North,* 2 June 2017. https://www.38north.org/2017/06/wgraham060217/

29 Joohee Cho, "North Korea Nears Completion of Electromagnetic Pulse Bomb," *ABC News,* 9 March 2011. https://abcnews.go.com/International/electronic-warfare-north-korea-nears-completion-electromagnetic-pulse/story?id=13081667

was "unintentionally transferred to North Korea." It is also suspected by South Korean intelligence that Russian scientists were assisting the North Koreans in developing an EMP weapon in 2009. And the Chinese military confirmed in 2013 that Pyongyang already possessed "Super-EMP nuclear weapons."[30]

In 2018, the North Korean government handed out propaganda documents to the country's farmers indicating that Pyongyang was still progressing toward the development and deployment of EMP weapons. These documents were obtained by a reporter for the *Daily NK*. Of course, it must be emphasized that, despite the fact that an independent source smuggled such documents out of the otherwise-closed North Korea, these documents were created by the Kim regime for propaganda purposes. The propaganda was intended to assure and remind North Korean peasants that, while Pyongyang had decided to engage in negotiations with the Trump administration in 2018, the regime was not relinquishing its perceived "right" to build weapons with which to threaten the United States, South Korea, and other American partners.

It should be noted that the North Korean document explicitly outlined the importance that EMP weapons play in the development of the overall North Korean nuclear weapons threat to the United States and its allies. As you read in the sections on the Soviet and American Cold War–era EMP tests, these weapons have the ability to totally destroy and disrupt the functioning of key satellites in orbit. Toward that end, the North Koreans have conducted four GPS jamming operations, directed against their archrivals in South Korea, since 2010. The most recent GPS jamming attack occurred in 2016. That North Korean GPS jamming attempt affected "hundreds of passenger jets, commuters, Uber drivers and commercial fishing boats while raising alarm bells in the Pentagon," according to a *Fox News* report.[31]

---

30 "Space Threat 2018: North Korea Assessment."

31 Staff, "Pentagon Concerned About North Korea Jamming GPS Signals, Officials Say," *Fox News*, 7 April 2016. https://www.foxnews.com/us/pentagon-concerned-about-north-korea-jamming-gps-signals-officials-say

As the *Daily NK* reports, "The [GPS jamming] attacks led to interference in communication lines used by military bases, airplanes, mobile phone relay stations, ships and individual mobile phones in Seoul and parts of Gyeonggi Province." A GPS jamming attack is considered to be a form of nonnuclear EMP attack, since its impact on the normal functions of a GPS satellite are the same as a direct nuclear EMP attack (and that's the point of such an attack, by the way). The aforementioned North Korean propaganda pamphlet from 2018 states that "EMP attacks lead to the destruction of electronic circuits in semiconductors, affecting the use of TVs, mobile phones and computers in the area. The widespread failure of South Korea's mass communications system due to an EMP attack and its effect on national infrastructure and military command systems could lead to widespread chaos."[32]

The North Koreans have spent decades building an unconventional military capability meant to preserve their bizarre and oppressive regime in the face of America's overwhelming conventional military power. To achieve this goal, Pyongyang has embraced nuclear weapons and has developed a space program designed to augment their threat to the United States. North Korea has a long history of brinksmanship directed against those they believe to be their enemies. What's more, as the opening of the Korean War showed, Pyongyang places great value on strategic surprise. Their nuclear weapons and EMP programs give Pyongyang a strategic edge over the United States that few Western analysts are willing to acknowledge. For all of the rhetoric about President Trump mitigating the North Korean nuclear threat, it must be understood that the Kim regime is existentially threatened by the United States. President Trump can be as diplomatic with Kim as possible, but there is still a very good chance that Kim will not abandon his pursuit of weapons, like nukes, that will keep the Americans from

---

32 Mun Dong Hui, "North Korean Propaganda Promotes EMP Attacks Using Nuclear Weapons," *Daily NK,* 23 November 2018. https://www.dailynk.com/english/north-korean-propaganda-promotes-emp-attacks-using-nuclear-weapons/

attacking North Korea. And just as with the Russians today, the chance for grave strategic miscalculations is very high, as Pyongyang presses ahead with their challenge to Washington, and Washington continues sending US military forces to the Korean Peninsula while attempting to negotiate with Kim Jong-un.

Should these negotiations fail, though, the reality will be that North Korea was given an additional eighteen months to develop their nuclear weapons and EMP threat to the United States.[33] Remember: even as the United States and North Korea engaged in negotiation over the last year, Pyongyang did not actually cease their nuclear weapons and ballistic missile testing.[34] Kim Jong-un merely quieted things down long enough to buy his nuclear weapons program time to develop better capabilities. Should the United States be forced into military conflict against North Korea, US troops may find an even deadlier opponent in North Korea that is more than willing and capable of pushing the American military juggernaut back.

What could North Korea do with space-based EMP weapons? Of course, they could outright launch a decapitating strike on the United States homeland that completely removes the threat of US military reprisals upon them. This would give North Korea the time it needed to invade South Korea. At the same time, a North Korean victory would not be assured—even under those battlefield conditions. The South Korean forces, as well as the American forces deployed to South Korea, would likely still attempt to mount an effective resistance against any North Korean invasion, although without their advanced technology, the South Korean and American forces might inevitably be overrun by the larger, North Korean military. Since Kim Jong-un comes across as

---

33 Brandon J. Weichert, "North Korea's Nuclear Program and the Cone of Uncertainty," *The Weichert Report,* 7 December 2017. https://theweichertreport.com/2017/12/07/north-koreas-nuclear-program-and-the-cone-of-uncertainty/

34 Elias Groll, "North Korean Missiles Just Keep Getting Better," *Foreign Policy,* 3 October 2019. https://foreignpolicy.com/2019/10/03/north-korean-missiles-just-keep-getting-better/

a villain from the Roger Moore era of the James Bond films, another action that he might take would be to hold the world ransom with the threat of his EMP weapons in orbit. Kim might demand that the world recognize Pyongyang's claim over South Korea or, more dangerously, that Washington permanently remove all of its forces from the Korean peninsula. If Washington did not comply, Kim might destroy the power grids of any Western countries he desires. The prospect of suffering an EMP strike is so caustic that an American leader—even one who perceives himself as a "tough guy," like President Trump—just might accede to North Korean demands rather than face the consequences of enduring a nationwide blackout that might last for two years and kill swaths of the American population over that time.

Space is key to North Korean ambitions. American leaders should understand this moving forward and develop necessary countermeasures against that North Korean space threat. The time to develop such counter-measures, though, is not *after* Pyongyang has fully deployed these terrible weapons against Americans in combat, such as EMP bombs. Instead, the United States must develop the necessary countermeasures now, *before* any crisis mounts. Diplomacy is never a bad choice, but diplomacy that is undergirded with the reassurance of reliable military defenses against an unpredictable foe, like North Korea, will be even better.

12

THE MAD MULLAHS LOOK TO

THE HEAVENS

IN MAY 1995, the current president of the Islamic Republic of Iran, Hassan Rouhani, a supposed "moderate" among Iran's political elite, stood before a group of pro-regime students, and intoned that "the beautiful cry of 'Death to America' unites our nation."[1] Later, in 2013, Rouhani repeated similar remarks while campaigning for Iran's presidency in the city of Karaj. This time, the future "moderate" Iranian

---

[1] Christopher Booker, "Iran's Moderate Leader Is Nothing of the Kind," *Telegraph* (UK), 23 January 2016. https://www.telegraph.co.uk/news/worldnews/middleeast/iran/12116458/Irans-moderate-leader-is-nothing-of-the-kind.html

leader spoke more explicitly about Iran's threat to the United States and its allies in the region. Rouhani explained to his entranced audience that "saying 'Death to America' is easy. We need to express 'Death to America' with action. Saying it is easy."[2] In fact, Iran's present regime was birthed in staunch opposition to what it viewed as America's unholy hegemony in the Middle East. From its very inception, Iran's leadership fought to undermine and ultimately destroy the American-led regional order at all costs. Saying that the United States and its allies—notably Israel—can get along with Iran is akin to saying that sheep can get along with wolves.

The statements from Iran's current president are a stark difference from how Iran entered the 1950s: as an unapologetic—modernizing, even—US client in the Middle East. Whereas today Washington laments the rise of Iran's nuclear weapons program, in the 1950s Washington helped to create that same program! Of course, when the Eisenhower administration provided Iran with technical assistance in developing nuclear energy reactors, purportedly for peaceful uses, under the US Atoms for Peace program, few envisioned that Iran would ever be anything but an American client state. Throughout the Cold War, Iran's leader, the monarchical shah, whose iron-gripped rule of Iran was assured by a joint covert US-British intelligence coup in 1953, was one of America's strongest allies. As the Cold War progressed, Iran under the shah's leadership became the lynchpin of America's Mideast defense strategy against the Soviet Union. So, giving Iran what was at the time advanced US military technology, boatloads of cash, and even nuclear technology made perfect sense. It was all done to keep the geostrategically vital Iran in America's camp during the ideological hurly-burly that was the Cold War.

---

2 Sohrab Ahmari, "About That New 'Moderate' Iranian Cabinet . . ." *Wall Street Journal,* 7 August 2013. https://www.wsj.com/articles/about-that-new-moderate-iranian-cabinet-1375917151?tesla=y

A TIMELINE OF UNFORTUNATE EVENTS:
IRAN'S NUCLEAR DEVELOPMENT

Throughout the 1970s, the shah's government had built "an impressive baseline capability in nuclear technologies."[3] This baseline of nuclear capabilities, thanks to their close relationship with the United States and other nuclear powers in the West, such as France, is the basis of Iran's present illicit nuclear weapons program. By 1979, America's nuclear venture with Iran would prove to be a disaster in the long term, as the shah was overthrown in a chaotic Shiite Islamist revolution that year. Virtually overnight, then, Iran went from being America's principal Mideast ally to its primary regional foe.

During their successful revolution in 1979, the shah of Iran and his wife fled to the United States. In retaliation, Iranian demonstrators stormed the US embassy in Tehran. The young Iranian radicals who invaded the embassy then held the US embassy staff hostage for more than four hundred hellish days. Throughout the ordeal, the Iranian invaders insisted that they would release their American hostages only if Washington agreed to return the aging shah and his wife to Iran. The United States did not submit to the humiliation that Iran's revolutionaries were imposing upon the country. Thankfully, the election of Ronald Reagan in 1980 ended the standoff. Yet, the Iranian threat was only rising at that point. In fact, the threat would persist and evolve for many more decades to come.

A few short years later, in 1982, Iranian-backed Hezbollah militants in nearby Lebanon blew up the US Marine Corps barracks, killing more than two hundred US Marines. Iran also kidnapped the CIA's station chief in Beirut, William Buckley, then tortured and murdered him. Iran's clerical leaders also plotted the overthrow of rival governments in neighboring Sunni Arab states. And as noted above, Tehran insisted on waging a campaign of terrorism against both Israel and the United States.[4]

---

3 "Iran," *Nuclear Threat Initiative,* May 2018. https://www.nti.org/learn/countries/iran/nuclear/

4 Brandon J. Weichert, "Ignoring Iran Will Not Help," *The Weichert Report,* 16 June 2019. https://theweichertreport.com/2019/06/16/ignoring-iran-will-not-help/

Saddam Hussein's Iraq eventually challenged the Islamist Iran in a gruesome war that persisted throughout the 1980s. Given Tehran's hostility toward the United States and its presence in the Middle East, Washington gave limited military and intelligence support to Saddam Hussein's forces in their decade-long struggle against Iran. Ultimately, the United States was dragged into the Iran-Iraq War when Iran attempted to blockade the vital Strait of Hormuz, through which nearly 30 percent of the world's oil supply passes daily.[5]

Relations between the United States and Iran only worsened from there.

In the 1990s, the Russians stepped in to facilitate the development of Iran's supposed civilian nuclear energy sector.[6] Just as with North Korea, US intelligence believes that Iran's civilian nuclear energy program is a cover for their illicit nuclear weapons program. By 2002, the Iranians had been caught red-handed, as the National Council of Resistance of Iran (NCRI) disseminated proof of the existence of previously undeclared Iranian nuclear research facilities, hidden behind the names of Iranian front companies to further complicate the ability of US intelligence to identify these covert nuclear research facilities.[7] Iran appeared to have a slight change of heart in 2003 shortly after the United States successfully toppled neighboring Saddam Hussein, when they agreed to suspend uranium enrichment, a key step on the path to a nuclear weapon. Yet, a year later, as it became clear that America's mission in Iraq was a failure, evidence surfaced that Iran was making their Shahab-3 missile able to carry a nuclear warhead. And by 2005, the entire denuclearization process

---

5 "FACTBOX-Strait of Hormuz: The World's Most Important Oil Artery," Reuters, 23 April 2019. https://www.reuters.com/article/usa-iran-oil-strait/factbox-strait-of-hormuz-the-worlds-most-important-oil-artery-idUSL5N2254EM

6 Eli Lake, "Iran's Nuclear Program Helped by China, Russia," *Washington Times,* 5 July 2011. https://www.washingtontimes.com/news/2011/jul/5/irans-nuclear-program-helped-by-china-russia/

7 "Remarks by Alireza Jafarzadeh on New Information on Top Secret Projects of the Iranian Regime's Nuclear Program," *Iran Watch,* 14 August 2002. https://www.iranwatch.org/library/ncri-new-information-top-secret-nuclear-projects-8–14–02

had broken down between Iran and the West after Tehran announced they'd resume uranium enrichment processes.

Today, most Western governments know that Iran's nuclear weapons program exists. Yet, between the consistent denials from Tehran, coupled with the many "intelligence gaps" that exist in the collection methods that US intelligence has for Iran, Tehran's nuclear weapons program can be said to exist and yet not exist at the same time in the minds of Western leaders. John R. Haines has likened this to the Schrödinger's cat paradox (in this case, Haines claims this is a case of "Schrödinger's Nuke").[8] What most understand is that Iran has rudimentary nuclear weapons capability—and Iran's nascent space program figures heavily into Tehran's quest for nuclear weapons, just as North Korea's space program plays heavily in North Korea's ambition for nuclear weapons.

Since 2004, Tehran has focused heavily on developing their missile technology. Beyond their relationship with Russia, Tehran maintains extremely close ties with their fellow rogue nuclear state, North Korea.[9] In fact, Iranian and North Korean scientists routinely share information on their developing nuclear programs. Specifically, North Korea has been selling missile technology to Iran for years. Former Israeli major general Amos Yadlin, who was the head of the Israeli Defense Forces Intelligence Branch, claimed that Tehran purchased missiles from North Korea that possibly exceeded the range of Iran's longest-range missile, the Shahab-3, which is based off North Korea's Nodong missile. The North Korean missile itself is based off the old Soviet SS-N-6 missile that the Soviets had sold to North Korea during the Cold War. What's

8 John R. Haines, "Schrödinger's Nuke: How Iran's Nuclear Weapons Program Exists—and Doesn't Exist—at the Same Time," *Foreign Policy Research Institute,* 20 May 2015. https://www.fpri.org/article/2015/05/schrodingers-nuke-how-irans-nuclear-weapons-program-exists-and-doesnt-exist-at-the-same-time/

9 Mintaro Oba, "Iran-North Korea Relationship Reflects Failed US Policies," *Atlantic Council,* 1 April 2019. https://www.atlanticcouncil.org/blogs/iransource/iran-north-korea-relationship-reflects-failed-us-policies/

more, Iranian leaders have publicly confirmed over the years that their missile program received copious assistance from North Korea. The ultimate goal behind such close cooperation with North Korea, Russia, and China is for Iran to be entirely self-sufficient in the production of missiles capable of launching nuclear weapons at the United States.[10]

Given their close reliance on North Korea for both technical and political support in their nuclear weapons and ballistic missile programs, it is likely that Iran's programs are not as sophisticated as North Korea's. Yet, this is not something to take comfort in. Because, as the timeline at the start of this chapter indicates, Iran has made steady progress in the development of their nuclear weapons and ballistic missile programs—and various international actors will continue to give these illicit Iranian programs support, despite the protestations of the United States. More important, just like Pyongyang, Tehran's obsession with nuclear weapons is firmly tied to their need to survive as a regime.

IRAN'S STRATEGIC GOALS

Since 1979, Iran has increasingly become isolated geopolitically. Their regime has employed terrorism and waged ceaseless unconventional warfare against those they believe are enemies of their regime. Iran's primary goal has been to spread their "Green Revolution" across the world, beginning with the Middle East. As I have noted in my writing for the *American Spectator*, Iran's ideological makeup is akin to that of the Soviet Union's communist ethos: both rivals historically worked to push their ideologies upon the world, much like how a contagion is spread. For Iran, terrorism and nuclear brinkmanship have been essential elements in their campaign of ideological warfare. In fact, the Iranian regime's ideological roots are often overlooked entirely or significantly

---

10 Paul Kerr, "Iran, North Korea Deepen Missile Cooperation," Arms Control Association, January 2007. https://www.armscontrol.org/act/2007–01/iran-nuclear-briefs/iran-north-korea-deepen-missile-cooperation

downplayed by Western liberals, who yearn for the days when Iran will no longer view the West as its existential foe.[11]

Or, as Kasra Aarabi argued in 2019:

> It is the entrenchment of a dogma, levelled against perceived enemies of Islam, that has been Iran's most potent resource. The events of 1979—not just the revolution but also the storming of Islam's holy city of Mecca by extremist insurgents and the Soviet invasion of Afghanistan—set major powers in the Middle East on a knee-jerk path to religious conservatism and away from modernisation [sic], altering the entire relationship between religion and politics in the Muslim world. The ideology of the Islamic Revolution bolstered Islamist groups and set an example that power through force was possible.[12]

The mullahs in Tehran have also gruesomely oppressed their own people for decades. Between their domestic policies and the onerous sanctions that Washington and its allies have imposed upon Tehran, the Iranian economy is collapsing—prompting protests against the regime from within Iran. There are many domestic problems that Iran's turgid, fundamentalist regime will face in the coming decade that most of its leaders are entirely unprepared to face. As David P. Goldman has assessed in his writing on the subject, Iran is facing demographic challenges, economic hardships, and increasingly unstable domestic politics.[13]

At the same time, Iran's population is majority Shiite Muslim in a region that is overwhelmingly Sunni Muslim. Historically, Iran has been at odds with their Sunni neighbors. Due to the internal division within

---

11 Brandon J. Weichert, "Trump Is Winning the Little Cold War with Iran," *American Spectator,* 23 June 2019. https://spectator.org/trump-is-winning-the-little-cold-war-with-iran/

12 Kasra Aarabi, "The Fundamentals of Iran's Islamic Revolution," Tony Blair Institute for Global Change, 11 February 2019. https://institute.global/insight/co-existence/fundamentals-irans-islamic-revolution

13 David P. Goldman, "The Norm Is NOT Democracy—The Norm Is Extinction," PJ Media, 3 January 2018. https://pjmedia.com/spengler/norm-not-democracy-norm-extinction/

the Islamic world between the majority Sunnis and the minority Shiites, Iran has long aspired to be a regional hegemon to preserve their way of life and to drive back the forces they believe are threatening their own survival. This not only includes the Sunni states, but also the mostly Jewish state of Israel, and the "Great Satan" that is the United States. Iran's nuclear weapons, ballistic missiles, and its space programs play heavily into Tehran's geostrategic ambitions. In the words of one Iranian general in 2006, Moshen Reza'i, "If we master nuclear technology, we will be transformed into a regional superpower and will dominate 17 Muslim countries in this neighborhood."

Much like the Russians, Chinese, and North Koreans, Iran's leaders understand that the US military is technologically superior and far more capable than that of Iran. Rather than buckle under the continuous pressure that Washington has subjected Tehran to over the decades, though, Iran's resistance quotient to American economic and diplomatic constraints imposed upon them has only increased. Iran is certainly not the first small country to challenge US global dominance. The aforementioned Saddam Hussein's Iraqi regime did that for decades. So, too, did Libya's Muammar Gaddafi. Of course, both of these regimes were toppled by American force of arms—Hussein's regime in 2003 and Gaddafi's by US and NATO forces in 2011. Both Tehran and Pyongyang assume that the real reason for America's willingness to use military force to replace the regimes of Iraq and Libya was due to the fact that neither regime was ultimately ever able to build a reliable nuclear weapons capability.

Neither Iran nor North Korea's leaders intend to make the same mistake that Saddam and Gaddafi did.

The presence of even a rudimentary nuclear weapons arsenal, then, could serve as a vital deterrent against the US military threat to Iran. Should the Americans or their regional allies insist upon resisting Iranian geopolitical ambitions, Tehran has developed a "mosaic defense" to complicate the ability of more advanced militaries from totally mollifying their national defenses.[14] In 2006, Iranian major general Mostafa Mohammad-Najjar said, "The armed forces of the Islamic Republic of Iran enjoy a unique superiority in asymmetrical defense in the entire region, relying on their own capabilities." Iranian brigadier general Mohammad Ali Jafari went even further that year when he assessed that "as the likely enemy [the United States] is far more advanced technologically than we are, we have been using what is called 'asymmetric warfare' methods . . . We have gone through the necessary exercises and our forces are now well prepared for [an American attack]."

The George W. Bush administration's ill-fated invasion of Iraq was intended to spread democracy throughout the region and replicate the success of the freedom movements that swept away the Soviet Union in eastern Europe at the end of the Cold War.[15] Yet, this was not the result of the Iraq War. Instead, it destabilized the region utterly. The Iraq War, coupled with the bizarre reaction of the Barack Obama administration to the events of the so-called Arab Spring, encouraged the further spread of Islamism throughout the region. All of these events diminished American power in the region and ensured the rise of Iran as a regional counterweight to the United States.[16] Next, President Obama's awful executive agreement with Iran over its nuclear weapons program went into effect in 2015. The agreement allowed for Iran to extend its influence and power throughout the region, because it removed most sanctions on Iran's regime and essentially gave Iran a safe path to acquiring nuclear weapons.[17]

---

14 Michael Connell, "Iran's Military Doctrine," *United States Institute of Peace: The Iran Primer,* accessed 28 October 2019. https://iranprimer.usip.org/resource/irans-military-doctrine

15 Christian Caryl, "The Democracy Boondoggle In Iraq," *Foreign Policy,* 6 March 2013. https://foreignpolicy.com/2013/03/06/the-democracy-boondoggle-in-iraq/

16 David P. Goldman, "Dumb and Dumber: When Neocons and Obama Liberals Agree," *Tablet Magazine,* 20 May 2013. https://www.tabletmag.com/jewish-news-and-politics/132459/dumb-and-dumber

17 Ray Takeyh, "The Nuclear Deal Is Iran's Legal Path to the Bomb," *Politico,* 22 September 2017. https://www.politico.com/magazine/story/2017/09/22/iran-nuclear-deal-bomb-215636

Today, Iran's geopolitical leverage is greater than it has ever been, as Iranian capabilities have increased thanks to the furthering of their nuclear weapons program.[18] What's more, Iran's ambitious foreign policy has kept the Americans off balance and allowed for Iran to dictate the course of its "Little Cold War" against the United States and Washington's clients throughout the Mideast.[19] Of course, under the Trump administration, Washington has embraced a "maximum pressure campaign" aimed at sapping Iran of its ability to threaten the United States or any of its regional allies. Many Western commentators and analysts have decried the Trump administration's revocation of the Obama administration's nuclear deal with Iran forged in 2015. These mostly liberal pundits worry that President Trump's actions will hasten war with Iran. Yet, Iran has maintained a "No-First-Use" policy regarding its nuclear weapons program, and Iran's leaders have striven to paint the United States as the aggressor against Iran. Therefore, Trump's actions are unlikely to militate Iran into attacking the way that many Western analysts fear. Iran's response, however, *will* be in the realm of unconventional war.

Recently, Iran engaged in a wave of violence directed against American-backed Saudi Arabia. Also, Iranian forces operating in Syria have attempted to increase their threat to US-backed Israel. In the case of the former, Iran purportedly used limpet mines to attack a Saudi Arabian oil tanker in international waters.[20] From there, it is suspected that Iranian commandos attempted to remove the mines to remove evidence of the Iranian mine attack. Yet, it is also possible that Iran wanted to appear as heroes in the international press. Tehran has maintained that they are not the ones who placed the limpet mines in international waters. Of

---

18 Staff Writer, "IRGC's Salami: Iran Today Is Stronger Than Its 'Enemies,'" *Al Arabiya English*, 25 October 2019. https://english.alarabiya.net/en/News/middle-east/2019/10/25/IRGC-s-Salami-Iran-today-is-stronger-than-its-enemies-

19 Weichert, "Trump Is Winning the Little Cold War with Iran."

20 Staff, "Iran News: U.S. Says Mines Used in Tanker Attacks Bear 'Striking Resemblance' to Weapons Touted by Tehran," *CBS News*, 19 June 2019. https://www.cbsnews.com/news/iran-news-us-shows-limpet-mine-parts-case-against-iran-in-tanker-attacks-today-2019–06–19/

course, Iran has consistently boasted of their limpet mine threat to the United States and international shipping.[21] Meanwhile, in 2018, Israel was prompted to launch devastating airstrikes against Iranian bases in Syria, as Israel had learned of Iran's intentions to attack Israel from those bases.[22]

Given this, Iran's intentions for regional conflict are clear: they want to become the regional powerhouse at America's expense. At the same time, though, Iran desires to win international support for their cause, which is why the Westerners who fear Trump's policies are wrong. Iran will avoid direct conventional war with the United States—and therefore Washington has enough leverage to focus on preventing an Iranian nuclear weapons breakout.[23]

IRAN'S SPACE PROGRAM: FAILURE TO LAUNCH (FOR NOW)

The prospect of an Iranian nuclear bomb has driven the fears of many American leaders. Various administrations have sanctioned Iran because of their unwillingness to give up their mad quest for nukes. Yet, one of the few areas that most sanctions have overlooked was Iran's civilian space program. This is why Iran's civilian space program has become a conduit for testing ballistic missile technology. Unlike North Korea's space program, though, Iran has endured a series of failures to launch. Most recently, in August 2019, Iran's launch of a purported weather satellite ended in a fiery failure. As Tom Rogan wrote in the *Washington Examiner,* "It's a good thing that Iran's satellite blew up on its launchpad."[24]

---

21 John Gambrell, "US Says Iran Took Mine off Tanker; Iran Denies Involvement," *Military Times,* 14 June 2019. https://www.militarytimes.com/news/your-military/2019/06/14/us-says-iran-removed-unexploded-mine-from-oil-tanker/

22 Lawahez Jabari and Linda Givetash, "Israel Claims It Thwarted Iranian Drone Attack with Syrian Airstrikes," *NBC News,* 25 August 2019. https://www.nbcnews.com/news/world/israel-claims-it-thwarted-iranian-drone-attack-syria-airstrikes-n1046151

23 Jed Babbin, "Shadowboxing with Iran," *American Spectator,* 8 July 2019. https://spectator.org/shadowboxing-with-iran/

24 Tom Rogan, "Be Glad Iran's Satellite Launch Failed," *Washington Examiner,* 30 August 2019. https://www.washingtonexaminer.com/opinion/be-glad-irans-satellite-launch-failed

That's the understatement of the decade.

Fearing the prospect of Iranian space program serving as a cheap cover for their ballistic missile program in 2005, Taylor Dinerman wrote at the *Space Review*:

> If Iran can build and test a nuclear weapon, and prove that it has the capability to build and launch a satellite, even a small one, it will join a new category of states that could be referred to as "mini-superpowers" . . . Having a satellite in orbit and a "bomb in the basement" gives a government options, and a certain amount of room to maneuver than states without that capability would have.[25]

Thankfully, the Trump administration recently imposed harsh sanctions directly against Iran's space program for the first time. President Trump has maintained that the series of failed rocket tests at Iran's space center "violated a United Nations Security Council resolution because the rockets incorporate the same technology used for ballistic missiles."[26] He is correct. But Iran's threat in space, like the threat of North Korea in space, is more than just ballistic missiles.

In 2009, Iran successfully placed its first satellite, a small device called "Omid" by its builders, atop Iran's first home-built satellite launch vehicle (SLV). Since that time, Iran has placed three additional satellites similar in size and technological capability to the Omid. Unfortunately for Iran, though, their rocket design has consistently proven insufficient to allow for more complex launches to occur. The SLV Iran used to launch their Omid satellite, the Safir, is based on the aforementioned Shahab-3 ballistic missile. Because of design limitations of the Safir rocket, for Iran to place larger, more complex satellites

---

25 Taylor Dinerman, "Iran's Satellite: A Look at the Implications," *Space Review,* 18 October 2004. http://www.thespacereview.com/article/250/1

26 Tal Axelrod, "Trump Administration Sanctions Iran's Space Agency," *The Hill,* 3 September 2019. https://thehill.com/policy/international/middle-east-north-africa/459764-trump-administration-sanctions-irans-space

into orbit, they must build a more powerful rocket. Iran's space agency ultimately produced such a rocket, the Simorgh. Despite its successful testing in 2016, though, the Simorgh program has failed to place a satellite in orbit. The August 2019 satellite launch failure means that Iran is entirely dependent on the Safir rocket, which is based on the medium-range Shahab-3 missile.[27]

The Iranians view space as a strategic domain, much like the other domains (land, sea, air, cyber). Iran's military can use space to conduct faster communications and better surveillance and to increase their reconnaissance capabilities. Space can also be used as a place to conduct strategic attacks against the United States and its allies. And here again, readers are shown the threat of EMP.

EMP: A SHARIAH-APPROVED NUCLEAR ATTACK

It has been noted that Iran maintains a "No-First-Strike" policy regarding nuclear weapons. The religious and true political leader of Iran, Ayatollah Khamenei, has argued against the embrace of a First Strike nuclear weapons policy because such a policy violates Islamic Shariah law. To the Ayatollah, the first strike would be immoral and would disproportionately murder innocent people overseas. Whether the Iranian regime is serious about this policy or is simply stating this to both cover up for the fact that their nuclear weapons program has not yet built a reliable nuclear weapons arsenal, as well as to lull the West into a false sense of security, is another matter entirely. There is one area, however, that the regime has not publicly announced its intentions with EMP weapons.

In fact, in 2015, the Defense Intelligence Agency (DIA) translated a textbook entitled *Passive Defense* that was written for Iran's general military staff in 2010. In that book, according to former CIA director James R. Woolsey and EMP Commission member Vincent Pry, "the

---

27 Shahryar Pasandideh, "Iran's Space Program Won't Get off the Ground While Under Sanctions," *World Politics Review*, 5 September 2019. https://www.worldpoliticsreview.com/articles/28167/iran-s-space-program-won-t-get-off-the-ground-while-under-sanctions

official Iranian textbook advocates a revolutionary new way of warfare that combines coordinated attacks by nuclear and non-nuclear EMP weapons, physical and cyber-attacks against electric grids to black out and collapse entire nations." Much like Russia and China today, Iran's military "makes no distinction between nuclear EMP weapons, non-nuclear radio frequency weapons and cyber-operations."[28] This is all in keeping with Iran's vision of asymmetric warfare, as well as their concept of mosaic defense.

Over the years, Americans have witnessed the way that Iran's military blends together the various aspects of statecraft to form a coordinated threat to US military power projection in the Middle East. For example, the United States and Iran have been engaged in a decade-long cyber war. Often, the Western press does not report this news because unconventional warfare is poorly understood by those outside of defense circles. Yet, Iran has increasingly viewed what was once the civilian realm of cyberspace as a conflict zone where their forces have advantages over the Americans. Similarly, Iran has boldly used terrorist proxies, notably Hezbollah, in its decades-long war against Israel and the United States to carry out its provocative foreign policy of regional dominance—while possessing barely plausible deniability. Meanwhile, Iran has used its limited military power to threaten international shipping, in an effort (mostly failed, thankfully) to undermine the economic stability of the United States, by spiking international oil prices. All of these are forms of what Thomas J. Wright would refer to as "all measures short of war." While the attacks are unconventional, they are explicitly meant to defeat America's perceived overwhelming conventional military threat to Iran. And the attacks, on some level, have worked.

An EMP weapon would allow Iran to fulfill its dream of pushing the United States out of the Middle East forever. The vacuum America

---

28 James Woolsey and Peter Pry, "A Shariah-Approved Nuclear Attack," *Washington Times,* 19 August 2015. https://www.washingtontimes.com/news/2015/aug/18/jams-woolsey-peter-pry-emp-a-shariah-approved-nucl/

left in the Mideast would, in the minds of Tehran's mad mullahs, be filled by Iran. Because the EMP "destroys electronics directly, but people indirectly, it is regarded by some as Shariah-compliant use of a nuclear weapon," according to Woolsey and Pry. The EMP Commission has assessed that Iran has practiced firing missiles and fusing warheads for a high-altitude EMP (HEMP) attack, similar to what many believe that North Korea is plotting. And while the aforementioned Iranian Omid satellites were too small and unsophisticated for their stated purposes of being a weather satellite, just as with North Korea's handful of satellites, they are capable of serving as EMP weapons in orbit.

THE SATELLITE THREAT

Even if the Iranian space program was entirely aboveboard (which it is not), Iranian use of space for satellite operations would itself be an indirect threat to the United States and its allies. As you've seen, satellites provide a country with incredible military and economic capabilities. Should Iran manage to get their larger Simorgh rocket working properly, Tehran could place a medley of weather, surveillance, communications, and reconnaissance satellites that would enhance the already-growing military capabilities of Iran. Iranian satellite constellations could allow for Iran's military to further threaten its regional foes.

Consider this: Iran has become a master in drone warfare technology. Iranian drones became such a threat to Israel in nearby Syria that Israel risked a conflict with Russia, whose forces are fighting alongside Iran in Syria, in order to blast an Iranian drone base. By creating a reliable space program, the Iranians could install advanced satellites in orbit that would allow for greater control—over farther distances—of increasingly sophisticated drones. Iran could also glean greater situational awareness in combat, thanks to the presence of weather satellites and surveillance satellites. Such awareness would allow Iranian forces to defend themselves against Israeli attack, which would, in turn, degrade Israel's ability to defend itself.

Oddly, there is little evidence to suggest that Iran has invested resources into developing ASAT weapons or space stalkers. Neither has

North Korea. Yet, given Iran's close relationships with both China and Russia—who do possess ASAT technology—garnering such capabilities would be relatively easy for either Tehran or Pyongyang. Rather than wait for the Iranian space threat to fully mature, the United States should act now to end the threat in its infancy. Iran possesses the strategic intent and desire to enhance its position relative to those of its US-backed neighbors. A reliable space program, not to mention its nuclear weapons, potential EMP, and ballistic missile programs, will further these malign Iranian strategic intentions.

The notion that Iran can be negotiated with in any meaningful way defies all common sense.[29] Much like the Soviet Union before them, Iran is an ideologically revolutionary power that seeks to export its dark ways to the world. Those who would dare resist Iran's revolutionary agenda, as the United States and its allies do, are made to endure ceaseless rounds of global terrorism, economic disruption, cyber warfare, and nuclear brinksmanship. Allowing Iran to develop a robust space program would only enhance the overall danger that tiny Iran poses to the United States. The Trump administration must remain steadfast in its commitment to stunt and stymie the growth of Iran's threat to the world.

---

29 Brandon J. Weichert, "Iran Keeps Asking for It," *American Spectator,* 15 September 2019. https://spectator.org/iran-keeps-asking-for-it/

13

NATO IN SPACE? ASK THE FRENCH

IT ISN'T JUST AMERICA'S RIVALS who are increasingly looking to space as a strategic domain. Traditional US allies, such as France, Japan, and Israel, are also seeking to exploit the bountiful economic and strategic opportunities that space development affords countries and corporations. This has been a long time coming, since the United States has spent the last thirty years dithering in space. It stands to reason that as hostile powers such as China, Russia, Iran, and North Korea expand their own space capabilities, other countries will begin taking matters into their own hands.

As the strategic orbits around Earth become increasingly crowded by many more players, thanks to America's apparent abdication of space, greater levels of investment will have to be made by both the US government and the private sector into the high-tech field to continue being the dominant force in orbit. At the same time, new partnerships will have to be forged between Washington and these other rising space powers. If Washington continues dithering in space, these rising space powers will inevitably join with the Chinese or Russians in space against the United States.

NATO SHOULD DO WHAT IT WAS CREATED TO DO

In the last few years, the United States has managed to elect some leaders who actually take space defense and overall space policy seriously. Yet, the situation for the United States has reached a critical stage in the strategic domain of space. As you've seen, US rivals are already exploiting strategic gaps that a succession of US leaders, from both political parties, have allowed to form. While it is essential for the United States to plug those strategic gaps, in the short term, the United States is going to need to take actions to protect its existing satellite constellations from enemy attack and disruption. Therefore, Washington should look to its existing alliances to do in space what those alliances have usually done on Earth: enhance American strategic power.

The North Atlantic Treaty Organization (NATO) has often been dubbed the most successful defensive military alliance in history. Yet, as has been noted, it is probable that post–Cold War NATO policies have helped to incite conflict with Russia rather than deter it. If NATO's purpose is to deter Russian aggression through Euro-American unity, then NATO has failed in its mission thus far. Despite these facts, though, there are many stakeholders throughout Europe and in the United States who simply refuse to allow for NATO to be fundamentally reformed or to simply end NATO.[1] As the Russian threat to the West

---

1 Brandon J. Weichert, "NATO Must Pay," *American Spectator,* 13 July 2018. https://spectator.org/nato-must-pay/

intensifies—specifically the Russian threat to American space systems—then, it stands to reason that the United States should put NATO to some good use. This is especially true since the NATO defense of Europe rests upon America's satellite constellations. Hopefully, giving NATO a new shared mission would reinvigorate its otherwise sclerotic bureaucracy into becoming the effective defensive military alliance everyone remembers it as.

Consider this: nearly 70 percent of the United States Army's systems requires satellites to function on some level.[2] Also, the main force of Americans that would be used to defend the Baltic states in Europe from any potential Russian invasion would be elements belonging to the US Army. These units depend upon the Wideband Global Satcom (WGS) satellite constellation to provide them with critical, instantaneous battlefield communications.[3] American forces forward deployed in Europe, alongside their NATO partners, are also heavily reliant on several other sensitive satellite constellations in geosynchronous orbit around Earth—everything from early missile warning systems to covert surveillance and reconnaissance satellites.

The bulk of the Russian forces arrayed against NATO are what's known as heavy infantry—think of armored vehicles, such as tanks, armored personnel carriers, and so on—as opposed to the mostly light infantry forces of NATO, which are more like the soldiers of the US Army Airborne Rangers. The NATO side is poorly matched against such Russian forces. Still, NATO forces have technological superiority and "home field" advantage (as they will be the defenders), and the NATO troops, specifically the American forces, are better trained than their Russian rivals. The bad news for the West is that their forces are,

---

2 Sandra Erwin, "U.S. Army 'Space Week' to Shine Light on Why Soldiers Care About Satellites," *Space News*, 24 November 2017. https://spacenews.com/u-s-army-space-week-to-shine-light-on-why-soldiers-care-about-satellites/

3 Brandon J. Weichert, "Washington Is Still Not Getting Space Force Right," *American Greatness*, 20 April 2019. https://amgreatness.com/2019/04/20/washington-is-still-not-getting-space-force-right/

at the very least, 70 percent dependent on satellites to effectively defend Europe from a Russian invasion using heavy infantry and asymmetrical warfare tactics—counterspace, cyberspace, EM spectrum attacks, and overall information warfare techniques.

The Russians are not stupid. Moscow understands that it cannot take the United States and its NATO partners in a straight-up fight. They are not interested in doing this. Should conditions between the United States and Russia deteriorate to even below where they are at present, any offensive action taken by Moscow will likely be done through unconventional means.[4] The Russians, being a potent space power themselves, understand that any NATO defense of Europe will be completely reliant upon satellites to provide key command, control, communications, intelligence, and reconnaissance. For example, should a comparably smaller NATO unit become pinned down by a faster-moving and larger Russian force in an invasion scenario under normal battlefield conditions, the NATO force could rely on precise airstrikes to winnow down their attackers' numbers. *What would happen, though, if those critical satellites in orbit were destroyed or their functions disrupted—even for a short period of time?*

STRATEGIC SURPRISE IN SPACE

The RAND Corporation has conducted a series of war games going back to 2016 in which they pit American forces against its great power rivals, such as Russia and China. Without getting lost in the details of the game, RAND found that the US military "gets its ass handed to it."[5] In fact, according to one analyst, "In these simulated fights, [Russia or China] often obliterate US stealth fighters on the runway, sends US

---

4 Massicot, "Anticipating a New Russian Military Doctrine In 2020: What It Might Contain and Why It Matters" (see chap. 1, n. 33).

5 Sydney J. Freedberg Jr., "US 'Gets Its Ass Handed to It in Wargames: Here's a $24 Billion Fix,'" *Breaking Defense,* 7 March 2019. https://breakingdefense.com/2019/03/us-gets-its-ass-handed-to-it-in-wargames-heres-a-24-billion-fix/

warships to the depths, destroys US bases, and takes out critical US military systems."[6] During a 2016 war game, RAND determined that Russian heavy infantry would punch through NATO defenses and push into the capitals of Latvia or Moldova within sixty hours of the start of hostilities before meeting any kind of stiff NATO resistance. By that time, countless NATO—American—light infantry would have been killed, captured, or totally cut off by the Russian blitzkrieg.[7]

A Russian "Space Pearl Harbor" on US satellite constellations would merely allow for Russia to have dominance in any engagement with the US and NATO forces opposed to them.[8]

THE FRENCH CONNECTION

One of the surest ways to continue inviting Russian aggression is by leaving US satellite networks vulnerable, particularly those sensitive and expensive systems that are in GEO. Although NATO has become a shadow of its former self—more of a bloated multinational bureaucracy rather than an effective multilateral military alliance—several of its members remain militarily competent. What's more, some, such as France, are keen to expand their own military operations in the strategic domain of space. In 2019, French president Emmanuel Macron proudly announced his country's intention to develop a space force of its own.[9] As the only indigenous continental nuclear weapons state, this is entirely in keeping with overall French strategic ambitions.

---

6 Ryan Pickrell, "The US Has Been Getting 'Its Ass Handed to It' in War Games Simulating Fights Against Russia and China," *Business Insider,* 8 March 2019. https://www.businessinsider.com/the-us-apparently-gets-its-ass-handed-to-it-in-war-games-2019-3

7 David A. Shlapak and Michael W. Johnson, "Reinforcing Deterrence on NATO's Eastern Flank," RAND Corporation, 2016. https://www.rand.org/pubs/research_reports/RR1253.html

8 Brandon J. Weichert, "Preparing for a Russian 'Space Pearl Harbor,'" *Orbis,* 23 August 2019. https://www.fpri.org/article/2019/08/preparing-for-a-russian-space-pearl-harbor/

9 Nicholas Wu, "French President Emmanuel Macron Announces Creation of French Space Force," *USA Today,* 13 July 2019. https://www.usatoday.com/story/news/world/2019/07/13/french-space-force-macron-announces-creation-space-force-command/1723998001/

More important, Paris is explicitly concentrating on defending French satellites from attack. According to one industry insider, the French Ministry of the Armed Forces displayed an exhibit at the 2019 Paris Air Show that depicted what military space operations could look like in a few short years. "In their scenario, a foreign country declared a space exclusion zone in the GEO belt. It then sent a satellite very close to a French [communications satellite] that was in this zone. The [communications satellite] was escorted by a notional 'bodyguard' small satellite, tasked with observing the space around it and taking pictures of any interference attempts."[10]

The bodyguard satellite that France's military leadership has spent 2019 referencing is a small satellite, what's known as a cube satellite, which travels alongside a sensitive satellite. This bodyguard satellite allows for greater situational awareness in defending what is usually an undefended satellite performing key functions. The bodyguard satellite would take hundreds of pictures and relay those images to ground controllers. Should a space stalker approach a sensitive satellite, then the bodyguard satellites could spring into action—working feverishly to identify the incoming interloper and, if necessary, to attack the space stalker. Failing that, the bodyguard satellite could move in to defend the primary satellite it was tasked with protecting. Further, the French calls for the creation of exclusive self-defense zones would be critical for better protecting sensitive satellite constellations in orbit.

Consider the Russian "Space Pearl Harbor" scenario again, this time with the presence of French bodyguard satellites protecting key systems. What's more, with the establishing of the exclusive self-defense zones in orbit, the presence of Russian co-orbital satellites would immediately raise red flags in the West, and decisive action could be taken against such space stalkers *before* they could get close enough to American or NATO partner space systems. *There would be no "Space Pearl Harbor."*

---

10 Gosnold, "Space Situational Awareness News From the Paris Airshow," SatelliteObservation.net, 20 June 2019. https://satelliteobservation.net/2019/06/20/space-situational-awareness-news-from-the-paris-airshow/

For the time being, at least, the critical gap in America's space defenses would have been plugged through burden sharing with traditional American partners, such as France. As my colleague, Brian G. Chow argued in 2019, "The United States should talk to NATO partners, especially France, about expanding France's bodyguard program into a NATO initiative, including an effort to create appropriate self-defense zones around vulnerable but critical satellites."[11]

FRANCE HAS ALWAYS BEEN A SPACE POWER

Even without the creation of a French space force, France has always been a potent force in space. Thanks to their nuclear weapons program developed independently from NATO, France established the capabilities and infrastructure to support a space program. What's more, France has been able to play a key role in the vital space launch services sector due to their holding of French Guiana in the Amazonian jungle. French Guiana is just a few degrees north of Earth's equator. Due to this, launches from this position are cheaper than elsewhere on Earth. The Earth spins the fastest at its equator, meaning that rockets launching from French Guiana are given an extra boost by Earth's natural spin. Consequently, less fuel is required for rockets to break free of Earth's gravity—while launching a payload at least two times heavier than what usual launches allow for—at a cheaper price.[12] From its South American redoubt, France has launched increasingly complex missions into space, namely the European Space Agency's (ESA) Galileo global positioning system, which is meant to both augment and rival the US-controlled GPS constellation.[13]

---

11 Brian G. Chow and Henry Sokoloski, "The United States Should Follow France's Lead in Space," *Space News,* 9 September 2019. https://spacenews.com/the-united-states-should-follow-frances-lead-in-space/

12 Robin McKie, "How France's Space Ambitions Took Off in French Guiana," *Guardian,* 22 October 2011. https://www.theguardian.com/world/2011/oct/22/france-space-french-guiana

13 Stephen Clark, "Ariane 5 Rocket, Galileo Satellites on Launch Pad in French Guiana," *Spaceflight Now,* 24 July 2018. https://spaceflightnow.com/2018/07/24/ariane-5-rocket-galileo-satellites-on-launch-pad-in-french-guiana/

France has spent tens of millions of euros to make its spaceport at French Guiana a top-tier launch facility. France's spaceport ensures that Europe has a space program independent of the United States or Russia as well as a program that is French-led. France also means for their spaceport to compete with the primary US spaceport in Cape Canaveral, Florida. Currently, many space companies and foreign countries pay the US government to use its conveniently located spaceport in Florida to launch their systems into orbit. French Guiana is even more strategically located for launches, meaning that France could potentially take away from the business that Cape Canaveral has benefited from.

The presence of the French spaceport has given France and the European Union a competitive advantage. In fact, NASA is using the French spaceport to launch its James Webb Space Telescope into orbit in March 2021—aboard a French Ariane rocket, no less. Rather than view the French spaceport as a threat, then, American policymakers should view it as a critical backup for and augmentation to US space launch capabilities.[14] A crisis may arise in the near future where US forces must quickly deploy satellites, either cubesats or replacement satellites for constellations that fall under attack in orbit. To cut down costs and increase launch efficiency, it might be necessary to use the French spaceport.

---

14 Leah Brennan, "James Webb Telescope Project Manager, a Columbia Resident, Plans for NASA Launch," *Baltimore Sun,* 24 July 2019. https://www.baltimoresun.com/maryland/howard/cng-ho-bill-ochs-james-webb-telescope-nasa-20190724-xnlflaxg4bh3nfgzaooaecgvhi-story.html

14

JAPAN BUILDS ITS SPACE MINING

PROGRAM (AT COST!)

IN THE COLD DEPTHS OF SPACE, the tiny Japanese *Hayabusa2* probe hovered silently above the rough surface of the gray, hulking asteroid known to astronomers as Ryugu. *Hayabusa2* was launched in 2014 by the Japanese Aerospace Exploration Agency (JAX). It was a follow-up mission to the highly successful *Hayabusa* mission that went to an asteroid called "Itokawa." When *Hayabusa2* had initially arrived at the Ryugu asteroid in 2014, the Japanese probe deployed two small rovers that could hop. A few weeks thereafter, the probe deployed a microwave-size lander, which lasted seventeen hours after conducting surface experiments.[1]

In the winter of 2019, the *Hayabusa2* probe prepared to fire a copper projectile at the surface of Ryugu. The goal was to kick up enough surface debris that the probe could swoop in and collect the samples. The samples will be returned to Earth in capsule in December 2020. In this instance, the Japanese mission is scientific: Japan wants to "better understand the solar system's early history and evolution, as well as the role that carbon-rich asteroids such as Ryugu may have played in life's emergence on Earth."[2] Yet the same techniques that can conduct cutting-edge scientific research on asteroids can also be used for resource harvesting operations on resource-rich asteroids. Japan is pioneering these methods and could one day use these methods to get rich from space.

It is also important to note that while Japan's efforts with the *Hayabusa* program are directed at scientific research and discovery, they are not entirely egalitarian. In fact, when the *Hayabusa2* was launched in 2014, several media outlets made a point of highlighting the economic potential that Japan's methods would have for future space mining missions.[3] Effectively, Japan was building the infrastructure and capabilities needed to not only conduct revolutionary scientific research in space, but more fundamentally, to corner the space mining market—at cost. After all, while the United States, China, and Russia spend gobs of their taxpayer dollars on an ever-growing array of space missions, Japan is spending a fraction of that on space missions that will actually have positive economic impacts for their country.

---

1 Loren Grush, "Japanese Spacecraft Successfully Touches Down on an Asteroid, Grabbing a Sample of Dust," *Inverse,* 21 February 2019. https://www.theverge.com/2019/2/21/18234782/jaxa-hayabusa-2-ryugu-asteroid-sample-return-mission

2 Mike Wall, "Japanese Spacecraft Successfully Snags Sample of Asteroid Ryugu," Space.com, 22 February 2019. https://www.space.com/japanese-asteroid-probe-lands-ryugu.html

3 Katie Forster, "Japan Is Launching an Asteroid Mining Space Program," *Business Insider,* 2 September 2014. https://www.businessinsider.com/japan-is-launching-an-asteroid-mining-space-program-2014–9

But space mining is not the only area of the space industry that Japan is looking to profit from. That's a more long-term prospect. There are still many legal hurdles that the various spacefaring nations and corporations must overcome before anyone can start claiming the natural resources of any celestial body (this will be discussed in a later chapter). Japan is also pioneering the low-cost launch business. As you've already observed, one of the biggest hurdles for any space program is cost. And one of the most onerous costs is the launching of objects into space. Larger countries, like the United States, can afford to do more launches than a smaller country, like Japan, can. Rather than be cut out entirely of the budding space launch business by bigger countries possessed of more resources, Japan opted to embrace its inherent advantages as a technologically innovative, though relatively small, country.

SMART ROCKETS, FEWER PEOPLE, LOWER COSTS

Around 2011, Japanese space policymakers realized that they needed a better rocket that was cheaper than their competition to launch a coterie of satellites into orbit. The Japanese invested heavily in the creation of the Epsilon-3 rocket, which could carry a larger payload than other rockets while also requiring fewer human beings to operate it from the ground.[4] The Japanese also included rudimentary artificial intelligence on the rocket. Some came to call this new Japanese rocket a "Smart Rocket."[5] The Epsilon-3 had fewer, though more advanced, components than traditional rockets. The presence of an artificial intelligence system also means that the Epsilon-3 was self-correcting. To be clear, the AI was composed of various sensors in the Epsilon launch vehicle. The central connection of the various sensors then disseminated commands to the rocket "body." The Epsilon-3, and now, Epsilon-4, also

---

4 Adam Hadhazy, "Artificially Intelligent Rockets Could Slash Launch Costs," Space.com, 21 March 2011. https://www.space.com/11181-rocket-launches-artificial-intelligence-japan.html

5 Elizabeth Howell, "Japan's Brainy Epsilon Rocket Launching on 1st Test Flight Tuesday," Space.com, 26 August 2013. https://www.space.com/22530-japan-epsilon-smart-rocket-test-launch.html

require fewer human controllers and, in fact, can be controlled from the ground by *just two personal computers*! All of this lowered overhead costs for Japan in the launch services sector.

In 2019, an Epsilon-4 rocket carried seven satellites built by ten different organizations. According to the *Japanese Times*, "With demand growing globally for small satellites, JAXA hopes to attract orders with its Epsilon rockets that are specially designed to carry such satellites at a lower cost." Presently, Japan hopes to cut the Epsilon's $50 million launch costs in half to around $30 million. Understand that the Epsilon rockets, which are more advanced than the standard Japanese rocket, the H-IIA, are already cheaper than any other, less advanced Japanese rocket used today.[6] By lowering their launch costs, investing in a truly advanced technological marvel like the Epsilon rocket, and trying to corner the market on international satellite launches, the Japanese will better compete against larger countries in space.

Once Japan's rockets become a key player in launching valuable objects in space, Tokyo will then be able to merge this capability with their growing space mining capability. In this way, then, we are witnessing the complete democratization of space exploration, defense, and economic development. After all, by cutting launch costs through better rockets, JAXA will then be able to deploy more sophisticated space mining equipment into the cosmos.

Given that Japan is a traditional US ally, as threatened—if not more—by China, Russia, and even North Korea, it stands to reason that the United States should share the burden more with Japan in space policy, just as it should with France. In the event of a foreign attack on sensitive American satellites, it will be necessary to have reliable launch partners, like Japan. Plus, for the Pentagon to lower overall costs of its

---

6 Staff, "Japan Launches Epsilon Rocket Carrying Seven Satellites, Including One Supposed to Generate Fake Meteor Shower," *Japan Times,* 18 January 2019. https://www.japantimes.co.jp/news/2019/01/18/national/japan-launches-epsilon-rocket-carrying-seven-satellites-including-one-supposed-generate-fake-meteor-shower/#.XbfZbi-ZPyU

space operations, contracting the launch services of an allied country like Japan might help the Pentagon ensure the survival of America's vital satellite constellations.

In terms of low-cost launch, Japan has successfully launched a total of eight critical unmanned cargo resupply missions to the ISS. The Japanese missions are critical for maintaining the aging space station. They are also a great way for burden sharing from Washington's perspective, as the Japanese have become masters of relatively low-cost launches. What's more, as other countries contribute to the space station, the United States can save its taxpayers some money while not losing access to space. This is also good for JAXA, since they are able to build up their capacities for maintaining low-cost access to space. As the new space race heats up, the United States will need allies.[7] Its relationship with Japan will be critical to defeat its geostrategic rivals, notably China, in the race to dominate the Moon's resources.

JAPANESE MOON MISSIONS

The Japanese have also begun bearing the fruits of their highly successful Kaguya lunar orbiter mission. Beginning in 2007, Kaguya spent two years in a polar orbit around the Moon, taking some of the most detailed photos of the Moon and of Earth from the Moon in human history.[8] Kaguya not only took some amazing photos for all of humanity to enjoy, but it also discovered a "cave stretching 50 km (30 miles) on the Moon that could potentially serve as a shelter for astronauts on future missions."[9] Therefore, this tiny craft just potentially located the most

---

7 Stephen Clark, "Japan Set to Launch Space Station Resupply Mission," *Spaceflight Now,* 9 September 2019. https://spaceflightnow.com/2019/09/09/japan-set-to-launch-space-station-resupply-mission/

8 Eric Berger, "Japan's HD Photos of the Moon Are the Coolest Thing You'll See Today," *Ars Technica,* 11 October 2016. https://arstechnica.com/science/2016/10/japans-space-agency-just-released-a-trove-of-jaw-dropping-moon-photos/

9 Jiji Kyodo, "Japan's Lunar Probe Discovers Moon Cave, Which May Be Optimal Base for Space Exploration," *Japan Times,* 18 October 2017. https://www.japantimes.co.jp/news/2017/10/18/national/science-health/japans-lunar-orbiter-discovers-moon-cave-potentially-suitable-use-shelter/#.XbhPFS-ZPyU

profitable real estate on the Moon. JAXA believes that the chasm is "structurally sound and its rocks may contain ice or water deposits that could be turned into fuel."[10] Most experts believe that the only way a country or company could ever successfully colonize the Moon would be if they could take the least amount of equipment and resources (such as water, materials for long-term habitats, etc.). This is because the more things one must lift into orbit from Earth, the more expensive that space mission will be. The more expensive that space operations become, the less likely they are to be undertaken. So, if astronauts heading to the Moon could take less equipment with them and instead have the materials needed to survive constructed from indigenous resources already on the Moon, the cost of the mission will be greatly reduced, and colonization becomes feasible.

Japan announced its intention to place a Japanese astronaut on the lunar surface for the first time in its history by 2030. Tokyo is open to those astronauts being part of an international mission. One thing is certain, though: there is a new space race at hand—and the United States is being left behind, even by its allies. JAXA is building off its discoveries of potentially resource-rich areas for colonization of the Moon by announcing plans to develop a manned lunar rover together with the Japanese company, Toyota. According to the *Hindu* (newspaper), "JAXA vice president Koichi Wakata said the aim is to launch a rover into space in 2020."[11] As Toyota representatives explained, "The rover will be used for missions to explore the Moon's polar regions, with the aim of both investigating the possibility of using the Moon's resources—such as frozen water—and of acquiring technologies that enable exploration of the surfaces of massive heavenly bodies."[12]

---

10 Justin McCurry, "Discovery of 50km Cave Raises Hopes for Human Colonisation of Moon," *Guardian,* 19 October 2017. https://www.theguardian.com/science/2017/oct/19/lunar-cave-discovery-raises-hopes-for-human-colonisation-of-moon

11 Hindu Staff, "Toyota Motor Corp. and Japan Space Agency to Develop Manned Lunar Rover," YouTube video, 0:24–1:01, 13 March 2019. https://www.youtube.com/watch?v=-OsEdVS-73Y

12 Mike Wall, "Toyota, Japan to Launch Huge Moon Rover for Astronauts in 2029," Space.com, 23 July 2019. https://www.space.com/toyota-japan-moon-rover-2029-timeline.html

The rover is designed to carry two people comfortably. Designers, however, say it can carry four people during an emergency situation. Given that NASA intends to send astronauts to the Moon's south pole by 2024 to establish a research outpost and that the European Space Agency desires to construct a "Moon village" at some point in the next century, JAXA and Toyota may be cornering the market on reliable rover technology. Plus, according to Mike Wall at Space.com, there are "a number of private companies [planning to] mine water ice near the lunar south pole, where the material is abundant on the floors of permanently shadowed craters. This water could be used for life support and, when broken into its constituent hydrogen and oxygen, as rocket fuel."[13]

By lowering the costs of space operations, as Tokyo has done, Japan is pushing their relatively small country to the front of the space development line. Access to space is being democratized and opened to a plethora of new actors, any of which *could* be used to augment American power in space or, more dangerously, roll back America's dominant position in space. Japan's space policy is yet another example that the United States should learn to emulate. What's more, the United States should look at potentially purchasing some of Japan's Epsilon rockets as a way of cutting down launch costs in the near term. In fact, NASA under the Trump administration has restated its "intent to cooperate on lunar exploration, including Japanese roles in the lunar Gateway [a proposed space station around the Moon] and human lunar landings."[14] Of course, the joint NASA-JAXA statement of cooperation was, in typical NASA fashion, noncommittal.

Japan will be moving ahead, at lower cost, regardless of whatever Washington decides to do. Furthermore, Japan is only starting to realize its space ambitions. Clearly, Tokyo's leaders will not allow their country to be left behind in the new global space race. The same, unfortunately, cannot be said about American leaders over the last thirty years.

---

13 Wall.

14 Jeff Foust, "NASA and JAXA Reaffirm Intent to Cooperate In Lunar Exploration," *Space News*, 25 September 2019. https://spacenews.com/nasa-and-jaxa-reaffirm-intent-to-cooperate-in-lunar-exploration/

15

ISRAEL IN SPACE: SMALL COUNTRY,

BIG DREAMS

JUST AS WITH OTHER RISING MIDDLE POWERS LIKE JAPAN, Israel dreams of being a space power. And just like Japan, Israel is very clear-eyed about the possibility their relatively small size might hamper such ambitions. But Israel is undeterred in its quest to do in space what it has accomplished on Earth: to make what many believed was an impossible dream into a reality. On Earth, Israel proved that an independent—*strong*—Jewish state could exist in the geopolitical cauldron that is the

Middle East. In space, perhaps, Israel could achieve similar dispropor-tionate results. This is especially true since Israel is a high-tech power.[1]

Unlike many Mideast states, Israel does not sit atop a treasure trove of oil and natural gas. To survive economically and geopolitically in the rough Middle East, Israel's leadership had to rely on the best resource of all to make their small country a global power: the people of Israel. Because of this, Israel has created a high-tech sector that has also served as the basis for their budding space program. And as you've seen throughout this book, the more high-technology capabilities that a country has at its disposal, the more likely they are to have a robust space program, able to compete with the world's best space programs. As Opher Doron, the head of Israel Aerospace Industries' (IAI) Space Division, told the *Jerusalem Post* in early 2019, "[Israeli kids] are talking about Mars nowadays and exploration, comets, landings. So space is exciting. It is the ultimate tech-nology. It brings together everything in tech—from physics, engineering and launchers and loaders, you name it."[2]

ISRAEL BECOMES A SATELLITE SUPERPOWER

In 2019, the IAI celebrated its thirtieth anniversary in space. IAI built the first satellite for the Israel Space Agency, which was created in 1983 by then prime minister Menachem Begin. Because of Israel's hostile Arab neighbors, the small country had to launch their first satellite to the west rather than the east. Thirty years later and Israel's space sector is a global leader.[3] In fact, Israel's satellites are so efficient that they've all outlived their intended life span.

---

1 David Rosenberg, "How Israel Is Turning Its High-Tech into Global Political Power," *Fathom,* November 2018. http://fathomjournal.org/how-israel-is-turning-its-high-tech-into-global-political-power/

2 Seth J. Frantzman, "Space: Israel's Final Frontier," *Jerusalem Post,* 21 February 2019. https://www.jpost.com/Magazine/Israels-final-frontier-567448

3 Ari Rabinovitch, "Israel Uses Military to Join Commercial space race," Reuters, 4 March 2015. https://www.reuters.com/article/us-israel-space/israel-uses-military-expertise-to-join-commercial-space-race-idUSKBN0M016Y20150304

Just as with America's space program, Israel's space program was birthed out of military needs. Specifically, Israel needed reliable early-warning and surveillance satellite capabilities. After all, Israel has been at war for most of its existence (the country was officially founded in 1947 out of the former British Mandate of Palestine). In June 1967, Israel was surprise attacked by its overwhelmingly large Arab neighbors during the Six-Day War. Again, Israel was surprise attacked in 1973 by its Arab neighbors during the Yom Kippur War.[4] Plus, Israel has had to contend with ceaseless terrorism waged by Iran's Hezbollah proxies in neighboring Lebanon and by the Sunni terror group Hamas to Israel's south.[5] Also, Iran has spent decades screaming, "Death to Israel!" meaning that Tel Aviv has had to constantly guard against another surprise attack.[6]

So, Israel's space program and its attendant private-sector industries have given Israel's intelligence and military services precisely the surveillance and reconnaissance satellite capabilities their relatively small force needs to prevent (or, to better prepare for, at least) another surprise attack. What's more, Israel's military is similar to the United States' Armed Forces in terms of their reliance on advanced technology. Just as with American forces today, Israeli forces are highly integrated and rely on capabilities, such as satellite linkages in space, to serve as force multipliers against their far more numerous, though lower-tech, adversaries.[7]

In fact, Israeli satellite capabilities played a decisive role in destroying an illegal North Korean nuclear reactor that had been built in the desert

---

4 "Israel's Wars," *Israel Ministry of Foreign Affairs,* 2013. https://mfa.gov.il/MFA/AboutIsrael/History/Pages/Israel-Wars.aspx

5 Caroline B. Glick, "Column One: Mowing the Lawn in Gaza," *Jerusalem Post,* 18 October 2018. https://www.jpost.com/Opinion/Column-One-Mowing-the-lawn-in-Gaza-569775

6 Agencies, "Crowds Chant 'Death to Israel' as Iran Marks 40 Years Since Islamic Revolution," *Times of Israel,* 29 October 2019. https://www.timesofisrael.com/iran-marks-40-years-since-islamic-revolution/

7 Jeffrey Caton, "Joint Warfare and Military Dependence on Space," Federation of American Scientists, accessed 29 October 2019. https://fas.org/spp/eprint/LSN3APP2.htm

of Syria in 2007. Because Israeli satellites could take high-quality images of countries like Iran or Syria every ninety minutes, Israeli intelligence managed to spot the illegal North Korean reactor being built. Soon, Israeli intelligence determined that not only was Syria allowing the North Koreans to build their illicit weapons site on their territory, but that it was a joint North Korean and Iranian program meant to enhance both countries' nuclear weapons program (Iran's nuclear program being a direct threat to Israel).[8]

Israel's satellites are so sought after that in 2016, one of Israel's best-known satellite fleet operators, Space Communications (Spacecom), shocked the world when it announced it was selling their company to a Beijing conglomerate. As Peter B. de Selding of *Space News* reported, the transaction was set to be finalized "pending the successful entry into service of Spacecom's Amos-6 telecommunications satellite, built by [IAI]." Under the agreement with the Chinese conglomerate, Spacecom's satellite fleet would have still been managed from Israeli territory.[9] Ultimately, however, the transaction collapsed, after a freak explosion destroyed the SpaceX Falcon 9 rocket that was supposed to lift Israel's Amos-6 satellite into orbit.

Even a great space power like Israel is not impervious to the dangers of space. After exploding on the launchpad in Cape Canaveral, Florida, in 2016, the Chinese conglomerate lost interest in purchasing Spacecom. Meanwhile, it set the space industry—specifically, the company of Spacecom—back. SpaceX was forced to answer hard questions from an already-skeptical US government (more on that in a later chapter); Facebook was affected, as they had purchased a large block of Amos-6's

---

8 David Makovsky, "The Silent Strike," *New Yorker,* 10 September 2012. https://www.newyorker.com/magazine/2012/09/17/the-silent-strike

9 Peter B. de Selding, "Chinese Group to Buy Israel's Spacecom Satellite Operator for $285 Million," *Space News,* 24 August 2016. https://spacenews.com/chinese-group-to-buy-israels-spacecom-satellite-operator-for-285-million/

operations.[10] The satellite was supposed to "provide Internet access to underserved regions of sub-Saharan Africa."[11] To quickly make up for the loss of such precious equipment, like the Amos-6 satellite, Spacecom quickly signed a deal with the Hong Kong–based AsiaSat, worth $88 billion, in 2017. This was a quick but necessary fix, allowing Spacecom to fill in some lost capabilities.[12] Although, it did give another Chinese firm access to the highly advanced Israeli satellite company—as you've seen in previous chapters, China desperately wants and needs access to the high-tech of other countries to add that technological distinctiveness to Beijing's own growing power.

The Spacecom deal with the Chinese conglomerate in 2016 raised many national security concerns in the West, particularly in Washington. In fact, a variety of conspiracy theories proliferated the internet in the immediate aftermath of the Amos-6 explosion. Some ranged from the hilarious (a flying saucer destroyed the Falcon 9 rocket) to very slightly more believable, a drone flew overhead and used a laser to damage the rocket just before it conducted its static test fire. Others said a sniper fired a .50 caliber sniper round at the rocket from a mile away, penetrating the rocket's fuel tanks just before launch.[13] While these are nothing more than wild conspiracy theories, the fact remains that

---

10 Rafi Letzter, "A Major SpaceX Explosion Just Destroyed Facebook's First Satellite," *Business Insider,* 1 September 2016. https://www.businessinsider.com/spacex-falcon9-explosion-facebook-satellite-amos6-2016-9

11 Brian B. Berger and Jeff Foust, "SpaceX Falcon 9 Rocket and Amos-6 Satellite Destroyed During Static-Fire Test," *Space News,* 1 September 2016. https://spacenews.com/developing-explosion-rocks-spacex-falcon-9-pad-at-cape-canaveral/

12 Celeb Henry, "Spacecom Begins Service with a Borrowed Satellite Rebranded Amos-7," *Space News,* 27 February 2017. https://spacenews.com/spacecom-begins-service-with-a-borrowed-satellite-rebranded-amos-7/

13 Eric Shear, "Conspiracy Theories Regarding Amos-6 Falcon 9 Explosion Not Based on Physics, Reality," *Spaceflight Insider,* 4 October 2016. https://www.spaceflightinsider.com/organizations/space-exploration-technologies/conspiracy-theories-regarding-amos-6-falcon-9-explosion-not-based-on-physics-reality/

the United States has been increasingly troubled by the willingness of Israel's high-tech sector—including its space sector—to do business with China. Washington fears Israeli technology transfers to China, as well as Israel's overall deepening ties with China, so much that in 2019 the United States Senate pushed for legislation "expressing concern about Israel over its deepening economic ties with China."[14] In the words of a recent RAND Corporation report, what China seeks most from Israel is to "learn from policies and practices that Israel has put in place regarding innovation and entrepreneurship as China attempts to shift its economy from one that is investment- and export-led to one that is consumption- and innovation-led." In the eyes of Beijing, "Israel represents a success story of how a 'start-up nation' with a small population can innovate at a high level and produce large multinational technology companies that are competitive in the global economy."[15] The RAND Corporation report explains that since 2005, Israel's leadership has become increasingly open to not only limited economic trade with China, but they also seek greater ties with China in the sharing of technology, such as space technology, that would be considered "dual-use" by Washington. This is, in part, because of "Chinese investment in Israeli technology companies and venture capital firms that invest in dual-use technology."[16] China is, therefore, becoming increasingly important for Israel's high-tech sector.

From a geostrategic perspective, China not only has ambitions to coopt as much advanced Israeli technology and brainpower to add to their own national distinctiveness, but they have also striven

---

14 Amir Tibon and Amos Harel, "U.S. Senate Warns Israel over Deepening Ties with China, Citing 'Serious Security Concerns," *Haaretz,* 14 January 2019. https://www.haaretz.com/us-news/.premium-u-s-senate-condemns-deepening-israel-china-ties-cites-serious-security-concerns-1.7368680

15 Shira Efron, et al., "The Evolving Israel-China Relationship," *RAND Corporation,* 2019, p. 27. https://www.rand.org/content/dam/rand/pubs/research_reports/RR2600/RR2641/RAND_RR2641.pdf

16 Efron et al., 23.

to incorporate Israel into their Belt and Road Initiative (BRI). Thus, during Israeli prime minister Benjamin Netanyahu's March 2017 visit to China, President Xi Jinping stated that China and Israel would "steadily advance major cooperative projects within the framework of jointly building the Silk Road Economic Belt and the 21st Century Maritime Silk Road." For his part, Netanyahu proudly proclaimed that his country was "ready to actively participate in infrastructure and other cooperation under the framework of the Silk Road Economic Belt and the 21st Century Maritime Silk Road." Remember, the point of China's BRI is less about entering into enriching trade relationships with other Eurasian powers than it is about depriving the United States of its ability to disrupt international trade headed to China during a future geopolitical crisis. China's BRI also aims to supplant American influence in key areas with Chinese influence. The ceaseless attacks on intellectual property, the pilfering of national security and trade secrets, the construction of the BRI itself, and China's robust strategy for space dominance all play a role in making China, not the United States, the center of a new geopolitical world order. Israel, being a key strategic US partner in the Middle East as well as being a technology "superpower," is a lucrative target for Chinese influence-builders.[17]

It must be noted, however, that not all Israeli leaders are supportive of closer trade with China. After all, they fully understand that China merely seeks to absorb their country into its wider collective. Plus, whatever long-term benefits Israel might gain from closer alliance with China, Israel's leaders know that they cannot rely on China to help defend them from Iran. Only their military alliance with the United States will do that. And that alliance is threatened if Israel becomes too close to China—particularly as tensions between Beijing and Washington intensify.

---

17 Efron et al., 38.

ISRAEL PIONEERS THE CUBESAT REVOLUTION

Israel has now parsed their decades of nearly perfect satellite operations into pioneering a new form of satellite: the cubesat. These tiny, rudimentary satellites are meant to increase the bandwidth available to countries that rely on space systems. Israel has been on the cutting edge of this basic though revolutionary technology. Cubesats significantly reduce the cost of launch while cheaply enhancing the space capabilities of the company or country using them. Cubesats are tiny—in some cases anywhere between one hundred kilograms and merely a few kilograms. This means that a space agency, like Israel's, could deploy an entire array of these tiny beauties into orbit and still maintain, and possibly expand, the bandwidth that modern societies require. Opher Doron of IAI believes that the cubesat movement could be "called the next frontier or 'new space.'"[18]

A former IAI employee, Raz Itzhaki-Tamir, co-founded an Israeli start-up called NSLComm, which many consider to be on the cutting edge of next-generation cubesat development. According to industry insiders, NSLComm "has developed a flexible and expandable pattern-shaping spacecraft antenna that self-corrects on deployment to enable small satellites to project directable high-capacity footprints of more than one Gbps of throughput onto the Earth's surface." This technology is extremely light in weight and low volume, meaning that its launch costs will be drastically reduced, even when compared to other cubesats. Plus, the "technology aims to supercharge bandwidth from smallsats previously limited to low Mbps in throughput to rates 100 times the previous limitations." As Jeffrey Hill of *Satellite Today* explained in 2018, NSLComm is "planning to disrupt the world of smallsats as we understand it today" with their new venture.[19]

---

18 Frantzman, "Space."

19 Jeffrey Hill, "Startup Space Spotlight: Israel's NSLComm Is Supercharging Smallsat Bandwidth," *Satellite Today*, 29 May 2018. https://www.satellitetoday.com/business/2018/05/29/startup-space-spotlight-israels-nslcomm-is-supercharging-smallsat-bandwidth/

NSLComm launched its first cubesat in July 2019 atop of a Russian *Soyuz* rocket. The company reports that the satellite in orbit has a "more than 90 percent success of [its] mission goals." The company is set to launch two more cubesats soon. These satellites, according to Raz Itzhaki, are "fully-funded" and have received funding from a "major multinational corporation" that Itzhaki will not publicly identify.[20] National security professionals must inquire if another Chinese conglomerate, like the one that attempted to purchase Israel's Spacecom in 2016, is the private benefactor. Only the NSLComm leadership knows the truth. Still, there can be little doubt that Israel has become a true space power—at cost—and that their "start-up nation" mentality is the reason for their unwillingness to accept failure of any kind.

ISRAEL DREAMS OF THE MOON, TOO

Lastly, Israel has planned to do more in space. Of course, as Israeli space experts have stated, Israel has accomplished much on what amounts to a shoestring budget. Considering how little Israel's space budget is compared to what other national space programs receive, it's a wonder that Israel has accomplished the feats it has. The reason that Israel has achieved greatness in space, notably in the satellite sector, is for the same reason that drove the United States and Soviet Union in the Cold War to look to the cosmos: for national defense purposes. While Israel's private space sector has accomplished much—especially in the satellite management business—even its budget is significantly smaller than the budgets of foreign private space companies.[21] NASA received its largest single budget increase in decades under the Trump administration. As of 2019, NASA's overall budget was $19.5 billion. Meanwhile, the national

---

20 Debra Werner, "NSLComm Reports 90 Percent Success on First Cubesat Mission," *Space News*, 24 October 2019. https://spacenews.com/nslcomm-mission-one/

21 Asaf Ronel, "Jerusalem, We Have a Problem: Why Israel's NASA Isn't Taking Off," *Haaretz*, 15 September 2018. https://www.haaretz.com/israel-news/.premium-israel-s-nasa-struggles-to-lift-off-amid-budgetary-issues-1.6471276

space budgets of China, the ESA, Russia, France, Germany, Japan, Italy, India, and the United Kingdom all made the top ten best funded space budgets in the world.[22] Still, Israel envisions having a robust national space program and even desires going to the Moon.

As the world marveled at the historic images that China's *Chang'e-4* lunar rover beamed back to Earth from the dark side of the Moon, a tiny Israeli spacecraft, built by an Israeli nonprofit, SpaceIL, in conjunction with IAI, sojourned to the lunar surface. This spacecraft, known as *Beresheet,* whose name is derived from the term "In the Beginning" found in the book of Genesis, was launched from Cape Canaveral on February 21, 2019. The tiny craft was planned to land on the north part of the Mare Serenitatis portion of the Moon in April 2019. The $100 million *Beresheet* was lost after suffering an engine failure eight minutes into its critical descent to the lunar surface.[23] Upon witnessing the crash of the craft on the lunar surface, Prime Minister Netanyahu vowed to fund another mission to the Moon within a year or two. Meanwhile, SpaceIL announced that the next *Beresheet* mission would not attempt to land on the Moon again. Instead, SpaceIL wants to send their next *Beresheet* to another celestial body.[24]

One thing is clear: Israel's space capabilities may be small, but they are reliable. And when Israel's space mission fails, as it did with its *Beresheet* mission, failure is not final. What's more, Israeli leaders recognize the inherent value that space affords any country or company willing to exploit space. The United States has witnessed an increasing interaction at the high-tech level between Israel and a strategic foe, like

---

22 "Space Agencies and Their Budgets," *Radio Free Europe, Radio Liberty,* 12 February 2019. https://www.rferl.org/a/space-agencies-and-their-budgets/29766044.html

23 Hanneke Weitering, "Israeli Moon Lander Suffered Engine Glitch Before Crash," Space.com, 12 April 2019. https://www.space.com/beresheet-moon-crash-engine-glitch.html

24 Natt Garun, "Israeli Team Will No Longer Send a Second Spacecraft to the Moon," *The Verge,* 25 June 2019. https://www.theverge.com/2019/6/25/18758794/spaceil-beresheet-2-moon-spacecraft-new-objective-challenge

China. Yet, Israel has often resisted entirely embracing a full technology development partnership with China explicitly because of Washington's concerns. Regardless of how the geopolitical dynamic between Israel and the United States has changed over the decades, the fact remains that the United States and Israel need each other. They have a common foe in Iran and are both threatened by bloodthirsty jihadist terrorism. Plus, the United States remains a far more compelling economic and technology development partner than China, despite China's vast investments in their own high-tech sector. Both Israel and the United States are democracies, and both countries stand to benefit from increased space exploration and exploitation.

Going forward, as America's legacy satellites are increasingly threatened by China and Russian counterspace capabilities, Washington should be looking to Israel for inspiration as well as partnership in using Israel's immense satellite capabilities. An array of Israeli-built cubesats might be the difference between the United States completely losing critical military communications during a crisis and being able to effectively resist a Russian or Chinese onslaught. Making US legacy satellites survivable in the near term is an essential mission for American space warriors to undertake. Therefore, Israeli views of satellite management and their development of cubesats must be front and center for any American push to make its satellite capabilities more survivable in a future space war.

16

FLYING INTO SPACE AS CHEAPLY

AS BUYING AN SUV

SPEAKING TO REPORTERS, Pallava Bagla, an expert observer of India's space industry, remarked that "the chicken-hearted can't do planetary missions. It's only the brave hearted space agencies that do missions to other planets."[1] To understand India's space program, then, is to understand the

---

1 Jamie Carter, "India On Mars? Despite Failed Moon Landing Expect Orbital Spaceflight and Missions to Venus and Mars," *Forbes*, 5 September 2019. https://www.forbes.com/sites/jamiecartereurope/2019/09/05/india-on-mars-historic-moon-landing-will-spur-human-spaceflight-and-missions-to-venus-and-mars/#57ae0ad16ae1

profound nationalist ethos that undergirds this space program.[2] Much like India's strategic rival, China or Russia, India is not interested in taking on missions to "better the world" or some other airy notion that most American and European liberal elites tend to bandy about. Instead, India seeks to make history in space for its own national prestige, economic benefit, and yes, national defense. As you will see, India means business in space, and it will take no guff from its geostrategic competitors, like China—or even its partners, like the United States.

India is set to become the most populous country in the world in less than a decade, at which time there will be approximately 1.45 billion people living within India's borders. Given its large population, India is the world's largest democracy.[3] India is also the fifth-largest economy in the world, displacing the United Kingdom. What's more, if trends persist, India will surpass Japan as the third-largest economy by 2025.[4] Interestingly, India was also the "fastest-growing trillion-dollar economy in the world."[5] This country is also highly integrated into the world economy and exists in a tough neighborhood, nestled as it is between nuclear-armed Pakistan and the power-hungry China. India's space program is an extension of its national power. It has accomplished much already, and it intends to accomplish still more in the coming decade, all in an effort to secure itself on Earth and to ensure that it enhances its national prestige as well as its economy in the rapidly growing space economic sector.

---

2 Brandon J. Weichert, "Nationalism Is the Key Ingredient for Space Exploration," *American Greatness,* 10 August 2019. https://amgreatness.com/2019/08/10/nationalism-is-the-key-ingredient-for-space-exploration/

3 Sam Miller, "Ten 'Big Facts' About India," BBC, 18 February 2014. https://www.bbc.com/news/world-asia-25881705

4 "India to Overtake Japan to Become 3rd Largest Economy In 2025," *Economic Times,* 12 July 2019. https://economictimes.indiatimes.com/news/economy/indicators/india-to-overtake-japan-to-become-3rd-largest-economy-in-2025/articleshow/70193869.cms

5 Calen Silver, "Top 20 Economies in the World," *Investopedia,* 7 June 2019. https://www.investopedia.com/insights/worlds-top-economies/

In 2021, India intends to place three of their astronauts—known in India as "gaganauts"—into low-Earth orbit by 2022. At the start of 2020, India's space agency announced that four officers from the Indian Air Force had been selected as the gaganauts who will embark upon India's first manned mission to space. These gaganauts would sojourn to nearby Russia and receive intensive training from the Russians.[6] Here again is messy geopolitics mucking up US foreign policy. After all, the more powerful India becomes in the strategic domain of space, and the closer they are to Russia relative to other space powers, the weaker those other powers on Earth become.

The leaders of India's space program plan to deploy balloons in the stormy skies of Venus. They also have designs for landing a rover on Mars. Furthermore, India plans on sending its gaganauts to the Moon in the next decade, potentially in 2024, the same time that both China and the United States plan on landing their people on the Moon.[7] With each advancement, India solidifies its place in a new geopolitical pecking order where the most innovative and powerful countries are those with great technologies and robust space capabilities, married to ambitious plans.

INDIA RACES AMERICA (AND CHINA) TO MARS!

In 2014, as the American people busied themselves in their daily routines, assured that, whatever may happen in the unseen orbits around Earth or on the nearby Moon, the United States was the only country to make it to Mars since the Cold War, this perception was shattered in a powerful way. This was the year that NASA's Mars Atmosphere and Volatile Evolution (MAVEN) orbiter was expected to arrive in orbit

---

6 Nolan Pinto, "4 Astronauts Identified for Gaganyaan Mission: ISRO Chief K Sivan," *India Today,* 1 January 2020. https://www.indiatoday.in/science/story/gaganyaan-4-astronauts-identified-isro-k-sivan-1633024–2020–01–01

7 Elizabeth Howell, "India Looks Beyond the Moon to Mars, Venus and Astronaut Missions, Space. com, 8 September 2019. https://www.space.com/india-moon-mars-venus-exploration-plans-beyond-chandrayaan-2.html

around Mars. Once there, MAVEN's mission would be to study the Martian atmosphere, to determine how the atmosphere lost its integrity and, presumably, ultimately vented the planet's water to space over the course of billions of years.[8] This was just a routine mission for many Americans and, therefore, nothing to get too excited about. Americans haven't really cared about space since the Moon race with the Soviets.[9] And as you've seen, once Americans made it to the Moon a few times, both the people and their elected leaders lost interest.[10]

Yet, the MAVEN mission was different. Shortly before the NASA probe was to enter Mars' orbit, India's government announced that its *Mangalyaan* Mars orbiter was hot on MAVEN's tail. Arriving two days after the MAVEN probe, the *Mangalyaan* became the first probe from an Asian country to arrive in Martian orbit.[11] The *Mangalyaan* mission was all the more impressive because India's first Martian orbiter cost a fraction of NASA's MAVEN orbiter. The Indian orbiter was a meager $74 million as opposed to the aforementioned $671 million NASA orbiter. At the time, Indian prime minister Narendra Modi even quipped that the *Mangalyaan* mission to Mars was "costing less than the make-believe Hollywood film *Gravity* [which cost $100 million to make in 2013]."[12]

Of course, the *Mangalyaan* satellite was nowhere near as sophisticated as the MAVEN orbiter. That was not the point of the mission,

---

8 Miriam Kramer, "MAVEN: NASA's Orbiter Mission to Mars—Mission Details," Space.com, 19 July 2018. https://www.space.com/23617-nasa-maven-mars-mission.html

9 Kristin Houser, "Americans Don't Really Care About Reaching Mars, Says Poll," *Futurism*, 21 June 2019. https://futurism.com/the-byte/americans-dont-care-mars-poll

10 Staff, "Why Americans Lost Interest in Putting Men on the Moon," BBC, 24 July 2014. https://www.bbc.com/news/av/world-us-canada-28450386/why-americans-lost-interest-in-putting-men-on-the-moon

11 Lisa Grossman, "US Probe Enters Mars Orbit, Indian Craft Close Behind," *New Scientist*, 22 September 2014. https://www.newscientist.com/article/dn26244-us-probe-enters-mars-orbit-indian-craft-close-behind/

12 Jonathan Amos, "Why India's Mars Mission Is So Cheap—and Thrilling," BBC, 24 September 2014. https://www.bbc.com/news/science-environment-29341850

though. Much like China's present space activities, India's intentions with the *Mangalyaan* mission was a proof-of-concept for the world to see. Further, the Indian move was designed to send a signal to the world's other powers: India was a major player on the world stage.

According to Jonathan Amos's reporting for the BBC, "*Mangalyaan* has gone equipped with an instrument that will try to measure methane in the atmosphere."[13] The Indian probe will help to determine whether or not there are "some methane-producing bugs, or methanogens" existing on Mars—possibly living "underground, away from the planet's harsh surface conditions."[14] This is something that the world's scientific community has striven to learn, but MAVEN, for all of its complexity, could not study. In fact, the *Mangalyaan* mission was designed to fill critical gaps in both NASA's MAVEN mission and European Space Agency's Mars Express mission—making India an indispensable partner in critical scientific exploration.[15] This will not only empower India's knowledge base, but it will also further increase India's space capabilities, while enhancing India's national prestige on the world stage.

INDIA'S MOON VEHICLE CRASHES HARD

Even before India's historic Martian orbit, India dreamed of making it beyond Earth's orbit. Like so many other countries today, India wanted to land on the Moon. In 2009, India's *Chandrayaan-1* spacecraft managed to enter into lunar orbit, where it "made more than 3,400 orbits around the Moon during its operational life of 312 days." In fact, *Chandrayaan-1* made a "path-breaking discovery with the detection of water in the vapour form on the lunar surface."[16] In 2019, the

---

13 Amos.

14 Amos.

15 Amos.

16 India Today Web Desk, "Remembering Chandrayaan-1 Launch: When India Touched Moon 11 Years Ago," *India Today,* 22 October 2019. https://www.indiatoday.in/science/story/remembering-chandrayaan-1-launch-when-india-touched-moon-11-years-ago-1611686-2019-10-22

Indian Science Research Organization (ISRO), strove to make a lunar landing with their *Chandrayaan-2* probe. Much like the Israeli experience with their *Beeresheet-1* lunar probe, the second Indian mission ended in disaster. In what ISRO leaders referred to as "a rapid dive to 15 minutes of terror," the *Chandrayaan-2*'s engines failed on its final approach to the lunar surface.[17] Due to this failure, *Chandrayaan-2* was lost in September 2019. However, according to Indian leaders, the *Chandrayaan-2* mission was not a total loss, as valuable data was gleaned even after the craft had crashed on the lunar surface.[18]

ISRO has vowed to return to the lunar surface in the wake of the *Chandrayaan-2* failure.[19] In fact, ISRO has promised to make *Chandrayaan-3* "bigger and better" by attempting to return soil samples from the Moon's south pole.[20] ISRO also reached out to Japan's space agency, JAXA, in order to enhance the capabilities and longevity of the robotic mission to the Moon (the previous *Chandrayaan* mission was done along with Russia).[21] This would be a classic cost-saving move that would not only allow India to enhance the success of their next mission to the Moon, but would also help India to establish linkages with another rising, advanced space power, like Japan. These two

---

17 Madhumathi D.S., "Chandrayaan-2: A Rapid Dive to 15 Minutes of Terror," *The Hindu*, 8 September 2019. https://www.thehindu.com/sci-tech/science/chandrayaan-2-a-rapid-dive-to-15-minutes-of-terror/article29369002.ece

18 Staff, "Chandrayaan-2: Was India's Moon Mission Actually a Success?" BBC, 30 September 2019. https://www.bbc.com/news/world-asia-india-49875897

19 "Chandrayaan 3 to Come Soon; ISRO to Prepare for New Moon Mission," *Kerala Kaumudi Online,* 10 September 2019. https://keralakaumudi.com/en/news/news.php?id=154181&u=chandrayaan-3-to-come-soon-%C2%A0isro-to-prepare-for-new-moon-mission

20 Jeff Foust, "India Confirms Plans for Second Lunar Lander Mission," *Space News*, 1 January 2020. https://spacenews.com/india-confirms-plans-for-second-lunar-lander-mission/?fbclid=IwAR3ikFkoMWW1wkjqAxMjcctVa97-wkQ9B1Qut6mLFkQjRp-NLPTV70Unc7s

21 Chethan Kumar, "India's Next Moon Shot Will Be Bigger, in Pact with Japan," *Times of India*, 8 September 2019. https://timesofindia.indiatimes.com/india/indias-next-moonn-shot-will-be-bigger-in-pact-with-japan/articleshow/71030437.cms

spacefaring countries are currently middle-tier powers on their own. When combined, though, they form a potent force in space. On New Year's Day of 2020, ISRO announced that it would be launching the *Chandrayaan-3* mission to the Moon, with little mention of JAXA involvement. However, ISRO did state that a "lunar mission is being discussed with [JAXA], but its elements have not been finalised."[22] It remains to be seen if the two countries will, in fact, coordinate on this important mission.

INDIA'S SPACE PROGRAM IS A MILITARY ENDEAVOR

Another way that India is outpacing its foreign competitors in the space sector is their satellite launch program. For example, ISRO successfully launched 104 nano-satellites (or cubesats) belonging to companies in multiple countries, including the United States, the United Arab Emirates, Israel, the Netherlands, and Kazakhstan. The 2017 launch set a record for the most satellites placed into orbit on a single launch. Before 2017, Russia had made history by placing thirty-seven satellites in orbit during a 2014 launch. India has dedicated much effort to become a primary mover in the multibillion-dollar-per-year space launch market.[23]

At its core, though, India's mission in space is largely based on national defense needs. India has placed the RISAT-2B satellite in orbit that will "watch over India's borders and the country's surrounding waters to monitor for foreign threats. For example, it will be able to provide views of jihadi terror camps in the Kashmir region, which has long been the subject of a territorial dispute between India and Pakistan."[24]

---

22 Special Correspondent, "Gaganyaan, Chandrayaan-3 in Mission Mode, Says ISRO," *Hindu,* 1 January 2020. https://www.thehindu.com/sci-tech/science/gaganyaan-chandrayaan-3-in-mission-mode-says-isro/article30449839.ece

23 Associated Press, "India Launches More Than 100 Satellites into Orbit," *Telegraph,* 15 February 2017. https://www.telegraph.co.uk/news/2017/02/15/india-launches-100-satellites-orbit/

24 Hanneke Weitering, "India Successfully Launches RISAT-2B Earth-Observation Satellite," Space.com, 22 May 2019. https://www.space.com/india-risat-2b-earth-satellite-launch-success.html

The RISAT-2B is the third imaging satellite in India's growing constellation of imaging satellites. India has invested heavily into these systems, as India has been locked in a decades-long struggle against its majority Sunni Muslim neighbor Pakistan for preeminence on both the Indian subcontinent and throughout south Asia. What's more, India has been locked in a geopolitical struggle with their northern neighbor, China.

India's advanced space program allows New Delhi's leaders to keep both threats in check. Pakistan is a rival nuclear power. Yet, India is far more technologically advanced than their Pakistani foes.[25] Presently, tensions are mounting again between India and Pakistan over the two powers in the aforementioned Kashmir region. Should a conflict ever erupt between Pakistan and India, it is quite likely that the Indians will have significant advantages over their Pakistani rivals. The space architecture that India is building not only ensures that they will have a more watchful eye over Pakistan, but it also ensures they will have significant strategic advantages over the Pakistanis in war.

As Raja Mansoor of the *Diplomat* argued in 2018:

> ISRO is mainly involved in launching commercial satellites, those dealing with the weather, space navigation, and communications. However, Pakistani authorities should be alarmed due to the multi-purpose nature of satellite. A satellite network provides India with a technological advantage on the ground and, in case of war, can be easily exploited for tactical and strategic gains. With a vast array of satellites, India can keep tabs on its borders through high-resolution imagery, intelligence gathering, navigation, and military communications—thus undermining Pakistan defenses. These satellites will also help India develop early warning systems specifically designed to detect ICBMs during different flight phases of incoming ballistic/cruise missiles.[26]

25 Zeba Siddiqui, "Factbox: India and Pakistan—Nuclear Arsenals and Strategies," Reuters, 1 March 2019. https://www.reuters.com/article/us-india-kashmir-pakistan-nuclear-factbo/factbox-india-and-pakistan-nuclear-arsenals-and-strategies-idUSKCN1QI405

26 Raja Mansoor, "Pakistan Is Losing the space race," *Diplomat*, 1 February 2018. https://thediplomat.com/2018/02/pakistan-is-losing-the-space-race/

Both India and China, meanwhile, have endured a tense relationship ever since the 1962 Sino-Indian War, which was waged over control of the mountain passes separating southern China from northern India. India has felt the pressure of China's geopolitical rise over the last forty years more than most other countries. The more powerful China becomes, the more aggressive along India's borders Beijing gets. In 2017, a conflict erupted between India and China over a tiny strip of mountains along the Indo-Chinese border, known as Doklam. India challenged China's claim to the territory.[27] Video from 2017 has since surfaced online of Chinese and Indian units facing off against each other along the Indo-Chinese border. In these videos, Indian and Chinese troops engage in sporadic physical altercations while their comrades film their encounters, as though they were partaking in some bizarre rip-off of the popular Brad Pitt and Ed Norton film, *Fight Club*.[28] New Delhi argued that India had won because they managed to restore the status quo. Beijing insisted that China was the victor, since the standoff at Doklam with India allowed China to permanently station its forces there whereas Indian forces withdrew.[29] In any event, this 2017 incident underscores the deep rancor and division the two nuclear-armed neighbors have toward each other.

Considering these threats to India, it should not be surprising how matters of defense and prestige have factored into the Indian government's choice of enhancing their space capabilities. It is true that India's space program is outwardly focused on commercial and scientific development. Yet, civilian space development allows for India's leaders

---

27 Joel Wuthnow, Satu Limaye, and Nilanthi Samaranayake, "Doklam, One Year Later: China's Long Game in the Himalayas," *War on the Rocks*, 7 June 2018. https://warontherocks.com/2018/06/doklam-one-year-later-chinas-long-game-in-the-himalayas/

28 WildFilmsIndia, "Troops at Indo-China Border Face Off Against Each Other," YouTube, 20 December 2017. https://www.youtube.com/watch?v=84b1PSOQAaU

29 M. Taylor Fravel, "Why India Did Not 'Win' the Standoff with China," *War on the Rocks*, 1 September 2017. https://warontherocks.com/2017/09/why-india-did-not-win-the-standoff-with-china/

to satisfy their own geopolitical ambitions and defend India's national security. It is very telling that ISRO conducted its first successful satellite launch in 2008 in response to the deadly Mumbai terrorist attacks. India wanted to create a reliable network of surveillance satellites that could do a better job of keeping an eye on the troubled Kashmir region since the Islamist terrorists who attacked Mumbai in 2008 belonged to Lashkar-e-Taiba, a Pakistani-supported terrorist group in Kashmir. Similarly, since Pakistan and India both have nuclear arsenals, India wants a technical and strategic edge over their Pakistani foes—such advantage has been provided for by the Indian space program.

In 2019, India shocked the world when it successfully tested an anti-satellite weapon of its own. Under the code name of "Mission Shakti," a derelict Indian weather satellite was targeted for destruction by India's military with an ASAT. Many of the Western governments decried India's action, as India's government, like the Chinese government in 2007, did not announce its ASAT weapons launch before doing it. Space policy analysts, such as Brian Weeden of the Secure the World Foundation, argued that India's launch was a "wake-up call" for the international community to come together and create better rules for operating in an increasingly "complex space domain."[30]

This is certainly a fair point. But the consternation over India's ASAT test is puzzling. Yes, India should have notified the world. And India certainly did create a dangerous debris field in orbit. Yet, unlike the Chinese ASAT test in 2007, the Indian ASAT test created a debris field that will only last for several weeks to months. The Chinese test, on the other hand, will last for years to come. The true purpose of the Indian ASAT test was to send a decisive signal to the Chinese, a common rival that India shares with the United States and Japan. India's leaders were conveying to China's leadership that they possess the same space

---

30 Brian Weeden and Victoria Samson, "Op-Ed: India's ASAT Test Is Wake-Up Call for Norms of Behavior in Space," *Space News,* 8 April 2019. https://spacenews.com/op-ed-indias-asat-test-is-wake-up-call-for-norms-of-behavior-in-space/

capabilities as China. In response to the Indian ASAT test, I wrote in the *American Spectator* that "the arms race [in space] began in 2007 and it was initiated by the People's Republic of China. India joining this new space arms race is a good thing. Together, the United States and India can counter the rising threat of China in space. Our two countries can bring their considerable resources to bear to prevent the Chinese from monopolizing space."[31] This is especially true as the United States attempts to organize a coalition of resistance against China's rising regional power in the form of the Quad Alliance—the United States, Japan, India, and Australia.[32]

Further, ISRO coexists with a lesser-known Indian group called the Defense Research and Development Organization (DSRO). This is a component of India's military. Another group belonging to the Indian military is the Defense Research and Development Service (DRDS). As their names suggest, each organization is involved in the development of weaponizing emerging technology. DSRO and DRDS have their hands in India's various missile programs; they are involved in the creation of new weapons for India's military, and they are heavily invested India's budding cyberwarfare and electronic warfare capabilities. In 2016, ISRO, DRDS, and DSRO all partook in the development of India's new Agni V ICBM, which can hit targets up to three thousand miles away.[33] All of this was made possible by India's small but consistent investment into the development of its space program.

Regarding the China threat to India, ISRO recently deployed the GSAT-7 satellite for the expressed purposes of enhancing their naval

31 Brandon J. Weichert, "The Arms Race in Space Is Here," *American Spectator,* 5 April 2019. https://spectator.org/the-arms-race-in-space-is-here/

32 Cary Huang, "US, Japan, India, Australia . . . Is Quad the First Step to an Asian Nato?" *South China Morning Post,* 25 November 2017. https://www.scmp.com/week-asia/opinion/article/2121474/us-japan-india-australia-quad-first-step-asian-nato

33 Franz-Stefan Gady, "India to Test Fire Missile Capable of Hitting China," *Diplomat,* 15 December 2016. https://thediplomat.com/2016/12/india-to-test-fire-nuclear-missile-capable-of-hitting-china/

capabilities. Presently, Beijing is intent on creating what's known as a "blue-water" navy capability.[34] In maritime parlance, there are three forms of naval power: the smallest, a "brown-water" navy basically means that your navy is operating on rivers and in local lakes within your territory.[35] The American PT Boat fighting in the Vietnam War, patrolling the rivers of South Vietnam, was a form of America's brown-water navy capability.[36] Next is the "green-water" navy. This is a navy that can only project power in its surrounding region that is possessed of "smaller, single-purpose [coastal and littoral] combatants."[37] China today is technically a green-water navy, as is India's navy. Both powers aspire to achieve a "blue-water" naval status—meaning that they have a global reach, with "multipurpose, expensive, heavily-manned [open-ocean] ships."[38]

What readers may not realize is that for a country to truly be a modern blue-water navy, they need a robust satellite architecture in orbit supporting those power projection activities globally. China has been making big moves to not only enhance its own naval capabilities, such as building modern destroyers and aircraft carriers, but Beijing has also used its newfound naval power to bully its neighbors, from Japan to Taiwan to Vietnam to the Philippines at sea. Meanwhile, Chinese naval strategists view the Indian Navy as a major strategic competitor, especially regarding Beijing's plans to become the dominant power not only in the western and southern Pacific Ocean but the Indian Ocean

---

34 Yoji Koda, "China's Blue Water Navy Strategy and Its Implications," Center for a New American Security, 20 March 2017. https://www.cnas.org/publications/reports/chinas-blue-water-navy-strategy-and-its-implications

35 "Building a Brown Water Navy," *American Battlefield Trust,* accessed 29 October 2019. https://www.battlefields.org/learn/articles/building-brown-water-navy

36 "Brown Water Navy in Vietnam," *Naval Historical Foundation,* 11 January 2012. https://www.navyhistory.org/2012/01/brown-water-navy-in-vietnam/

37 Wayne P. Hughes Jr., "Build a Green-Water Fleet," *Proceedings,* June 2018. https://www.usni.org/magazines/proceedings/2018/june/build-green-water-fleet

38 Hughes.

as well. India's recent GSAT-7 and similar satellites will support the Indian Navy in its mission to beat back China's unwanted influence in the Indian Ocean—and perhaps challenge China in waters closer to their shores.

Writing in 2013, when the GSAT-7 was launched atop a French rocket, Ajey Lele of *Space News* said:

> This dedicated satellite is expected to provide the Indian navy with an approximately 3,500-to-4,000-kilometer footprint over the Indian Ocean and enable real-time networking of all its operational assets in the water (and land). It also will help the navy to operate in a network-centric atmosphere. The Indian peninsula is an extremely tricky region for operations because of its geographic location. One of the deadliest operations on Indian soil, the infamous 2008 Mumbai attack, was launched using the Arabian Sea route. It is expected that GSAT-7 will be helpful for gathering communications and electronic intelligence in respect to moving platforms in the sea, particularly through its [ultra-high frequency] facility. GSAT-7 also will help the navy monitor activities over both the Arabian Sea and the Bay of Bengal region. Broadly, India's strategic area of interest extends from the Persian Gulf to the Malacca Strait, and now a significant portion of this region will be covered by the navy's satellite.[39]

Everything that India has done in space has a direct impact on its geopolitical strategy and position on Earth. Thanks to its mastery of civilian satellite launches, India has integrated itself into the global business community. Everyone from foreign countries, like France or even the United States, to major technology firms, like Google, are coming to rely upon the relatively cheap and safe Indian model for space development. These are connections that Beijing desires to make for its own purposes. India is, in many respects, edging China, Inc. out of

---

39 Ajey Lele, "Commentary: GSAT-7: India's Strategic Satellite," *Space News,* 9 September 2013. https://spacenews.com/37142gsat-7-indias-strategic-satellite/

the space launch sector—which will have larger ramifications geopolitically, should the trend persist. India has been quietly building a reliable
military satellite network as well that enhances New Delhi's ability to
defend itself against both China and Pakistan.

CHINA FEELS THE HEAT FROM INDIA

In fact, India is so hot and heavy into the new global space race that
China has announced new rocket plans meant to challenge India's
reputed dominance of low-cost satellite launches. These new Chinese
rockets are seen as Beijing's response to New Delhi's challenge in the
civilian space launch market. When India conducted its historic single
launch of 104 satellites into Earth orbit, the Chinese state-run *Global
Times* ran an article warning China's leadership that "China's space
industry is lagging behind that of India in commercial space industry."[40]
The *Global Times* article went on to argue that China needed to lower
its commercial space costs to better compete with India. China's new
rocket, the *Tenglong*, is a liquid rocket that could "peg the launch cost
for each kilogram of payload to within USD 5,000."[41] The Chinese
rocket is expected to make its maiden launch in 2021.[42]

Taken together, India's space program is a net positive for the
United States. Washington must increase its partnerships with India's
defense industry, specifically its industries related to space. Already US
companies use India's cheaper launch system to deploy commercial

---

40 Sutirtho Patranobis, "ISRO Record: Chinese State Media Says Beijing Can Learn Lessons
from India," *Hindustan Times,* 20 February 2017. https://www.hindustantimes.com/india-news/
isro-launch-china-could-learn-some-lessons-from-india-on-space-commerce-says-state-media/story-
hUXI9C9ReS0KmTMOnwZYEM.html

41 "China Unveils New Commercial Carrier Rockets to Compete with India for Global Market,"
Economic Times, 21 October 2019. https://economictimes.indiatimes.com/news/science/china-unveils-
new-commercial-carrier-rockets-to-compete-with-india-for-global-market/articleshow/71684151.cms

42 Andrew Jones, "Chinese Commercial Launch Companies Are Preparing Second-Generation
Rockets," *Space News,* 24 October 2019. https://spacenews.com/chinese-commercial-launch-
companies-are-preparing-second-generation-rockets/

satellites into orbit. The Pentagon may want to get in on that action as Washington's leaders explore innovative ways to make America's precious satellite constellations more survivable. Meanwhile, rather than join the great international pile-on of India for their ASAT test, the United States should be giving India pointers for how best to move forward and to enhance their strategic threat to China. Mutual space development between the United States and India is a natural fit—it only enhances US dominance of space rather than weakens it. Washington must act boldly, though, to woo Delhi out of Russia's orbit in matters of its space program and toward the West. To paraphrase former US President Lyndon B. Johnson, I'd rather have India inside America's strategic tent pissing out than India outside of America's tent, pissing into it. India is a real space power, and they are not going away anytime soon.

17

BRINGING BACK BRAZIL

THE UNITED STATES has a long history with Latin America, a region that the United States shares a massive land border with as well as several maritime passes. Washington has always had an outsized presence in Latin America. Because of this proximity and the fact that Washington has long favored a hegemonic foreign policy—particularly close to its borders—its relationship with Latin American states has historically been negative. Things got progressively worse during the Cold War. The region became a proxy battleground between Communist insurgents, with support from the Soviet Union, fighting to ensure that the region embraced Communism. When the Cold War ended, many of the negative ideological aspects of Communism remained. Brazil was no exception.

Brazil is the fifth most populous country in the world.[1] Brazil has the most powerful military in South America.[2] It also forms the "B" in what Goldman Sachs referred to as the "BRIC" nations (Brazil, Russia, India, and China) in 2001. These were the cluster of rising states that were moving rapidly from developing countries to modern developed ones. Basically, economists think "these four nations will become dominant suppliers of manufactured goods, services, and raw materials by the year 2050."[3] When Goldman Sachs first posited this acronym in 2001, Brazil and the other BRICs were seen as the countries of the future.[4] Today, Brazil has fallen on some relatively hard economic times.[5] Yet, the fundamentals of Brazil make this country, so geographically close to the United States, an enticing potential partner for Washington.[6] Still, the relationship between Brasilia and Washington has been frosty.

From 2003 until 2013, a succession of grotesquely corrupt socialist governments in Brazil ensured that there was little growth in what should have been a profitable and amicable US-Brazil relationship.[7] Finally, though, events in Brazilian politics took a drastic turn as the Brazilian voters, tired of the endless corruption and absolute degradation of their

---

1 "Brazil," *Encyclopedia Britannica,* 28 October 2019. https://www.britannica.com/place/Brazil

2 "South American Powers Ranked by Military Strength," *Global Firepower,* accessed 30 October 2019. https://www.globalfirepower.com/countries-listing-south-america.asp

3 Christina Majaski, "Brazil, Russia, India, and China (BRIC)," *Investopedia,* 22 August 2019. https://www.investopedia.com/terms/b/bric.asp

4 Jim O'Neill, "Building Better Global Economic BRICs," Goldman Sachs, 30 November 2001. https://www.goldmansachs.com/insights/archive/archive-pdfs/build-better-brics.pdf

5 Heather Timmons, "The BRICs Era Is Over, Even at Goldman Sachs," *QZ,* 9 November 2015. https://qz.com/544410/the-brics-era-is-over-even-at-goldman-sachs/

6 Marcos Degaut, "Why Washington Doesn't Get Brazil," *The National Interest,* 21 November 2016. https://nationalinterest.org/feature/why-washington-doesnt-get-brazil-18473

7 Paula J. Dobriansky, "When Trump Supporter Bolsonaro Visits D.C., It Will Reset Brazil's Relationship with U.S.," *Miami Herald,* 11 March 2019. https://www.miamiherald.com/opinion/op-ed/article227429304.html

country—at the hands of domestic socialists—elected a right-wing government to power for the first time in more than a decade. The sixty-three-year-old former Brazilian Army captain, Jair Bolsonaro, was elected to the presidency in 2019 by winning about 75 percent of the Brazilian vote.

Speaking for every Brazilian voter, Bolsonaro told the legislators of Brazil's Congress to take up "the mission of rebuilding our nation, freeing it of the yoke of corruption, crime, economic irresponsibility, ideological submission." He vowed to work "toward sustainable public accounts, pursuing free-market principles and efficiency without an ideological bias." Bolsonaro is known as a lover of free-market capitalism and has a policy bent toward a smaller government that is far less regulated than it was under the decade-long reign of corrupt leftists. He is also the most overtly pro-American Brazilian leader in decades.

Bolsonaro's "stripped-down" presidential cabinet consists of an "astronaut, an anti-corruption czar, retired military officers, and a group of liberal economists."[8] It is the first man nominated to Bolsonaro's cabinet that draws my attention for this work: the selection of Brazil's first man into space, Marcos Pontes, as the minister of science, technology, and innovation.[9] Bolsonaro, like President Donald Trump in the United States today, is hated by Brazil's political establishment—and anyone associated with Bolsonaro has endured ceaseless attacks on their character. Thus, when Marcos Pontes was nominated, and days later, he accepted the nomination to head Brazil's Science, Technology, and Innovation Ministry, Pontes was endlessly lampooned by establishment figures on both the left and the right.[10]

---

8 Mario Sergio Lima and Flavia Said, "Bolsonaro Takes the Reins in Brazil In Nationalist Surge," *Bloomberg*, 1 January 2019. https://www.bloomberg.com/news/articles/2019–01–01/former-army-captain-takes-brazil-s-reins-in-surge-of-nationalism

9 Rodrigo Viga Gaier, "Brazil's Bolsonaro Taps Astronaut, Courts Judge for Cabinet," Reuters, 31 October 2018. https://www.reuters.com/article/us-brazil-politics/brazils-bolsonaro-taps-astronaut-courts-judge-for-cabinet-idUSKCN1N52Q3

10 Evanildo de Silveira, "Um Ministro em Órbita," *Apublica*, 18 March 2019. https://apublica.org/2019/03/um-ministro-em-orbita/

Still, Pontes's nomination indicates that the Bolsonaro administration of Brazil will look to the cosmos as a place to enhance Brazil's flagging international prestige to better its economic position in the lucrative civilian space economy and, potentially, to bolster its national defense.[11] After all, it must now contend with a collapsing Venezuela to its north; the presence of Iranian, Russian, and Chinese elements throughout the region; and a constantly hectoring Cuba. Now that Brazil's government has decisively shifted to the right and taken a loud pro-American stance on most issues, Brazil will face threats in its near future that it has not had to deal with for some time. As I have argued in previous chapters, Brazil's space program could be a place to build real strategic linkages with the United States while bettering Brazil's indigenous capabilities and wealth.

BRAZIL AND CHINA COOPERATE IN SPACE

Under the previous left-wing governments, Brazil moved inextricably closer to both China and Russia. Brazil's Space Agency, in particular, worked in tandem with China to develop and launch the Chinese-Brazil Earth Resources Satellite (CBERS). The program between China and Brazil was created in 1984. Its first satellite was launched in 1999. Three of its satellites have since been retired, and another was lost in an accident during launch aboard a Chinese *Long March 4B* rocket. One remains today in orbit and two more are currently being designed. The CBERS satellite program was designed "for Earth observation for use in areas including environment monitoring, meteorology and map making."[12] This program was a major strategic boon for Beijing, as it got China into a technological relationship with a major power in the Western Hemisphere.

---

11 Staff, "Bolsonaro Says Brazil 'Liberated From Socialism' at Inaugural Ceremony," *France 24*, 1 January 2019. https://www.france24.com/en/20190101-live-jair-bolsonaro-inauguration-oath-president-brazil

12 Andrew Jones, "New China-Brazil Earth Resources Satellite in Launch In H2 2019," *GB Times*, 23 November 2018. https://gbtimes.com/new-china-brazil-earth-resources-satellite-in-launch-in-h2–2019

More importantly, this relationship, through the CBERS program, would allow for Beijing's exploitation of vital natural resources there.

CBERS allows China to seek out new potential resource spots in Latin America. As you've seen, China is resource hungry. In 2017, for instance, "a major new infrastructure fund backed by China" was announced. The focus of the fund was to create the "infrastructure to speed resource development across Brazil."[13] China is making major moves toward dominating the resource sector of South America. Beijing is especially interested in tapping into Brazil's oil market.[14] The CBERS satellite constellation, which Brazil is still required to provide support for, will help both China and Brazil discover new resources for exploitation in Brazil and the wider Latin American region.

As the administrator for the Brazilian Space Agency (Agência Espacial Brasileira, or AEB), Jose Raimundo Braga Coelho, said to Chinese state-run media in 2018:

> For all countries, especially when they use certain a [sic] kind of satellite like CBERS, that is good for everybody, especially Asian countries and also South American countries. The importance of our cooperation is not to develop satellites, but it's to use the satellite data to develop applications that are good for every country, many countries.[15]

Even with the removal of Brazil's socialist government and its replacement by a right-wing nationalist government, as led by President Bolsonaro, Brazil remains interested in working on infrastructure development with the Chinese. Brazilian vice president Hamilton Mourao

---

13 Dave Forest, "This Region Is China's Next Target for Resource Deals," *Oil Price,* 28 June 2017. https://oilprice.com/Metals/Commodities/This-Region-Is-Chinas-Next-Target-For-Resource-Deals.html

14 Peter Millard, "Why Billions of Barrels of Oil Go Untapped in Brazil," *Bloomberg,* 3 March 2019. https://www.bloomberg.com/news/articles/2019–03–03/why-billions-of-barrels-of-oil-go-untapped-in-brazil-quicktake

15 Staff, "China, Brazil to Launch New Earth Resource Satellite Next Year," *Xinhua,* 22 November 2018. http://www.xinhuanet.com/english/2018–11/22/c_137624776.htm

met with China's leader, Xi Jinping, in May 2019, when Mourao insisted that "a Chinese company cannot arrive [in Brazil] and bring 100,000 Chinese people to work in Brazil."[16] Yet, Mourao also reaffirmed his country's interest in joint infrastructure development projects in Brazil. Despite both the flowery language of Jose Coehlo and the purported pragmatism of Hamilton Mourao, I believe that the Sino-Brazilian relationship is lopsided. The country that most benefits from satellites aimed at identifying exploitable natural resources in South America is China. Jose Coelho was also, unfortunately, correct when he stated to Chinese state-run media in 2018 that the CBERS program was not intended to develop satellites. It truly is about the data. What do you think Beijing will do with that data? Sure, China will use the data CBERS collects to "develop applications." But those applications are specifically to be used for acquiring the untapped natural resources of Latin America—first and foremost to benefit Beijing! Whatever benefits Brazil enjoys will be ancillary compared to the windfall China reaps from these development deals.

And because Beijing entered into a cooperative agreement with Brasilia to build, launch, and maintain these satellites, this partnership gives China's blatant hunt for resources in Latin America an air of legitimacy that it would otherwise lack. In fact, Chinese mining and infrastructure investment rarely benefit the local countries where Chinese overseas mining and infrastructure projects occur. Just ask the Kenyans.[17] Apparently, the Bolsonaro government has made inquiries into this matter, as Vice President Mourao's comments in 2019 suggest. By insisting that any infrastructure deal with Beijing disallow scores of Chinese workers from moving into Brazil to staff these projects,

---

16 Ricardo Brito and Lisandra Paraguassu, "Brazil Wants China to Invest in Its Infrastructure: Vice President," Reuters, 13 June 2019. https://www.reuters.com/article/us-brazil-china/brazil-wants-china-to-invest-in-its-infrastructure-vice-president-idUSKCN1TE2YH

17 Joseph Goldstein, "Kenyans Say Chinese Investment Brings Racism and Discrimination," *New York Times*, 15 October 2018. https://www.nytimes.com/2018/10/15/world/africa/kenya-china-racism.html

the Bolsonaro regime obviously thinks it can avoid Africa's horrible experience with China's neocolonialism. China needs the resources of countries like Brazil and the wider Latin American region. Therefore, Beijing will play great lip service to respecting the national sovereignty and rights of countries, like Brazil. But for Beijing, Brazil is nothing but food with which to feed China's insatiable hunger for natural resources. It is another resource to assimilate into its collective. This relationship also provides Red China with a strategic foothold in a part of the world that Washington has long considered to be its proverbial backyard. No amount of diplomatic legerdemain on the part of Brazil's leadership will prevent potential infrastructure development deals with China from going as badly for them as they've gone for African countries that have accepted Chinese development deals.

Of course, there are *some* benefits for Brazil as well. Satellites, like CBERS, can help Brazil improve the management and use of resources. They can also assist local governments in better protecting against natural disasters and disaster cleanup. Data provided by the CBERS could help Brazil analyze soil erosion, for example.[18] Yet, the Chinese have a long history of not treating their partners fairly.[19] China usually views its partners in countries like Kenya or Brazil as mere vassals to China's burgeoning Middle Kingdom.

Let us consider China's ongoing strategic development of Africa, as well as China's growing Belt-and-Road Initiative and apply these lessons to other parts of the world that Beijing seeks to develop.[20] In fact, there

---

18 Ray A. Williamson, "The Growth of Global Space Capabilities: What's Happening and Why It Matters," Secure the World Foundation, 19 November 2009. https://swfound.org/media/36582/swf-testimony%20to%20the%20us%20congress%2019%20nov%202009.pdf

19 David Smith, "Workers Claim Abuse as China Adds Zimbabwe to Its Scramble for Africa," *Guardian,* 2 January 2012. https://www.theguardian.com/world/2012/jan/02/china-zimbabwe-workers-abuse

20 Reid Standish, "China's Path Forward Is Getting Bumpy," *Atlantic,* 1 October 2019. https://www.theatlantic.com/international/archive/2019/10/china-belt-road-initiative-problems-kazakhstan/597853/

was a famous *Economist* cover in 2008 that depicted the Chinese looking like the old British colonizers from the nineteenth century, aptly titled, "The New Colonialists."[21] Neocolonialism and mercantilism are precisely the Chinese worldview.[22] Given this, Beijing is not interested in helping the poorer and developing nations of the world rise from poverty into modernity. They will gladly make investments in infrastructure, so long as Chinese companies benefit from them. So, something as small as a satellite program, like CBERS, could completely upend Brazil's ability to exploit its own natural resources for itself. This is especially true considering Chinese capital and firms form the basis for any exploitation of these natural resources, or those of the infrastructure projects meant to move those resources from the region onto the market. This is the pernicious threat of China's increasing influence globally.

If Brazil wants to have assistance in analyzing soil erosion or enhancing crop yields through satellite imaging or in exploiting their natural resources, Brasilia does not need Beijing's assistance to do this. The United States is a large, economically dynamic, and technologically advanced neighbor much closer to Brazil than China. While the US, unfortunately, does have a mostly negative history of interaction with Latin America, the United States has fundamentally changed over the last century since the dark days of colonial imperialism.[23] Whereas the Chinese appear to be in the process of reenacting the worst excesses of the colonial era, the United States and its Western partners have moved beyond these practices. Washington is interested in preserving what it considers to be its strategic interests, yes, but Washington is also committed to facilitating the development of other countries, such as Brazil.

---

21 "The New Colonialists," *Economist,* 13 March 2008. https://www.economist.com/leaders/2008/03/13/the-new-colonialists

22 Spencer P. Morrison, "America: China's Mercantile Resource Colony," *American Greatness,* 11 December 2018. https://amgreatness.com/2018/12/11/america-chinas-mercantile-resource-colony/

23 Christopher Hitchens, "Imperialism: Superpower Dominance, Malignant and Benign," *Slate,* 10 December 2002. https://slate.com/news-and-politics/2002/12/american-imperialism-then-and-now.html

Just ask the South Koreans about how being a strategic partner of the United States differs vastly from being viewed as a vassal state by a foreign empire like China.

RUSSIA AND BRAZIL FORMED A SOUTHERN CROSS (FOR A WHILE)

Another relationship that the previous socialist government of Brazil was interested in cultivating was with Russia. Since the Cold War, Moscow has long sought access to Latin America as a way of placing strategic pressure on the United States. In the 1990s, even as Russia was reeling from the collapse of the Soviet Union, Moscow insisted on developing its relationship with Brazil. As you've seen in the previous chapters detailing Russian space plans, theirs is far more interested in the direct military applications of alliances in distant lands, like Brazil. Specifically, Brazil is a place that is very attractive for the space launch services sector. In fact, Brazil hosts the Alcântara Launch Center, which provides a convenient location for companies or countries seeking to place objects in low-Earth orbit—at cost.[24]

As David Logsdon and Reeve Wolford assessed in 2018, after they were part of a delegation composed of US space executives to the Alcântara Launch Center:

> Brazil has a prime launch site at Alcântara which is only 2 degrees south of the equator. Launches from Brazil take advantage of the increased rotational velocity of the earth for an "extra push" into equatorial orbits, translating to about a 30 percent increase in the amount of payload a rocket can lift as compared to Cape Canaveral. This would be a major efficiency gain for US launch companies.
>
> An equatorial launch site is even more important for launches into geostationary orbit, as it avoids the need to lift considerable amounts

---

24 Imanuela Ionescu, "Brazil-Russia Military-Technical Cooperation," *Army University Press,* December 2018. https://www.armyupress.army.mil/Journals/Military-Review/English-Edition-Archives/November-December-2018/Ionescu-Brazil-Russia/

of fuel and instead burns much of it in order to change an inclined orbit to an equatorial one. Approximately half of US medium and large satellite launches go into geostationary orbits. If US companies were operating from Brazil, they could potentially increase US geostationary launches by 30 percent at the expense of our competitors.[25]

This was precisely what interested Russia when they made their initial agreements with Brazil's socialist regime in the 1990s. Remember, back then the Russians were in the process of receiving copious assistance from the United States to keep their space program going. Following the Cold War, when Russia was fighting for its very survival, the well-preserved Russian space program was looking to expand its outreach to other countries, in some cases, even at America's expense. By the 2000s, as Russia's economic and political situation stabilized, Moscow wanted to enhance its profitability in the private space sector.

Few people know that, until the construction of the Vostochny Cosmodrome in the Amur Province of Russia's Far East (very near to China's border), the Russian Federation did not have a spaceport in its own territory.[26] The Russians have used the Baikonur Cosmodrome. Baikonur was made famous during the Cold War for being the place from where Yuri Gagarin became the first human being to enter Earth's orbit, although, the Baikonur Cosmodrome ceased being on Moscow-controlled territory when the Soviet Union collapsed. Instead, it became part of present-day Kazakhstan. To mitigate the loss of Baikonur to Kazakhstan, Moscow decided to lease the facility from the fickle Kazakh government. This was always a dubious prospect for Moscow, though, considering there was much political opposition to Russia's presence at

25 David Logsdon and Reeve Wolford, "Op-Ed: U.S., Brazil Should Act Now to Forge a Partnership in Space," *Space News,* 28 February 2018. https://spacenews.com/op-ed-u-s-brazil-should-act-now-to-forge-a-partnership-in-space/

26 Matthew Bodner, "The Long Road to Vostochny: Inside Russia's Newest Launch Facility," *Space News,* 30 January 2019. https://spacenews.com/the-long-road-to-vostochny-inside-russias-newest-launch-facility/

Baikonur in the post–Cold War era.[27] Today, Baikonur is falling apart and a continual reminder of how damaging and humiliating the fall of the Soviet Union was for Moscow.[28]

There was a period of time when Moscow did not think they could have all of their space launch eggs in one basket. Russia helped France build its launch center in French Guiana, but France was ultimately a NATO member, making them unreliable for Moscow's purposes. In 2007, Russia started building Vostochny. Even that project, though, has suffered terrible delays, as evidenced by the fact that more than a decade later it is only partly complete. So, Moscow looked to Brazil's Alcântara launch center. Russia was working in tandem with the previous socialist government of Brazil to enmesh their defense technology and space launch sector in Moscow's orbit rather than Washington's. Not only did Russia want to use Brazil as a spaceport, but they also wanted to ensure that the most powerful country in South America was a Russian ally as opposed to an American one.[29]

Brazil may have fortuitous geography for affordable space launches, but it needed the larger rockets of countries like Russia and the United States to place more complex systems in orbit. Brazil, like India and the other rising space powers mentioned in this work, sees the strategic and economic value in making their space program as self-reliant as possible. By partnering with other countries that have more robust space capabilities, Brazil intends to learn how to replicate certain capabilities that they do not yet have. In 2005, the Russian government proposed the

---

27 Staff, "Kazakhstan Mulls Ending Russia's Cosmodrome Lease," *Moscow Times,* 10 December 2012. https://www.themoscowtimes.com/2012/12/10/kazakhstan-mulls-ending-russias-cosmodrome-lease-a20035

28 Patrick Reevell, "Russia's Crumbling Baikonur Spaceport Is Earth's Only Launch Pad for Manned Flights," *ABC News,* 9 December 2018. https://abcnews.go.com/International/russias-crumbling-baikonur-spaceport-earths-launchpad-manned-flights/story?id=59677739

29 Jill Aitoro, "Brazil Defense Minister: Space Partnership with US Not Dead Yet," *Defense News,* 20 December 2017. https://www.defensenews.com/space/2017/12/20/brazil-defense-minister-space-launch-partnership-with-the-us-not-dead-yet/

construction of the joint-Russian and Brazilian family rockets, artfully called the "Southern Cross." Yet again, the advanced Russian rocket technology that the United States had helped to preserve following the Cold War was being used by Moscow as a strategic cudgel against the United States. And to add insult to injury, Moscow was doing this in America's geopolitical backyard. Few American leaders from either party seemed to care about this. What's more, Brazil's succession of socialist leaders openly encouraged the creation of the Southern Cross family of rockets as a proverbial poke in the eye of America. After all, most socialists hate the United States and everything it stands for.

Alas, the corruption and incompetency of the socialists in Brazil led to the collapse of the Russo-Brazilian rocket deal. Brazil simply did not have the money to support even its commitments to Russia, thanks to the terrible economic mismanagement that the socialists had undertaken. Instead, Brasilia rebranded its Alcântara Launch Center as a place from which to launch lower-cost, smaller satellites into low-Earth orbit. Brazil's space sector languished as the Russians lost interest and the Brazilians went broke. We can all thank the inherent ignorance of socialists for the great strategic opportunity that now exists between the United States and Brazil today. The leaders in both Brazil and the United States should not waste this unique opportunity to enhance cooperation, particularly as China makes overtures to partner with Brazil's government in developing the Alcântara Launch Center, effectively replacing the Russian threat with the greater Chinese threat to the United States.

BRAZIL EMBRACES SPACE NATIONALISM, AND SIDES WITH AMERICA

The Brazilian Space Agency was a turgid entity for several years until the rise of Jair Bolsonaro. The United States should have been a natural ally for Brazil. In fact, fifty-four years ago, Brazil's first foray into space from the Alcântara Launch Center was done with explicit American assistance. Together, the Brazilians and Americans launched the Nike

Apache rocket. After that, the political situation between the Western Hemisphere's two most important countries changed, and the relationship never truly restarted until the rise of two conservative nationalists, Donald J. Trump of the United States and Jair Bolsonaro. Seemingly overnight, Brazil performed an about-face and began moving into the future, arm in arm, with the United States. There was even some loose talk about Bolsonaro's Brazil being made a member of NATO in early 2019.[30] While that would have been an extreme move, Brazil did become an official NATO partner country, proving that the renewed US-Brazil relationship had real geopolitical staying power.[31]

Happily, the Trump and Bolsonaro administrations have signed a historic agreement that will allow for the mutual development of space. This agreement is a major setback for the socialist forces within Brazil's political establishment. More important, it is a strategic defeat for both the Chinese and the Russians, who have long sought to have influence in Brazil at America's expense. The Technology Safeguards Agreement (shortened to the unfortunate acronym of TSA) "establishes protections for American space technologies exported to Brazil for space launches." You see, the TSA has been around for a while—waiting for Brazil's former socialist government to ratify it. Brazil's previous government refused because they argued that the agreement impinged on Brazil's sovereignty. In fact, the United States, knowing that the socialist regime there had entered into tight technology and space development deals with both Beijing and Moscow, could not have allowed such an agreement to go into effect: Washington rightly feared that its proprietary technology would be transferred from Brazil to either Russia or China. With the Bolsonaro government, everything about the TSA is different.

---

30 Jan Jekielek, "Exclusive: On Brazil Joining NATO and Defending the Soul of the West—Foreign Minister Ernesto Araujo," *Epoch Times,* 5 April 2019. https://www.theepochtimes.com/exclusive-on-brazil-joining-nato-and-defending-the-soul-of-the-west-foreign-minister-ernesto-araujo_2848997.html

31 Brandon J. Weichert, "Brazil Does Not Need to Join NATO," *American Spectator,* 22 March 2019. https://spectator.org/brazil-does-not-need-to-join-nato/

Brazil's new leadership allowed for changes to be made to the original TSA, first proposed as far back as 2000, so that American technological secrets could be protected in Brazil. At the same time, the renegotiated Technology Safeguards Agreement will finally allow Brazil to be more integrated into the global supply chain for high-tech and space development.[32] However, Washington needs to move with greater alacrity in securing the allegiance of Brazil. Should Chinese firms partner with Brazil to develop the Alcântara Launch Center, American technological secrets will not be protected.

The executive secretary of the National Security Council, Scott Pace, reminded audiences in March 2019 that, while Brazil's fortuitous position so near the equator is that country's most attractive feature for space development projects, it is not the most important. In fact, according to Pace, "It's going to be these other, broader aerospace activities as Brazil becomes more integrated into the global supply chain for aerospace products and as they start to step up and do more work on the satellite side." Toward that end, at the 35th Space Symposium hosted in Colorado Springs, Colorado, in 2019, Brazil's air force executive vice-president major general José Vagner Vital indicated his government's intention to "have a backup for the Geostationary Defense and Strategic Communications Satellite, SGDC-1."[33] Thus, the proposed SGDC-2 will "have an open procurement for the world's satellite manufacturers," meaning that, because of their newfound relationship, the United States will likely benefit from this and other developments in Brazil's promising space sector.[34]

Even if US satellite manufacturers don't directly benefit from this

---

32 Livia Peres Milani, "Brazil's Space Program: Finally Taking Off?" *Wilson Center,* 24 October 2019. https://www.wilsoncenter.org/blog-post/brazils-space-program-finally-taking

33 Caleb Henry, "Brazil to Order Second Dual Civil-Military Communications Satellite," *Space News,* 10 April 2019. https://spacenews.com/brazil-to-order-second-dual-civil-military-communications-satellite/

34 Henry.

development, the US military will. As the US and Brazilian militaries grow closer over the coming years, it will be essential for Brazil's forces to operate in tandem with America's, the way that NATO and South Korean forces do. The SGDC satellites are advanced military communications satellites, much like the aforementioned Wideband Global Satellite Communications constellation that the US Army manages and that Russia likely seeks to disable in the event of a conflict with NATO.

By building up a space power with fortuitous launch geography, a very friendly government, and a modern military—in the same hemisphere as the United States, no less—Washington is potentially creating a partner in Brazil with real capabilities that could augment US capabilities in the event of a crisis. Just as the United States should partner in space much more closely with France, Japan, and India, so too must Washington encourage greater relations with Jair Bolsonaro's Brazil. After all, America's world dominance is predicated largely on strong alliances based on shared strategic and economic interests. American leaders should never again allow either China or Russia to get as close to coopting such a potent country like Brazil into their growing anti-American alliance network.[35]

If America is going to win the new space race, it must be able to share some burdens with allies. Otherwise, the combined technological powers of China and Russia—and even Iran and North Korea—will leapfrog the United States at the strategic level. More dangerously, these powers will gain dominance in the strategic high ground of space—threatening everything and everyone below.

---

35 Pepe Escobar, "Trump Will Try to Smash the China-Russia-Iran Triangle . . . Here's Why He Will Fail," *South China Morning Post,* 22 January 2017. https://www.scmp.com/week-asia/opinion/article/2064005/trump-will-try-smash-china-russia-iran-triangle-heres-why-he-will

18

SPACE NATIONALISM

A RECENT STUDY asked children ages eight to twelve in the United States, the United Kingdom, and China what they wanted to be when they grew up. The children in both the US and the UK overwhelmingly said they wanted to be a "Vlogger/YouTuber" when they reached adulthood, whereas in China, a whopping *50 percent* of children there said they wanted to become an astronaut.[1] Say what you will about the

---

1 Paige Leskin, "American Kids Want to Be Famous on YouTube, and Kids in China Want to Go to Space: Survey," *Business Insider,* 17 July 2019. https://www.businessinsider.com/american-kids-youtube-star-astronauts-survey-2019–7?fbclid=IwAR35XokQkveZAkPTDE2eXNFinpbOEWTrc4rE klYUUqZ8KWxKz_DXI8QRkBY

tyranny of statistics and their ability to mislead, but this Harris Poll survey, done to commemorate the *Apollo* Moon landing's fiftieth anniversary, is a perfect snapshot of where Western and Confucian cultures are. And if you're living in the United States or the United Kingdom, judging from this Harris Poll, the future seems bleak. After all, a society is only as successful as it believes it can be, and the will of most societies throughout history are reflected in the dreams as well as the aspirations of their people. American kids used to dream of being astronauts. Spaceflight was the pinnacle of societal achievement. It united us. When I was growing up in Florida, every time a space shuttle or satellite launched from Cape Canaveral, my neighbors and I—Republicans and Democrats, young and old, gay and straight alike—would gather in our shared backyards on the water and marvel at the glories of that technological feat. Of course, even at that time in the late 1990s and early 2000s, most Americans were ambivalent about space launches. They had become routine. When the shuttle program went away in 2011, no one really cared. Our once confident and proud culture of empiricism and strength was replaced by a lax sense of nihilism with a twinge of ambivalence.[2] Everything went into the narcissistic pursuits of texting, Facebooking, tweeting, and YouTubing—all in an effort to get famous. And why would one seek fame? Possibly for wealth. But even today, fame does not necessarily equal the fortunes it once did. It truly is about being identifiable to strangers. That, more than anything, drives Americans today: to be known more than your fellow citizen. What few understand, I believe, is that this is all an attempt to belong. Americans today have been deprived of their shared sense of culture—and it shows. Many are now desperate to fill that void with anything they can.

AMBIVALENT TO WONDER

In this morass, all of the wonders of modernity are taken for granted. Little thought is given to just how fragile our modern society is, to how

---

2 Encounter Books, "Victor Davis Hanson on The Fate of the West, Trump, and the Resistance," YouTube, 20 June 2018. https://www.youtube.com/watch?v=Uyui7MMxs8A

much work is put into making sure everything operates as it should, by a few unfamous people holed away in a control center somewhere in the Pentagon or in a cubicle at some tech firm. As is often the case, an American comedian, the now-disgraced Louis C.K., captured the cultural malaise well when he went on the Conan O'Brien late-night program in 2007 and riffed:

> Now, we live in an amazing, amazing world, and it's wasted on the crappiest generation of just spoiled idiots that don't care. Because this is what people are like now, they're like, "Ugh! [my smartphone isn't working]!" Give it a second! [Its signal is] going *to space*! Can you give [the signal] a second to get back from space? I was on an airplane, and there was high-speed internet on the airplane—that's the newest thing that I know exists—and I'm sitting on the plane, and they're going, "Open up your laptop; you can go on the internet." And it's fast and I'm watching YouTube clips—*and I'm on an airplane*! And then it breaks down, and they apologize about the internet not working, and then the guy next to me goes, "Pfft! Bullshit!" Like, how quickly the world owes him something he knew existed only ten seconds ago![3]

In the recent past, Americans used to look up at the stars and dream of what might be a better tomorrow. Western children used to want to be Buzz Aldrin. Now, they want to be Justin Bieber.[4] The kids today, the adults of the future, are clearly *not* all right. It's the Chinese kids who do the dreaming. Remember the *Field of Dreams* mentality: if you build it, they will come. Kevin Costner's character first had a dream and then the urge to "build it." Man cannot build something unless he first envisions—dreams about—it first. When one's culture is subsumed in narcissism and nihilism, such as the West is today, how can the West, particularly the United States, do the great things it used to do?

---

3 Louis C.K., "Louis CK Everything Is Amazing and Nobody Is Happy," YouTube, 1:30–2:32, 16 November 2016. https://www.youtube.com/watch?v=nUBtKNzoKZ4&list=RDnUBtKNzoKZ4&start_radio=1

4 Desiree Adib, "Pop Star Justin Bieber Is on the Brink of Superstardom," *ABC News,* 14 November 2009. https://abcnews.go.com/GMA/Weekend/teen-pop-star-justin-bieber-discovered-youtube/story?id=9068403

The answer, of course, is that America cannot achieve much greatness today. Writing in *American Greatness* in 2018, Matthew J. Peterson observed that "over the last century, the world has witnessed the slow but steady depoliticization of America and Western Europe in favor of the rule of the apolitical 'expert.' The great challenge of our era is to foster the return of actual political life between the people and their elected officials."[5] The American people have allowed their leaders to create giant bureaucracies in their name, those of both the government and major corporations, to regulate and control our lives.[6] Americans have ceded increasing amounts of power to these central bureaucracies because most Americans assume that they cannot get through daily existence without the "expert" help of these organizations. As my colleague Steven F. Hayward observed in 2014, "Bureaucratic rule nowadays permeates every level of government right down to many small-town councils and county commissions."[7]

What does the United States have to show for these bureaucratic expansions? It's a country that doesn't *dream* anymore. We have no aspirations beyond feeding our own selfish desires.[8] Americans today worry—*about everything.*[9] And when one is constantly anxious about

---

5 Matthew J. Peterson, "Thank God Trump Isn't a Foreign Policy Expert," *American Greatness,* 12 June 2018. https://amgreatness.com/2018/06/12/thank-god-trump-isnt-a-foreign-policy-expert/

6 Timothy M. Gill, "Why the Power Elite Continues to Dominate American Politics," *Washington Post,* 24 December 2018. https://www.washingtonpost.com/outlook/2018/12/24/why-power-elite-continues-dominate-american-politics/

7 Steven F. Hayward, "Bureaucracy in America Now Goes All the Way Down," *Forbes,* 20 January 2014. https://www.forbes.com/sites/stevenhayward/2014/01/20/bureaucracy-in-america-now-goes-all-the-way-down/#1bb9d72db1c0

8 Celia Cole, "Overconsumption Is Costing Us the Earth and Human Happiness," *Guardian,* 21 June 2010. https://www.theguardian.com/environment/2010/jun/21/overconsumption-environment-relationships-annie-leonard

9 Peter Dockrill, "America Really Is in the Midst of a Rising Anxiety Epidemic," *Science Alert,* 9 May 2018. https://www.sciencealert.com/americans-are-in-the-midst-of-an-anxiety-epidemic-stress-increase

everything, one rarely accomplishes anything.[10] The more bureaucracy we've created to make us feel safer or to assist us just prevents us from achieving our dreams. As we fail to achieve those dreams—much less to even have them—the less we've been fulfilled, and the greater the cultural malaise.

Meanwhile, because of these trends, our society has become more atomized. The combination of narcissism and consumerism has merged with the Left's insane quest to divide us according to race, religion, sexual orientation, and so on. Identity politics is merely an outgrowth of the terrible way the United States has developed since the 1960s. For children today, the one ubiquitous component of their shared community is social media—the glorifying of self above all else. This translates outwardly to the whole country over time. It explains why we don't accomplish great things as a nation (other than blowing up a random jihadist in a desert somewhere).

Don't listen to what the democratic globalists say: the development of high technology or a country's space sector is *not* an exercise in globalism. *There is no human mission in space.* There are, however, *national* aspirations for space. There is a need for the military and economic development of space, but these are for national priorities, not global yearnings. What's more, the idea that all nations in the world can be subordinated to the supposedly "common goal" of space exploration is ridiculous. Despite what the globalists desire, humanity has *never* fully progressed to the point where we are willing to shake off the traditional bonds of family, town, country, and civilization in favor of a global bureaucracy in some distant city. Why would any rational person want to do that?

How many times have astronauts bearing the hideous blue flag of the United Nations landed on the Moon? How many satellites has the UN launched into orbit? The answer is: none. What's more, it's likely that the UN will never land on the Moon or go to Mars—at least not anytime soon. There is no unified priority because there is no unified mankind.

---

10 Daniel Smith, *Monkey Mind: A Memoir of Anxiety* (New York: Simon and Schuster, 2013), 1–9.

IT'S *ALWAYS* ABOUT THE CULTURE

Humans are community-loving beings. The community is not found in some amorphous international governing body, such as the United Nations. The community is best represented in the nation-state. NASA may have left a plaque on the Moon that reads, "We come in peace for all mankind," but we went to the Moon for the United States, and it was the shared Western values of most Americans that propelled us to the forbidding vacuum of space. That we blasted men to the Moon is part of the unfolding story of *American history and Western civilization.* American politics and its policies, whether they be on the domestic or international fronts, therefore, are extensions of our interests—which are themselves prime exhibits of *who we are as a community within the human race.* Nationalism compels countries to take up the daunting challenge of space development and defense.

The late conservative activist Andrew Breitbart was fond of telling audiences that "politics lies downstream of culture."[11] He was correct. Politics is merely a component of the shared civilizational construct. It is an expression, not the cause, of our country and every action it takes.

As Lawrence Meyers wrote in 2011:

> Our lives—indeed, our very species—has storytelling wound into our DNA. From the earliest cave drawings, man has expressed himself in terms of story. Ancient civilizations understood that stories are vital to understanding our place in the world, so much so they codified storytelling and found base rules that form it . . . Stories instill moral and ethical values. They place joy and tragedy in context. They preserve cultures. At their best, they deliver the secrets and meanings of life.[12]

---

11 Leif Weatherby, "Politics Is Downstream from Culture, Part 1: Right Turn to Narrative," *The Hedgehog Review,* 22 February 2017. https://hedgehogreview.com/blog/infernal-machine/posts/politics-is-downstream-from-culture-part-1-right-turn-to-narrative

12 Lawrence Meyers, "Politics Really Is Downstream from Culture," *Breitbart,* 22 August 2011. https://www.breitbart.com/entertainment/2011/08/22/politics-really-is-downstream-from-culture/

What's the more appealing story here for us or our children? That we go into space for "all mankind," when, as you've seen in the preceding chapters, *no one else is going into space for the species*? Or, to claim that *your country* is developing high technology and blasting off into the cosmos—and that it is doing things other countries could only ever dream of? I think most Americans would say they favor the latter narrative. More than anything, the *Apollo* missions are a story of an *American* success. They are examples of *American* exceptionalism. As you've seen in this work, the Soviets took to the stars for their revolutionary ideology that had become their culture. America did the same for its values and culture.

Today, the United States' once-dominant position in space has eroded and allowed for other, less scrupulous powers, to rise in that key domain precisely because American leaders and our people generally stopped caring about space. In fact, the United States has generally stopped caring about many things.[13] This was partly because of the post-1960s cultural movement from the Left, which had created a national sense of stifling narcissism and nihilism. After forty years (and counting) of this left-wing ethos prevailing in American society, everything from winning wars to developing timely treatments of diseases to space exploration has been stunted.

At times like this, I am reminded of the famous monologue that Jeff Daniels delivered in the first episode of Aaron Sorkin's otherwise turgid HBO series, *Newsroom*. Daniels's character, Will McAvoy, was asked by an ignorant college girl what makes America so great. During her question, McAvoy snapped. In a five-minute monologue, McAvoy explicates the feeling that I, and so many people in Middle America, have today about the state of our culture and country. At one point, the Will McAvoy character laments the loss of American prestige and capabilities by saying:

---

13 Robert Kagan, "The Ambivalent Superpower," *Politico*, March/April 2014. https://www.politico.com/magazine/story/2014/02/united-states-ambivalent-superpower-103860

We built great big things, made ungodly technological advances, explored the universe, cured diseases, and cultivated the world's greatest artists and the world's greatest economy. We reached for the stars, and we acted like men. We aspired to intelligence; we didn't belittle it; it didn't make us feel inferior. We didn't identify ourselves by who we voted for in the last election, and we didn't scare so easy.[14]

Just remember what Ye Peijian, the head of China's lunar program, said in 2018, "If we don't go [to space] now even though we're capable of doing so, then we will be blamed by our descendants. If others go there, they will take over, and you won't be able to go even if you want to."[15] Most public intellectuals focused heavily on the first half of Ye's remarks wherein he compared the universe to an ocean, the Moon to the Senkaku islands, and Mars to Huangyan Island. What few people analyzed was the last part of Ye's remarks.

Yes, China's leaders rightly view space in geopolitical terms. But geopolitics itself is merely an outgrowth of culture. The great British geostrategist Sir Halford Mackinder at the turn of the twentieth century separated the world into two camps: countries belonging to land-locked, continental powers, such as Russia and China, and those belonging to maritime civilizations, such as the United Kingdom or the United States. Mackinder observed that continental powers tended to be more autocratic and centralized whereas maritime powers, by their nature, are expeditious and democratic.[16] The geography influenced culture, and the culture influenced grand strategy and foreign policy. Culture matters more than anything else in technological development and space development.

14 "Jeff Daniels (written by Aaron Sorkin): 'America Is Not the Greatest Country in the World Anymore,' The Newsroom—2012," *Speakola*, 24 June 2012. https://speakola.com/movie/jeff-daniels-sorkin-newsroom-2012

15 Malcolm Davis, "China, the US and the Race for Space," *The Strategist*, 12 July 2018. https://www.aspistrategist.org.au/china-the-us-and-the-race-for-space/

16 Colin Dueck, "The Return of Geopolitics," *Real Clear World*, 27 July 2013. https://www.realclearworld.com/articles/2013/07/27/the_return_of_geopolitics_105345.html

When Ye made his comments in 2018, he was not talking about China's space development plans in airy terms of one-worldism, as American leaders so often do. Instead, he was proclaiming Beijing's intent to blast forth into the stars, in order to make *Chinese children* proud, all to ensure China's place in this new and wonderful strategic domain *before* China's rivals could deprive Beijing access to it. These are cultural and national, not the stilted arguments of globalists. Of course, as has always been the case, these cultural and national arguments have immense global ramifications. When President Donald Trump told UN audiences that the future belongs to "patriots, not globalists" in 2019, he was correct. Space, like the future, belongs to national patriots rather than blinkered globalists.[17]

In fact, the nation-state, which is the basis of all international politics, is itself a Western construct. It was a product of the Thirty Years' War in Europe, which is considered one of the bloodiest wars in European history. What came out of that conflict was the Treaty of Westphalia, which is the basis for the modern international order. Subsequent great conflicts, such as the world wars, created additions to this construct—namely, multilateral, globalist institutions, such as the United Nations. Although, even the name "United *Nations*" indicates a truth that its globalist founders could not ignore: the unifying political force for any global order is based on the nation-state.

The same is true, by the way, of the European Union, which fancies itself as a supranational body but is nothing but a giant common market with a huge external tariff—most power within the EU is granted to the EU's governing authority. And within the EU, power remains firmly in the hands of a select few powerful nation-states. When the EU got too big for its own britches, the political backlash within the national polities of many of its members has been profoundly anti-EU. Due

17 Kathryn Watson, "Trump Says Future Belongs to 'Patriots,' Not 'Globalists,' In U.N. General Assembly Speech," *CBS News,* 24 September 2019. https://www.cbsnews.com/live-news/trump-un-speech-future-belongs-to-patriots-not-globalists-united-nations-general-assembly-today/

to this backlash, the United Kingdom is undergoing an exit from the EU. This effectively proves that the nation-state, not multinational or globalist institutions, like the UN, is the basis of human affairs—and it has been an enduring feature of human existence since at least 1649, when the Treaty of Westphalia was signed.

THE WORLD OF BITS VERSUS THE WORLD OF ATOMS

The noted tech investor Peter Thiel has an excellent tagline for his venture capital firm, Founders Fund: "We wanted flying cars, instead we got 140 characters."[18] It's a pithy tagline that embodies much truth. Thiel laments that Silicon Valley has spent much time investing in the world of "bits." In the world of "atoms," unfortunately, the American tech sector has been lagging behind. Space development would classify as belonging to the universe of "atoms" whereas things like social media platforms belong to the world of "bits."[19] It's another example of how small-thinking the United States has become in its aspirations. The onerous regulatory environment, according to Peter Thiel, is the single greatest threat to innovation in the world of "atoms."

Here are Thiel's own words on the matter:

> Things have been surprisingly stagnant [in the West], and there's this weird question of "what's going on?" You have Moore's Law on the one hand; if I had to sort of simplify it, I'd say you had incremental but relentless progress on the computer side. And on the other hand, you've basically no progress on [anything else].

Thiel went on to argue that many other industries—in the atom world, such as the energy sector, or the space sector—has undergone

---

18 Daniel Weisfield, "Peter Thiel at Yale: We Wanted Flying Cars, Instead We Got 140 Characters," *Yale School of Management,* 27 April 2013. https://som.yale.edu/blog/peter-thiel-at-yale-we-wanted-flying-cars-instead-we-got-140-characters

19 Mark Sullivan, "VC Peter Thiel: You Can Either Invest in 'Bits' or 'Atoms," *Venture Beat,* 8 September 2014. https://venturebeat.com/2014/09/08/vc-peter-thiel-you-can-either-invest-in-bits-or-atoms/

little real innovation since the 1970s. Innovation and economic growth have been achieved through the "relentless progress" in the world of bits. Thiel argues that "we've basically outlawed everything having to do with the world of stuff. And the only thing you're allowed to do is in the world of bits. And that's why we've had a lot of progress in the world of computers and finance. Those were the two areas where there was enormous innovation in the last forty years. It looks like finance is in the process of getting outlawed, so it looks like the only thing that will be left is in the area of computers."[20]

America's failure to believe in itself—the imbuing of this nihilistic and ambivalent ethos that the Left has foisted upon the country since the 1960s—has played heavily in the development of massive regulations in fields that require as little regulations as possible to spark innovation. These government-created hurdles, coupled with the government's decline in federal research and development, has meant that private capital is not very much interested or able to take the risk needed to invest in things, such as the private space sector.

NATIONALISM IN CHINA

In China, however, as you've seen, this is not a problem. There is but one regulation imposed upon advanced foreign companies entering into this billion-person market: you must cede a degree of your corporate sovereignty over to Beijing. And as you've read in previous chapters on the matter, most Western firms have little compunction about doing this. With the exception of forcing American companies to *kowtow* to the Chinese leadership, there is much in China that lures foreign companies and entrepreneurs to their shores. In China, entrepreneurship is encouraged, as it is seen as a net positive for China's growth as a major power.[21]

---

20 Fluxspiel, "2012—Eric Schmidt and Peter Thiel," YouTube, 11 March 2015. 9:45–11:38. https://www.youtube.com/watch?v=PsXFwy6gG_4

21 Edward Tse, "The Rise of Entrepreneurship in China," *Forbes,* 5 April 2016. https://www.forbes.com/sites/tseedward/2016/04/05/the-rise-of-entrepreneurship-in-china/#1c3af2513efc

Until the recent wave of protests in Hong Kong (and the repre-hensible crackdowns that Beijing is imposing on that great city), Hong Kong was consistently rated as having a higher ranking than even the United States in the respected Economic Freedom Index.[22] Meanwhile, cities nearby in China, such as Shenzhen, Guangzhou, and Shanghai, are all becoming booming metropolises because of their embrace of low regulations and low taxes in their business sector.[23] This, along with China's massive investments in the cultivation of human capital over the last thirty years, has created a renaissance in that country. How is it that Red China understands economic development better than the United States, the land of the automobile and airplane?

It's because the Chinese are not hindered by the kind of cultural gutting that the United States has been subjected to since the late 1960s. You see, the Chinese already had their cultural revolution under Mao Zedong. It was an utter failure. Most of China's leaders from Deng Xiaoping, the man who succeeded Mao, all suffered on some level during Mao's idiotic Cultural Revolution.

Mao feared that he was losing his power in the 1960s. In order to maintain control, Mao embarked on an ideological campaign meant to root out any members of the country's citizenry who were not "suf-ficiently revolutionary." Mao turned to China's youth, specifically those from the urban population centers. He indoctrinated them in Maoism, his particular brand of Communism, and unleashed his youthful "Red Guards" upon the country. From 1966 to 1976, every single aspect of China was uprooted, disassembled, and subjected to extreme condi-tions for perceived ideological slights or variances—anything that might challenge Mao's power.[24] Most of China's post-Mao leaders viewed this

---

22 "Hong Kong," *2019 Index of Economic Freedom,* accessed 11 November 2019. https://www.heritage.org/index/country/hongkong

23 "China Plans to Make Shenzhen a 'Better Place' Than Hong Kong," *Straits Times,* 20 August 2019. https://www.straitstimes.com/asia/east-asia/china-plans-to-make-shenzhen-a-better-place-than-hong-kong

24 "A Brief Overview of China's Cultural Revolution," *Encyclopedia Britannica,* accessed 1 November 2019. https://www.britannica.com/story/chinas-cultural-revolution

decade, the Cultural Revolution, as highly destructive to China's course and Mao's stated goals in 1949 of "catching up" to the United States and overtaking them.

Singapore's former leader Lee Kuan Yew famously told audiences in 2013 that "this generation [the Baby Boomers and Gen-X'ers in China] has been through hell: the Great Leap Forward, hunger, starvation, near collision with the Russians—the Cultural Revolution gone mad."[25] It was precisely the fact that nearly every major player in China's post-Mao government experienced the excesses of Mao's Cultural Revolution that they helped build China into the juggernaut that it is today. China's leaders continued their human rights abuses, but they allowed enough economic freedom to enrich and empower the entire country.

Xi Jinping, however, is an oddity in this pattern. Like so many of those who suffered under the Cultural Revolution, Xi's own father was removed from his office and sent to work in a factory. From that point on, Xi's family was scattered across the vast Chinese countryside for being considered by the Red Guards as ideological "deviationists." Yet, Xi was also a young man whose formative years in school and in his local community were all informed by the radical ideological politics of the Cultural Revolution. He appears to have incorporated these ideas into his worldview today, despite the personal toll it took on his own family.

Here's how one observer wrote about Xi and his revolutionary politics:

> There are only two types of people [according to Xi]: Those who support the [Communist] party and those who do not. Like Mao in 1966, Xi believes that his power hinges on making all Chinese—government officials and ordinary citizens alike—loyal and obedient

---

25 Graham Allison and Robert Blackwill, "Interview: Lee Kuan Yew on the Future of U.S.-China Relations," *Atlantic*, 5 March 2013. https://www.theatlantic.com/china/archive/2013/03/interview-lee-kuan-yew-on-the-future-of-us-china-relations/273657/

through any means possible. Power is founded on the repression of opponents, such as Nobel Peace Prize winner Liu Xiaobo and the tens of thousands of other jailed authors and scholars.[26]

These notions of power, combined with Xi's obsession to "Make China Great Again" through his China Dream, make China a potent force going forward. Recently, Xi reiterated his commitment to make China a technological innovation hub by 2035. As noted earlier in this work, China plans on becoming a dominant space power through Xi's "Space Dream" by 2049, the one-hundredth anniversary of the founding of the Chinese Communist Party. And while Xi is clearly the leader most similar to Mao since his death, the institutions, culture, and population of China are unlikely to allow for the same level of excesses and internal destruction that Mao's insane ideological Cultural Revolution wrought for China—if only because enough people living in China today still remember the abject misery of the whole affair.

NATIONALISM IN RUSSIA

Similar trends exist in Russia today. A country that was ground zero for the Communist revolution, Russia has been laid low. In many respects, it will never fully recover. Vladimir Putin, who many believe to be a mere extension of the bad old Communist days, has repeatedly decried Communism. In fact, Putin remarked that Lenin was possessed of "bad ideas" in 2005. Many Western observers point out that Putin publicly stated that the collapse of the Soviet Union was the greatest "geopolitical disaster" of the twentieth century. President Putin was not saying he wanted to yet again raise the Iron Curtain of Communism in Europe. Instead, he was specifically referencing the loss of hundreds of miles of Russian territory to new countries.

Since taking office, Putin has also worked hard—as have China's leaders—to reconnect Russians with their past beyond the Soviet era.

26 Ma Jian, "Xi Jinping—a Son of the Cultural Revolution," *Japan Times*, 15 May 2016. https://www.japantimes.co.jp/opinion/2016/05/15/commentary/world-commentary/xi-jinping-son-cultural-revolution/#.Xbu7vy-ZPyU

Whereas the Communists sought to create a new entity in the Soviet Union by erasing the vast and rich history of Russia and its surrounding countries, Putin has fought to rehabilitate former symbols of Russian power and cultural majesty. Whether sincere or not, Putin has encouraged the return of the Orthodox Christian Church to its pre-Bolshevik place as a cultural weight in Russia. Putin has also reopened the old Cossack military schools for young Russians after Lenin himself shuttered the schools during his bloody revolution. The Putin regime has embraced pro-natalist policies, meant to encourage family formation in Russia while ensuring that those families have more than one child. All of these actions are direct refutations of the old Communist order. Putin rightly believes that Soviet Communism eviscerated Russia. While Putin mourns—and seeks a return of—the geopolitical power that the Soviet Union once enjoyed, neither Putin nor his cadre desire a return of the self-destructive, nihilistic Communist order that gutted Russian culture and deprived it of decades of true political stability, economic growth, technological innovation, and peace.

Vladimir Putin is both a nationalist and an imperialist, seeking to return Russia to its former glory in what Moscow arrogantly refers to as its "Near-Abroad." But just as China's leaders want to make China great again for the Chinese people, so too, does Putin want to make Russia great again for his people. Under Putin, Russians have been given a reason to have hope, however small. Today, Russia is staring down a demographic disaster, economic instability, and political uncertainty in the long run. So long as Putin reigns, though, Russians will continue to have a reason to stay optimistic. Again, the Russian space program is an outgrowth of the nationalist-inspired optimism.

NATIONALISM FOR ALL

Both Iran and North Korea have been decimated by their history of internal political revolution. Today, Iran is suffering the hardest, as its revolutionary Islamist regime is still in place and still imposing its antedated laws on the country. Yet, there are immediate strategic reasons

Iran's leadership is pursuing the strategy that it is against the United States—and why they are not only appealing to the country's Islamic history, but also to its history of being a great nation of the Middle East (when it was known as Persia). North Korea, as well, has been gutted by the strange cult of personality that has dominated there since the rise of the Kim family to power after the Second World War. Kim Jong-un, though, must make an appeal to nationalism in order to keep his own military in line and to ensure that the people do not lose their fear of him while he seeks negotiations with President Trump.

France, Israel, and Japan all use the rhetoric of nationalism and the national interest to expand their roles in space. After all, space is merely an outgrowth of the nation-state that is going into space. So, too, do India and Brazil use nationalism as opposed to internationalism or democratic globalism to invest in and expand their space programs. Weak-kneed globalism does not carry countries to greatness. Strong countries are moved to greatness through the shared cultural worldview of their citizens and leaders. Most of these countries I've surveyed in this book all experienced some kind of cultural revolution. And for the most part, their societies recoiled.

After experiencing the terrible excesses, nearly all of them (the cases of Iran and North Korea remain an open question) gave up on their quest for radical social reengineering and, in the words of the sonar officer from the 1989 hit film *The Hunt for Red October*, when nations get confused, they kind of run "home to mama." Mama, in this case, is their original culture. And working backward, then, many of these countries appeal to nationalism as a means of reestablishing the lost links to their culture and history. All of these countries are reacting in big ways to the excesses of the revolutionary politics that ensnared them in the twentieth century.

AMERICA RUNS HOME TO ITS NATIONALIST MAMA

No such reaction happened in the United States until the rise of Donald J. Trump as president. Since the countercultural revolution of the 1960s, the United States has stagnated in all ways. This is most notable in America's deteriorating place in space—and its relative lack of vision in that strategic domain. Just as with other countries that endured usually left-wing revolutions of one kind or another, the United States has been subjected to one ideological excess after another. Not content to spit on Vietnam War veterans returning from that terrible conflict, the Left agitated for Marxist "reforms" to all aspects of our nation, anything to tear down the edifices of America's cultural heritage. Today, everything that can be deconstructed for political purposes has been. Things are so bad now that math and science, the building blocks of advanced countries and economies, have been deemed "racist" by America's own (blessedly less violent) form of Red Guards, the social justice warriors.[27] Unlike the other countries that I've analyzed, the United States has still not had to come to terms with the terrible excesses of its cultural revolution fifty years ago. But America will soon have its own cultural reckoning. Given current trends in American domestic politics, it is likely already underway.

Evoking Peter Thiel once more, 1969 was a seminal year. It was the year that American astronauts first landed on the Moon. Three weeks after that event, the Woodstock festival began. It was a fork in the road of the nation's development. Americans should have chosen the Moon landings. By 1969, however, the cultural revolution in the United States was in full force. Most Americans chose Woodstock.[28] We are paying the price for that awful choice today.

Nationalism allows countries to see the world in realistic terms and to craft similarly realistic strategies—based on who they are, which itself is informed by the history and culture of a given nation. It forces

27 Lee Ohanian, "Seattle Schools Propose to Teach That Math Education Is Racist—Will California Be Far Behind?" Hoover Institution, 29 October 2019. https://www.hoover.org/research/seattle-schools-propose-teach-math-education-racist-will-california-be-far-behindseattle

28 Jay Yarow, "Peter Thiel: Progress Ended After We Landed on the Moon and 'Hippies Took Over the Country,' *Business Insider*, 23 October 2014. https://www.businessinsider.com/peter-thiel-on-progress-2014—10

leaders to balance their own political goals with what can be achieved. And when something must be achieved that is considered to be an impossibility, nationalism and the national interest will inspire more people in a country to commit to that dream more than any paean to globalism or internationalism will. We saw this during the Cold War, when Americans feared losing the strategic high ground of space to the Soviet Union and took decisive action—at the national level—to prevent the Soviets from beating the United States in space.

Until Washington changes its policies and the laws governing the United States' use and development of space as a strategic and economic asset, the world's other rising powers, friends and foes alike, will continue rising to challenge America's dominant position in space. But in order to make the necessary reforms, one must first acknowledge that there is such a thing as space nationalism and that the national, rather than the global, interest is served best by the United States committing itself to building up a robust presence in space at all costs. David P. Goldman has argued that a national investment of $1 trillion into expansive technological research and development over a decade would ameliorate the losses America has suffered in its technological development and space program relative to its rivals. Whatever the number, the national commitment to achieving such a goal over a prolonged period of time will be key. Without that national commitment to accomplishing such an endeavor, the United States will eventually be made a second-tier power. The first example of America's new second-class world status will be made in space, where technology is taken and tested to its operational extreme.

19

DEREGULATING SPACE

WHEN NASA TERMINATED ITS MOON PROGRAM, it dismantled its heavy-lift Saturn V rockets. Washington then channeled its energies into the space shuttle program, which, as you've seen, did have some strategic applications in the context of the Cold War. However, it proved to be a proverbial dead end in America's space development. Meanwhile, the United States had signed a series of international agreements that essentially hamstrung both the private sector development and the complete weaponization of space. The most important treaty the United States signed regarding the use of space was the Outer Space Treaty. This was the international agreement that purportedly ensured space would remain a "weapons-free sanctuary."

Here are the bullet points from that treaty, as they appears on the UN website:

- The exploration and use of outer space shall be carried out for the benefit and in the interests of all countries and shall be the province of all mankind;

- Outer space shall be free for exploration and use by all states;

- Outer space is not subject to national appropriation by claim of sovereignty, by means of use or occupation, or by any other means;

- States shall not place nuclear weapons or other weapons of mass destruction in orbit or on celestial bodies or station them in outer space in any other manner;

- The Moon and other celestial bodies shall be used exclusively for peaceful purposes;

- Astronauts shall be regarded as the envoys of mankind;

- States shall be responsible for national space activities whether carried out by governmental or nongovernmental entities;

- States shall be liable for damage caused by their space objects; and

- States shall avoid harmful contamination of space and celestial bodies.[1]

Of course, at the time of this treaty, the technology in space was very rudimentary. It was also very expensive to conduct space missions. Had it not been for the fact that the Soviet Union had reached space

---

1 "Treaty on the Principles Governing the Activities of States in the Exploration and Use of Outer Space, Including the Moon and Other Celestial Bodies," *United Nations Office for Outer Space Affairs,* accessed 1 November 2019. https://www.unoosa.org/oosa/en/ourwork/spacelaw/treaties/introouterspacetreaty.html

first, or that nuclear weapons ultimately traversed space on their way down to their targets below, it is likely that space would have remained a distant dream for most people. But the military implications of one power coming to dominate the strategic high ground of space made space development an essential national security mission for a country, like the United States. The Outer Space Treaty was an attempt to hem in the two superpowers, who at the time were clearly marching toward conflict in the cosmos. The Outer Space Treaty was signed following the devastation that the high-orbit nuclear tests wrought (which eventuated in the aforementioned Partial Test Ban Treaty of 1962). It was also crafted a few years after the tense moments of the Cuban missile crisis. All of this played into the thinking of those who created the treaty.

Let's get real, though: the Soviets and Americans entered into the treaty not out of some commitment to fanciful notions of belonging to a shared global community. Instead, both sides were simply buying time for their technology to mature in order to better exploit space. In fact, even as representatives from Washington and Moscow spoke of their dreams of a sanctuary for humanity in space, both sides were at different times merely jockeying for more leverage over each other in space. At first, in the 1960s, the Soviets were still considered ascendant in their development of space technology. This was why Moscow focused its attention on the development of space weapons, like the aforementioned military space station program. This fact also explains why the USSR spent so much time and money developing anti-satellite weapons. The United States, too, continued looking at space—at least partly—as a strategic domain, which explains why American forces in Vietnam began using satellites as a strategic asset during that terrible conflict.

AMERICA MASTERS THE ART OF SELF-DETERRENCE

Yet, at the same time, Western attitudes toward space were slowly changing. On the one hand, there was the onerous cost of continuing space operations—particularly manned spaceflight. In one instance, the United States Air Force had developed its own manned spaceflight program, known as the Manned Orbital Laboratory (MOL). First conceptualized in the early 1960s as manned satellites, the program was lavishly funded during the Lyndon B. Johnson administration.[2] The Pentagon had desired to place a constellation of manned space stations in orbit, similar to what would become the Soviet Union's Almaz military space stations or the *Tiangong* space stations that China would eventually place in orbit in the 2000s. Much of the equipment had been assembled while numerous US Air Force and Naval officers were chosen to perform missions aboard these exclusively military space stations. The MOL Program had even developed what I think would have been iconic blue spacesuits—as opposed to the NASA white and silver Gemini spacesuits—with a revolutionary helmet that allowed for total situational awareness in space. Ultimately, the program was cut due to cost overruns.[3] The Richard M. Nixon administration instead favored the creation of cheaper, unmanned satellites.

In many respects, the Nixon administration made the right call: the optical gear and communications packages on unmanned satellites today are more sophisticated than they would have been on large, clunky manned orbital platforms. And elements of the MOL Program were inevitably folded into NASA's successful *Skylab* space station program.[4] What was lost, though, was the military capability of manned spaceflight. The MOL program ultimately called for the creation of an advanced, manned national command space station in high geosynchronous orbit that would have functioned as a sort of orbiting NORAD in the event of a nuclear war between the Soviet Union and the United States.

---

2 Al Hallonquist, "The MOL-Men Come into the Light," *Air and Space Magazine,* 4 December 2015. https://www.airspacemag.com/daily-planet/mol-men-come-light-180957353/

3 Leonard David, "Declassified: US Military's Secret Cold War Space Project Revealed," Space. com, 30 December 2015. https://www.space.com/31470-manned-orbiting-laboratory-military-space-station.html

4 "50 Years Ago: NASA Benefits from Manned Orbiting Laboratory Cancellation," *NASA,* 10 June 2019. https://www.nasa.gov/feature/50-years-ago-nasa-benefits-from-mol-cancellation

More importantly, the technological "ecosystem" for a permanent and robust manned American presence in space would have been established decades ago by the MOL program. Due to the cutting of this and similar military space programs in favor of NASA, many capabilities were lost, and the United States was needlessly restrained in the race to develop space.[5]

It should be noted at this point that the Soviets rarely heeled to a treaty unless it benefited them. For example, the United States signed the Anti-Ballistic Missile (ABM) Treaty with the Soviet Union in 1972. The treaty was meant to prevent either side from developing even rudimentary defenses against nuclear weapons, out of fear that such defensive systems would destabilize the carefully crafted "balance of terror" between Washington and Moscow. At that time, the Cold War was governed by an insane ethos that argued for parity in the creation of both American and Soviet nuclear weapons arsenals, to allow for one side to hold the other side hostage in the Cold War. In so doing, it was believed, the risk of "mutual assured destruction" (MAD) would prevent the Cold War from turning into a hot war. With the exception of a few instances, like the Cuban missile, this theory was never fully put to the test. Yet, one thing is clear: the Soviets did not respect MAD. In fact, the writings of prominent Soviet general V. D. Sokolovsky prove that the Soviet Union believed it could fight—and win—a nuclear war against the West, only if Moscow had the right technology to defend themselves from an American nuclear counterattack. If Moscow possessed such weapons, then the Soviets could then initiate a conflict (the Reds favored a nuclear surprise attack in Europe). To achieve these aims, the Soviets had unapologetically developed the rudimentary *Galosh* antiballistic missile system that was meant to better protect Moscow from American nuclear attack.

---

5 Joseph Trevithick, "Revealed—the Pentagon's Scheme for a Spy Outpost in Space," *War Is Boring*, 23 October 2015. https://warisboring.com/the-u-s-military-had-plans-for-a-spy-outpost-in-space/

Derek Leebaert assessed that the Soviet Union, like America's rivals today, were incredulous about the American penchant toward pacifism and restraint in the face of their continual provocations. Here's what Leebaert, a national security expert and technology investor argued in 2002:

> The Soviet national missile defense system that emerged was not intended to create a perfect, or impermeable, "box," particularly since such defenses would be swamped by a US attack against the [Soviet ICBM force]. The purpose of the ABMs was instead to provide a degree of effectiveness, after a successful Soviet preemptive strike, against a ragged US retaliation. The Soviets knew these dual-purpose systems (the SA-5s and their radars, for starters) were relatively ineffective, but deployed them on a massive scale anyway. The United States accepted revised explanations that they were not intended to be used against missiles. In sum, there was no reason at all to believe that the [Soviets] who had fought from Moscow to Berlin [in the Second World War] had recently been socialized to the hypermodulated [sic] contingencies of US professors.[6]

As it turns out, the Soviets were not wooed into pacifism by their ideologically blinkered American foes in the 1970s. As Soviet general V. M. Milstein told Derek Leebaert, "Of course [the Soviet Union has] a 'war-winning' doctrine! What do you expect it to be, a *war-losing* one?"[7] Similarly, Chinese, Russian, Iranian, and North Korean leaders have crafted what they believe are war-winning strategies to use against the United States today—and they almost always resort to this concept of asymmetrical, high-tech warfare. Neither the Soviets of yesteryear nor America's foes today will play by the lofty notions of America's increasingly detached and gelded political class. When people on the

---

6 Derek Leebaert, *The Fifty-Year Wound: How America's Cold War Victory Shapes Our World* (New York: Back Bay Books, 2002), 390.

7 Leebaert, 391.

right were warning of the dangers of engaging the Soviets in these arms control treaties; about how the Soviets routinely lied and cheated, their concerns were often ignored. Or worse, the people issuing the warnings about the Soviet Union's intentions and capabilities were ridiculed.

Former CIA director and future president George H. W. Bush was tasked with assembling a group of contrarian analysts during the Gerald R. Ford administration in 1976. This was known as the "Team B" exercise wherein several people, like Richard Pipes, who would go on to serve during President Ronald Reagan's time in office, argued that the West was completely inaccurate about the strategic intentions and capabilities of the Soviets. The Team B exercises concluded that CIA assessments had been wrong about "the accuracy of MIRV-equipped Soviet missiles; the scope of Soviet civil defense, including the existence of entire underground cities, as in Sevastopol [in the Crimean Peninsula of Ukraine], the ABM violations, and certainly, the emerging Soviet capability to wipe out US ICBMs in a first strike. Indeed, sometimes Team B underestimated [the Soviets]."[8]

The United States had invested in ABM interceptors of their own. But after the ABM Treaty was signed in 1972, there was a pause in what was then a very promising American ABM system.[9] The Soviets, on the other hand, took the pause in America's development of ABM systems as a green light to invest even more money in their own ABM technology. The Americans had created a strategic gap that the Soviets hoped to shamelessly exploit. Such a strategic imbalance would have potentially allowed for Moscow to win an engagement with the West, as their cities and critical infrastructure might have been defended against any Western nuclear retaliation. What's more, the Soviets expanded their nuclear arsenals and enhanced the capabilities of their nuclear weapons to better overwhelm their Western foes in a first strike.

---

8 Leebaert, 491.

9 "Missile Wars: Missile Defense, 1944–2002," *PBS Frontline*, September 2002. https://www.pbs.org/wgbh/pages/frontline/shows/missile/etc/cron.html

Whatever remained of America's nuclear retaliatory capability would have been degraded further by the Soviet ABM defenses, thereby giving Moscow the hope that it might beat the Americans in a nuclear war.

America's rivals, whether the Soviets of yesteryear or the Chinese, Russians, Iranians, or even the North Koreans today, rarely respect treaties they enter into with the United States. Each time that the United States enters into a treaty, such as the ABM or the Outer Space Treaty, it tries to comport itself with the dictates of that agreement—even to its own detriment. Washington has mastered the unfortunate art of self-deterrence. A stronger commitment to an American nationalist ethos that placed the US national interest at the forefront of American politics, rather than some amorphous international concern, would prevent such misguided, foolish policies from being attempted in the first place.

UTOPIANISM AS POLICY

What's more, during the time of the passage of the Outer Space Treaty, Western culture had shifted away from one of progress and productivity to the aforementioned condition of nihilism and narcissism. Again, I remind you of Peter Thiel's postulation that the United States in 1969 confronted a choice between the values of Woodstock or those of the Apollo Moon landings. In the words of that old knight at the end of *Indiana Jones and the Last Crusade*, America chose poorly by favoring Woodstock. During this time period, academia and the wider policy-making community had started to embrace the soft-headed notions of one-worldism. The space policy community, in a sense, became ground zero for such fuzzy thinking. In an attempt to capture the zeitgeist of the 1960s, Western space policy advocates came to believe that space would not only always be a weapons-free sanctuary, much like Antarctica, but that it *should* be a weapons-free sanctuary. These well-connected policy advocates worked very hard to ensure that the United States remained limited in the strategic domain of space indefinitely, while Moscow strove to leapfrog the Americans. Inevitably, these views trickled across the government, and have influenced many policies in space since that time.

It is a shame that America's foreign rivals do not share the soft-headedness of American Leftists.

Matthew Mowthorpe provides an excellent working definition of the sanctuary view of space:

> The sanctuary view of space doctrine believes that the realm of space should not be weaponized. The intrinsic value space provides for national security is that satellites can be used to examine within the boundaries of states, since there is no prohibited overflight for satellites as there is for aircraft. This enables arms limitation treaties to be verified by satellites in space serving as a national technical means of treaty verification. Early warning satellites serve to strengthen strategic stability since they provide surveillance of missile launches which increases the survivability of retaliatory strategic forces. The sanctuary school argues that such is the importance of the functions of these space systems that space must be kept free from weapons, and antisatellite weapons must be prohibited, since they would threaten the space systems providing these capabilities.[10]

The taint of the Cold War era is still felt today, as Mowthorpe's definition suggests. One current and influential space policy analyst, Joan Johnson-Freese, had this to say about the weaponization of space in 2016:

> The Pentagon's recent aggressive focus on fighting and winning a space war may make for great news stories and science fiction, but it's lousy policy. While the situation in space has changed, in many ways for the worse, moving to an offensive strategy is a dangerous overreaction [on the part of the US military].[11]

---

10 Matthew Mowthorpe, *The Militarization and Weaponization of Space* (Lanham: Lexington Books, 2004), 12.

11 Joan Johnson-Freese, "Stopping the Slide Towards a War in Space: The Sky's Not Falling, Part 2," *Breaking Defense,* 28 December 2016. https://breakingdefense.com/2016/12/stopping-the-slide-towards-a-war-in-space-the-skys-not-falling-part-2/

Just as with those American policymakers who held the sanctuary view during the tough times of the Cold War, Freese and those like her embedded within the space policy community lament that the "aggressive approach undertaken by the military rests on incomplete premises and carries a significant chance of being counterproductive if space stability is the goal."[12] Yet, again, we are forced to ask the question that has pervaded US national security policy debates since 9/11: How is "stability" achieved when the forces involved are so lopsided? I have spent previous chapters informing you on how the United States is challenged by unconventional, asymmetrical warfare. Thus, if "stability" in space is to be secured, it will *not* be through passive policies that maintain the status quo. Our enemies—and even our friends, as you've seen—simply won't abide by the status quo any longer. It will be through an unapologetic expansion—and, if necessary, use—of overwhelming American power in space that maintains the status quo.

By the way, that 2016 article quoted above by Joan Johnson-Freese was written in response to the 2014 Space Strategic Portfolio Review (SPR), which argued for a more militarized approach to defend America's vulnerable space assets.[13] The 2014 SPR, in itself, was created in response to a May 2013 Chinese launch of what Beijing said was a "science mission" to geosynchronous orbit. As you've seen, GEO is where the bulk of America's most sensitive military satellites are housed, because GEO was usually believed to be a place where few other countries could reach—and therefore threaten those critical satellites. Clearly, Beijing never got the message about space purportedly being a sanctuary. The Pentagon was right to be concerned, considering how every US rival, going back to the Soviet Union, has unapologetically embraced "dual-use" technology to, at first, lull Americans into passivity and then to attack those forces in what we've come to call a "grey rhino" event.

---

12 Johnson-Freese.

13 Maximilian Betmann, "A Counterspace Awakening? (Part 1)," *Space Review*, 22 May 2017. http://www.thespacereview.com/article/3247/1

Maybe the Chinese 2013 GEO launch was scientific in nature. Yet, whatever lessons Beijing gleaned from that launch, it can now apply to a surprise ASAT attack on a future date.

Johnson-Freese's fellow utopian globalist thinkers in the space policy community, Helen Caldicott and Craig Eisendrath, claim that in the world today

> each person is a neighbor. We cannot anymore pretend that the fate of people in other countries can be outside our moral consciousness. If someone is starving in Nigeria, or suffering from AIDS, he or she is our moral neighbor, and his or her fate is part of our collective responsibility. For the history of humankind, this responsibility was limited by knowledge, distance, and communication. Today, those limits have disappeared, and we are part of a human family. If this responsibility has not been recognized by our governments, it is our duty to insist that, with the knowledge, communication, and accessibility possible, we are indeed responsible and need to act.[14]

At no point in their ode to globalism do these space policy analysts advocate for defending, much less defining, what is in America's national interest. Their argument sounds as if it should be murmured in a drum circle as opposed to legitimate national security policy circles. Alas, this is what decades of ceaseless cultural war at all levels of American society have wrought. In fact, utopian globalists are downright hostile to doing anything in space that smacks of base nationalism.

These are the people who have most influenced US space policy over the last thirty years. They are the ones who consistently stunt and stymie any move toward a nationalistic space program, out of fear that better protecting our satellites will somehow, through the butterfly effect, risk the populations of some poor African country . . . or something. It's all very strange and hugely self-destructive for a country that purports

---

14 Helen Caldicott and Craig Eisendrath, *War in Heaven: The Arms Race in Outer Space* (New York: New Press, 2007), 24.

to be the world's superpower. For what is a space program if not an extension of a nation's will? Such wrongheaded utopian assumptions undergird our present space policies, but more importantly, they form the bedrock of the international laws that outline how space can be used by nation-states—which explains why America is so vulnerable in space and, therefore, on Earth.

Interestingly, by 1970, when the world's powers sought a new space treaty, this time governing the use of the Moon, it was the so-called Third World nations that stood opposed to the United States *and* the Soviet Union. When the Moon Agreement of 1979 was being forged, both the Soviets and Americans were competing for greater levels of access to space, in keeping with the rapid development of space technology. Recognizing the vast resource potential that the Moon provided, Moscow wanted to ensure that the parts of the Outer Space Treaty that limited a country's—or a private-sector space company's—ability to claim and exploit space would not be extended to the Moon. Washington, on the other hand, preferred for the private sector to exploit the Moon. Meanwhile, the developing world, notably led by countries like Brazil, was significantly opposed to the idea of allowing either the United States or the Soviets to use their much greater power and wealth to further exploit space—thereby depriving the smaller, developing countries their chance to exploit space. The United Nations, in the throes of its mad love affair with global socialism, embraced this concept since the countries pushing it were themselves under socialism's dangerous enchantment.

At that time, Brazil and other developing states were not necessarily opposed to the exploitation of space resources, so long as they were given a cut of the loot. Of course, both Washington and Moscow had issue with that concept. If either the United States or Soviet Union made the investment and provided the manpower as well as the equipment, why would a third party that had nothing to do with that transaction suddenly be given a slice of the money and resources extracted from space mining? It was a ludicrous suggestion, but the Third World countries were understandably concerned that they were going to get cut out of access to the promising new domain of space. Yet, even the USSR believed in the capitalist notion that "to the victor go the spoils."

Due to the controversial aspects of the Moon Agreement, the United States Senate never ratified the treaty. Despite this fact, the federal government has gone to great lengths to ensure that the United States comports with the tenets of the Moon Agreement as best as it can. The logic is that the United States will inevitably seek to officially join this odious and inherently un-American document at some point. This is precisely what America's permanent administrative state did regarding the United Nations Convention on the Law of the Sea for years.[15] The idea is that these unelected, lifelong bureaucrats in Washington know best, and, inevitably they will convince some unsuspecting American president to embrace a treaty that is widely supported globally, but in fact, harms US national interests and sovereignty. So, the administrative state gets the proverbial ball rolling by enacting as many regulations as it can at the administrative level to make the treaty's inevitable ratification that much easier. It sounds like an efficient move. Of course, if a treaty, like the Moon Agreement, is ratified, one can kiss the entire notion of *American* development of space goodbye.[16]

Does one believe that either China or Russia will sit idly by and comport their behavior with these treaties? In the case of China, when have they followed a treaty or not exploited a loophole to their advantage? This is silliness masquerading as serious policy. There's at least a $1 trillion space mining economy waiting to be exploited. Who among us believes that either Beijing or Moscow or Tokyo, for that matter, will simply bow before the gods of globalist utopianism, when there are mouths to feed and strategic interests to be gained?

---

15 "U.N. Convention on the Law of the Sea: Living Resources Provisions," *Every CRS Report,* 7 January 2004. https://www.everycrsreport.com/reports/RL32185.html

16 Thomas Gangale, *The Development of Outer Space: Sovereignty and Property Rights in International Space Law* (Santa Barbara: Praeger, 2009), 68–73.

Under the present legal conditions, the United States can neither defend nor prosper from the strategic domain of space. The language of the Outer Space Treaty, in particular, makes some investors twitchy about the legality of their plans to develop space. When the treaty was written, the primary components of human activity in space resorted to the use of satellites and the manned space programs of both the United States and the Soviet Union. This is why both satellite usage and manned spaceflight functions have remained the primary elements of human activity in space. This is also why the United States is starting to lag behind its competitors in space. Additionally, the very fact that the Outer Space Treaty and the unratified Moon Agreement of 1979 explicitly prevent both the private sector development and the weaponization of space, which leaves space open only to national governments and those businesses in the satellite and space launch sector. The Soviets of yesteryear and most of the other spacefaring powers today are happy to keep this formulation in place, as most countries utilize some variation of the state-owned enterprise (SOE) model. Public and private endeavors are officially fused in these countries, meaning that countries possessed of SOE models have a comparative advantage in space development over the United States, with its relatively strict separation of the public and private sectors, as the law is currently written.

Peter Thiel was correct when he argued with Eric Schmidt about the fact that the world of atoms has gone stagnant in terms of innovation, whereas the world of "bits" continues to expand. The reason is simple: the world of stuff is usually easier for government regulators to understand, whereas the world of bits is tougher for lawmakers with legal or liberal arts backgrounds. This is yet another reason why the internet and computer sector were relatively deregulated for as long as they were—and why the tech tycoons today are working hard to lobby lawmakers to go easy on their industry.[17] The Outer Space Treaty was ratified by the United States Senate, and therefore all relevant laws in the United States dealing with space were made to comport with the language of that treaty. It's a disaster for the country, since American rivals have no qualms about violating or simply ignoring the treaty.[18]

One thing is clear, though: the United States cannot afford to simply leave things as they are with the legal regime governing the use of space. The world will move on without the United States at the helm in orbit—and beyond—if Washington continues dithering on this matter. The United States would go from a superpower to a middle power, stuck on Earth, looking up at all of the progress and wonders of China, feeling threatened by the peon states of Iran or North Korea, and constantly worrying about what Russia might do to the US in orbit. Meanwhile, the lucrative space sector, which "could generate revenue of more than $1 trillion by 2040, up from its current revenue of $350 billion," according to the investment experts at *Morgan Stanley,* would be left to American rivals. There are some reservations from investors, though, as the United States experienced a similar interest in the private space sector in 1990s. That craze was short-lived, unfortunately, as various satellite and launch companies floundered.[19] Yet, it is important to understand that space technology had still not matured to what it is today. Further, there was little capability in other countries to challenge the United States in this critical domain. The reality has changed. It's time for our laws to reflect this altered paradigm.

---

17 Lorelei Kelly and Robert Bjarnason, "Our 'Modern' Congress Doesn't Understand 21st Century Technology," *Tech Crunch,* 6 May 2018. https://techcrunch.com/2018/05/06/our-modern-congress-doesnt-understand-21st-century-technology/

18 Brandon J. Weichert, "Deregulate Space," *American Greatness,* 18 October 2017. https://amgreatness.com/2017/10/18/deregulate-space/

19 "Space: Investing in the Final Frontier," *Morgan Stanley,* 2 July 2019. https://www.morganstanley.com/ideas/investing-in-space

MAKING LAWS REFLECT REALITY

Besides, when I worked on Capitol Hill, the House passed the US Commercial Space Launch Competitiveness Act in 2015. The bill basically legalized space mining. Of course, given the bizarre presence of the Outer Space Treaty in US legal code, the House Resolution was written in a way that basically did "not allow ownership of an asteroid, or a swath of the Moon [*sic*], or any other section of extraterrestrial real estate—just the resources extracted from such a body." In essence, "space mining will thus be analogous to fishing in international waters."[20] According to Sagi Kfir, general counsel of asteroid mining company Deep Space Industries, the US Commercial Space Launch Competitiveness Act will allow "capital markets to take a closer look at the space resource utilization industry, now that we have a legal framework for operations."[21] It remains unclear whether the 2015 space mining law is in violation of the Outer Space Treaty. The lack of clarity on this matter, though, might stymie enough investors to slow the pace of development in the private space sector yet again.[22]

The investor community needs greater legal clarity to reduce risk on their potential investment. It is time for the United States to lead the world in getting the great spacefaring powers to meet and work out the differences between what the various countries around the world want to do in space and what the Outer Space Treaty allows. Brian Weeden of the Secure the World Foundation and other space policy experts believe that diplomacy, rather than militarism, in space is needed. While we can carp about the details of his stance, Weeden's supposition

---

20 Mike Wall, "New Space Mining Legislation Is 'History in the Making,'" Space.com, 20 November 2015. https://www.space.com/31177-space-mining-commercial-spaceflight-congress.html

21 Jeff Foust, "House Passes Commercial Space Bill," Space.com, 18 November 2015. https://www.space.com/31147-house-passes-commercial-space-bill.html

22 Peter B. de Selding, "New U.S. Space Mining Law's Treaty Compliance Depend on Implementation," *Space News,* 9 December 2015. https://spacenews.com/u-s-commercial-space-acts-treaty-compliance-may-depend-on-implementation/

is correct but it will need to be framed differently to the world. I argue that Washington should tell the world's spacefaring powers: either agree to make the national and private sector exploitation of space a reality by changing key parts of the existing Outer Space Treaty, or the United States will unilaterally abrogate the treaty—likely destroying whatever legal regime exists for space.[23]

In the past, I've argued in favor of simply discontinuing our involvement in this treaty, as we have done in the Kyoto Protocol, the ABM Treaty, the Paris Accords, and the INF Treaty. Yet, some space lawyers—yes, that is a real profession—believe that if the United States were to unilaterally abrogate the Outer Space Treaty, Washington would still be required to follow its rulings, under the legal concept of *jus cogens,* or "compelling law." As Thomas Gangale warns, "If the United States withdrew from the Outer Space Treaty, the other parties of the treaty would act to defend it and could make life difficult for US nationals who conducted activities that those parties regarded as inconsistent with the treaty."[24] This would, according to Gangale and others, inject uncertainty in the private space sector. Such a move, as Gangale argues, would exponentially increase risk to private capital too. All of this would likely drive potential investors away from the private space market, setting the dream of space development back significantly.

However, the United States *has* abrogated important treaties before. In 2002, it terminated its involvement in the US-Russia Anti-Ballistic Missile (ABM) Treaty. The George W. Bush administration did this to allow for the creation of a reliable missile defense system against a potential nuclear weapons attack from a rogue state. At the time, Moscow opposed this and continues decrying this move twenty years later. Former President Bush encouraged the end of the ABM Treaty because he believed it might finalize the construction of a space-based

---

23 Weeden and Samson, "Op-Ed."

24 Gangale, *The Development of Outer Space*, 53.

missile defense system initially envisaged by Ronald Reagan. The Trump administration followed this move with the termination of the Intermediate Range Nuclear Forces (INF) Treaty. Again, Russia balked. The sky did not fall. The dirty little secret with international law is that, since there is no independent enforcement body, only a coalition of self-interested nations, all international "laws" are really subject to the power imbalances that each country has with each other. If the powerful countries decide to prevent the development of space, they will. Their actions indicate otherwise, despite the presence of the Outer Space Treaty. Again, nationalism and the pursuit of perceived national interests are far more important here than any sense of international obligation.

Ideally, the United States could get the other spacefaring powers to craft more sensible laws governing the use of space and the celestial bodies therein. To be fair to America's rivals, both the Russians and Chinese sought to craft a better version of the Outer Space Treaty. Of course, this 2008 Russo-Chinese proposal would have negated America's advantages in space by explicitly banning all weapons in space. But as you've seen, the Chinese and Russians have been consistent in their commitment to building out dual-use weapons for space. While the Americans routinely try to live up to their treaty obligations, meaning that had the previous Obama administration embraced the Russian and Chinese proposal, the United States would have been more vulnerable than it already is to a Russian or Chinese strategic surprise in space.[25] The Chinese and Russian proposal, however, was an opening for the United States to gather together *all* of the spacefaring powers—including American friends, like France or India—to craft a real proposal that would have acknowledged how significantly times in space have changed since the original Outer Space Treaty was crafted

---

25 Bill Gertz, "U.S. Opposes China-Russia Space Arms Treaty," *Washington Times,* 2 December 2015. https://www.washingtontimes.com/news/2015/dec/2/inside-the-ring-us-opposes-china-russia-space-arms/

in 1967. After all, it would be in the long-term strategic interests of all of the spacefaring powers to encourage dynamism in their space policies: to accept that the economic development of space, but also the securing of national strategic interests in space, are essential for all space powers going forward.

Alas, not even this was tried by the former Obama administration. Instead, Russia and China were simply left to stew while Washington did very little to secure itself in space. This dithering sent a chilling signal to the other space powers: the United States would not work with them on shared national and economic interests. Yet, Washington refused to fully embrace a weaponized view of space and to totally encourage the private sector development of space—before the various other space powers could engage in similar activities. Indecision and lack of moral clarity on the part of America's leadership has most contributed to America's vulnerability in space over the decades.

If the United States is to win—or even survive—a space war, or if it wants to win the second space race, then it must fundamentally change the way it views space. Either the United States deregulates space and allows for the private sector development of space, or it will fall behind its competitors. In nearly every other high-tech industry, the United States has figured out the right admixture of deregulation coupled with copious federal research and development. American leaders in generations past understood the stakes of ignoring the rising threats in a new domain, like space. Today, it remains unclear that they understand the danger. For all of the Trump administration's hoopla about creating a space force to better defend US satellite constellations from enemy disruption and destruction, little thought has apparently been given by anyone to the fact that weaponizing space even for defensive purposes or harvesting the resources of space is mostly forbidden under current international space law. Unless the United States intends to simply ignore this reality, then real steps must be taken to fundamentally alter the legal framework.

20

BEING CHEAP AND EASY IN

SPACE IS GOOD

"IF YOU'RE A START-UP, you almost certainly won't get a chance to
sell your product or service to the US government," according to a report
by the national nonprofit consumer advocacy organization known as
Public Citizen. As *Vice News* reported in 2015, "Derek Khanna, one
of the report's four authors, says that most government contracts are
awarded to firmly entrenched, old-line companies based on the fact
that they employ large portions of the 'consulting class.' In other words,
[legacy contracting firms] employ lobbyists who can get legislators to
steer business their way. Start-ups tend not to have this advantage, which

makes it extremely difficult for them to compete for the $6.2 trillion [of government contracts per year]."[1]

Given the opaque nature and secrecy of the Department of Defense, many Pentagon leaders are averse to using smaller start-ups, like Elon Musk's SpaceX to launch sensitive satellites into orbit—even though Musk's company has a proven track record of drastically reducing the obnoxiously high costs of military space operations.[2] In fact, former Air Force chief of staff General Norton Schwartz openly expressed his favoring of the Boeing and Lockheed Martin conglomeration, known as the United Launch Alliance (ULA), over SpaceX. In Schwartz's words, he "didn't want to trust the 'crown jewels' [the Air Force's classified payloads]" to a start-up, like SpaceX.[3] More to the point, had it not been that SpaceX was more of a hobby for the billionaire, Musk, it is unlikely that another start-up would have managed to make it as far in the DoD contracting process as SpaceX has gotten. After all, Musk had to initially sue the Pentagon in 2014 after it had awarded SpaceX a series of launch contracts and then, inexplicably, revoked those contracts and handed them off to the ULA.[4] Remember: not only should lowering the cost of space operations be a strategic priority for the US military, but it also should be a moral necessity for the US government, since the money they spend comes from taxpayers like you and me.

---

1 Justin Rohrlich, "Why Can't Startup Companies Get US Government Contracts?" *Vice News,* 24 September 2015. https://www.vice.com/en_us/article/d3978k/why-cant-startup-companies-get-us-government-contracts

2 Wendy Whitman Cobb, "How SpaceX Lowered Costs and Reduced Barriers to Space," *The Conversation,* 1 March 2019. http://theconversation.com/how-spacex-lowered-costs-and-reduced-barriers-to-space-112586

3 Jed Babbin, "Elon Musk: King of the Crony Capitalists," *American Spectator,* 8 October 2018. https://spectator.org/elon-musk-king-of-the-crony-capitalists/

4 Christian Davenport, "Elon Musk's SpaceX Settles Lawsuit Against Air Force," *Washington Post,* 23 June 2015. https://www.washingtonpost.com/business/economy/elon-musks-spacex-to-drop-lawsuit-against-air-force/2015/01/23/c5e8ff80-a34c-11e4-9f89-561284a573f8_story.html

It's not surprising, though, that America's so-called deep state would refuse to engage with start-ups. Most major defense contractors, like Lockheed Martin and Boeing, all have an incestuous relationship both with the Pentagon's leadership as well as the United States Congress.[5] Elected officials and former defense leaders inevitably land on the boards of several defense companies that they used to regulate. For that old revolving door between government and private sector to keep spinning, then, those who are charged with oversight of the defense sector won't peer too deeply into the practices of firms whose boards they hope to sit on after leaving public service.[6]

Musk and those small start-ups in the space sector seeking access to defense contracts are disruptors. And the permanent bureaucracy dislikes disruption. Initially, the Pentagon, along with the bigger defense contractors, fought Musk's entry into the defense market; they and their ULA partners were especially hard on Musk's firm when SpaceX first got access to Pentagon launches—far harder and less forgiving than the Pentagon has ever been with the retinue of cost overruns and massive failures that so many programs built by the major defense contractors have encountered.[7] Just think how much more the Pentagon could do in space if its leaders simply allowed for disruption and innovation to occur within the sclerotic defense sector, particularly in the military's space launch services sector. Disruption can be a force for good when seeking greater efficiency. Greater efficiency allows for costs to decrease,

5 Mandy Smithberger, "Brass Parachutes: The Problem of the Pentagon Revolving Door," *POGO*, 5 November 2018. https://www.pogo.org/report/2018/11/brass-parachutes/

6 Daniel Cebul, "New Watchdog Report Decries 'Revolving Door' Between the Pentagon and Defense Contractors," *Defense News*, 6 November 2018. https://www.defensenews.com/industry/2018/11/06/new-watchdog-report-decries-revolving-door-between-the-pentagon-and-defense-contractors/

7 Christian Davenport, "ULA Chief Accuses Elon Musk's SpaceX of Trying to 'Cut Corners,'" *Washington Post*, 18 June 2014. https://www.washingtonpost.com/business/economy/ula-chief-accuses-elon-musks-spacex-of-trying-to-cut-corners/2014/06/18/a7ca0850-f70d-11e3–8aa9-dad2ec039789_story.html

which, in turn, increases overall innovation—all things that the military space program must favor.

One of the biggest political hurdles US military leaders must overcome regarding the implementation of more robust plans for space is the cost factor. Lowering costs, by embracing a more competitive launch services capability, would drastically increase efficiency and allow for the military to accomplish that which it seeks in space. Since 2005, the Department of Defense, in particular, has struggled with keeping costs down in the military space sector to allow for more satellite launches. This was one reason why the DoD made Lockheed Martin and Boeing coalesce into the United Launch Alliance to begin with. This was also a factor behind why Elon Musk's (at that time) newly minted SpaceX was competing for contracts with the Air Force. Get costs down to accomplish more missions in space—that should be the Pentagon's mantra. As you've seen, most other spacefaring powers have either already figured out or are in the process of learning how to drastically reduce the costs of space operations. By lowering those costs, they are increasing innovation while ensuring greater levels of access to space as well as a more reliable defense of satellites. The United States is falling behind.

When I was earning my master's degree, the senior vice president of the Atlantic Council, Barry Pavel, sojourned to our little graduate program in the heart of Washington, D.C. and gave an informal talk about emerging threats, specifically in the high-tech arena. My professor, a colleague of Pavel's, intimated to him that I had become ensconced in the study of the national security space policy. He prompted me to speak with Pavel on the subject. What stuck with me was something that Pavel had mentioned in passing during our two-hour conversation. Pavel had said that he had worked hard during the early 2000s to get the cyberwar policy people to do a better job of integrating themselves into the wider national security policy debate. Pavel said that the cyber folks tended to keep to themselves, which made understanding the threat America faced from this new domain hard for policymakers to grasp. The more difficulty in understanding the complex area of cyberspace that policymakers had, the less likely they would be to either fund or place resources into defending and developing that strategic domain. After a decade, though, everyone came to understand the importance of cyberwar and the significant threats faced by the United States in cyberspace. The space policy community, despite having been a strategic domain accessible to humans far longer than cyberspace, is as isolated away from the larger national security dialogue as the cyberspace policy people were in the early 2000s.

In fact, the space policy community has been isolated from the larger national security policy conversation in much the same way that the counterterrorism policy folks were in the decade leading into the 1990s. After all, it took 9/11 for Washington to fully understand the extent and nature of the threat America faced from global terrorism. Unlike counterterrorism, however, US policymakers still have difficulty in understanding cyberspace.[8] This, despite the fact that nearly every single American today, regardless of age, uses cyberspace on a minute-to-minute basis, every day of every year. Space is just as important and technically complex as cyberspace. And while everyone relies on the signals that pass through our satellites in space, few people understand how critical these systems are.

What's more, when one tells American lawmakers the baseline price for making national investments into defending these vital satellite constellations, those lawmakers balk. Obviously, they would prefer to wait for a Space Pearl Harbor to happen before they decide to make the necessary investment into protecting American assets in orbit. Of course, by that point, it'll be far costlier for Americans. After such an attack, not only would new satellites have to be built and launched, which would take time, but Washington would also have to fund greater defenses for these systems. Further, following a debilitating attack in

---

8 "5 Tensions Between Cybersecurity and Other Public Policy Concerns," *At the Nexus of Cybersecurity and Public Policy: Some Basic Concepts and Issues,* National Research Council (Washington, D.C.: The National Academies Press, 2014). https://www.nap.edu/read/18749/chapter/7

space, there is no promise that the United States would be able to reclaim a dominant position there, depending on the level of damage inflicted upon American satellites. Besides, for the military to truly be able to operate effectively in the new contested and congested strategic domain of space, it will need a fundamental change in the institutional culture of the Pentagon and its outlook on space.

BUREAUCRACY KILLS

In 2011, retired Air Force brigadier general Thomas D. Taverney wrote about his experience in satellite acquisitions for the Pentagon. Satellites are expensive. Military satellites are more expensive than even the most sophisticated civilian satellite. Taverney observed that onerous government regulations and stifling bureaucracy prevented greater levels of innovations and did little to ameliorate the high cost of launching military satellites into orbit. In fact, very often, the government standards for developing and deploying military satellites caused significant cost overruns. The cost overruns themselves pose a significant national security threat, as they usually make legislators jittery about funding such projects in the future. And evoking Taverney, such budgetary constraints mean that there are few spares available to replace an important military satellite should it be damaged or destroyed. The lack of replacement satellites means that critical capabilities will be denied to American forces responding to a crisis. Therefore, an American rival will have a significant strategic advantage over American forces.

Until the rise of Donald J. Trump and his nationalist agenda to the presidency, few in the isolated military space policy wing of the Pentagon could formulate innovative ways to allow for the survival of key American systems in orbit, in the event of a space war. The United States was losing the military side of the second space race because of this. Taverney referred to the situation that designers of and managers for military satellite programs as the "Space Acquisition Vicious Circle." According to Taverney, "each attempt to resolve problems in the acquisition process creates new problems." Because space launches are an expensive undertaking for the Department of Defense, the military tends to build and maintain a handful of satellite constellations in orbit. Due to this, in the words of Taverney, "it needs to get everything it possibly can out of each. This has meant, within current systems architecting approaches, bundling capabilities. With bundling, the government imposes as many requirements as possible onto a single mission. This is called *aggregation,* and is often used to build stronger cross-mission advocacy."[9]

Taverney has been at the forefront of resolving the problem of launch costs for the US military's space policy. He insists that the DoD stop doing the incremental adjustments that it has been doing to its acquisition of satellites. Instead, he calls for boldness. Since he wrote about the "Space Acquisition Vicious Circle" in 2011, the Trump administration has moved more toward the direction that Taverney and other military space reformers have been advocating. These needlessly complex systems are far too expensive for the Pentagon to afford, given current budgetary restraints, and how much more money are taxpayers realistically going to give the Pentagon? Taverney believes that "disaggregation, mixed constellations, and shorter life systems [for military satellites]" will help to reduce costs significantly.[10]

Disaggregation basically means making satellites simpler, so that their up-front costs are cheaper, meaning that the Pentagon will not only get more proverbial bang for the taxpayers' buck, but that Congress would be more willing to fund US military space endeavors, since it would be cheaper than it presently is to conduct missions in space. The mixed constellation suggestion is an interesting one, basically mixing the more advanced and expensive satellites with less advanced and cheaper satellites to create a satellite constellation of varying layers of capability. Cheaper, less complex military satellites, though, would

---

9 Thomas D. Taverney, "Resilient, Disaggregated, and Mixed Constellations," *Space Review,* 29 August 2011. http://www.thespacereview.com/article/1918/1

10 Taverney.

force the government to repair and replace these systems sooner than they normally would. That's all right. Under these new guidelines the overall cost of doing business in space would be cheaper. Thus, the US military could repair and replace these less sophisticated satellites with greater ease and diligence than they currently do. What's more, they could offload some of the functions onto other satellites until repairs could be made or a replacement launched.[11]

Presently, many of the military's satellites are too unwieldy to "ride share" with civilian systems on civilian rockets. The DoD must rely on specialized, expensive rockets, usually provided by the ULA, half of which run on the Russian-built RD-180 rocket engine. While the ULA has lowered the overall cost for launches from what they were before 2005, when the ULA was created, Elon Musk has proven that military space launches can be done at 40 percent less the cost of what the United Launch Alliance can do! This explains why the entrenched bureaucracy fought Musk when he first began bidding on Air Force launch contracts.[12] It also further proves that the system has been rigged against smaller companies. After all, the military technology sector is one of the most lucrative sectors in the US economy today. Space launches for the military are among the most expensive undertaking in the military technology field. Disrupting the market by radically reducing costs, the way that Musk, an outsider to Washington, D.C., has done, is a direct threat to the unnatural order that has arisen in the iron triangle of interest groups, the permanent bureaucracy, and congressional Committees.[13]

---

11 Taverney.

12 David Z. Morris, "Is SpaceX Undercutting the Competition More Than Anyone Thought?" *Fortune,* 17 June 2017. https://fortune.com/2017/06/17/spacex-launch-cost-competition/

13 Gordon Adams, "The Iron Triangle: The Politics of Defense Contracting," *Foreign Affairs,* Spring 1982. https://www.foreignaffairs.com/reviews/capsule-review/1982-03-01/iron-triangle-politics-defense-contracting

SPACE IS WHAT YOU MAKE OF IT

"Space is hard," said one national security analyst who opposes a space force at a briefing I delivered to elements of the Defense Department in 2018. This individual was a senior civilian at the Pentagon. "Well, it's hard because of the costs involved," I retorted. "Lower that, and everything for you becomes much easier in space," I added. The individual rolled his eyes skeptically. Another person at the meeting insisted that weaponizing space—or even defending our satellites—would constitute a violation of one of our country's hallowed diplomatic agreements, the Outer Space Treaty. Yet, America's enemies are not blinkered by such "deep" thoughts as many of America's professional "strategists" apparently are, having found ways to lower the cost of their space activities. Lowering costs allowed for innovation, which opened entirely new modes of attack on the United States from space. Due to this, US rivals are weaponizing space anyway with the supposed "dual-use" technologies, such as co-orbital inspection satellites that can be converted into the dreaded space stalkers.

The cost of space operations is high because, until very recently, no one in the Pentagon, in Congress, or in the defense contracting community was willing to address this simple but critical issue. And why should they have been made to? The Pentagon is one of the largest bureaucracies in the world, with an opaque budget that few in Congress understand. Initially, the government wanted to have a handful of defense companies, such as Lockheed Martin, that it could contract its functions out to. And because the DoD is involved in top-secret technology development, the government wanted to have these few preferred contractors enmeshed in the national security state that was established at the end of the Second World War, with the National Security Act of 1947. This was done to increase efficiency and still maintain secrecy during the Cold War. It made sense to do this—at first.[14]

---

14 William D. Hartung, "Pentagon Socialism: The Pentagon's Plan to Dominate the Economy," *Salon,* 12 November 2018. https://www.salon.com/2018/11/12/pentagon-socialism_partner/

Over time, however, human nature kicked in, and, and the inefficiency and corruption that comes with most government programs increased as the budget expanded. As those budgets remained high throughout the Cold War, the more opportunities there were for the preferred contractors to make increasing amounts of money. Under current conditions, the defense contracting sector is one of the least capitalistic systems in America today. Whereas competition is an *esprit de corps* in a free market, competition is not in any way a factor in the defense contracting community. And the defense contracting community has not only grown in size, but it has also been consolidated under only a handful of major defense contractors. Think about it this way: even when a small start-up does appear in the defense contracting community, these firms are usually bought out by the larger firms in order to keep competition from drastically lowering prices in the defense industry.[15]

As an entrepreneur who was already wealthy by the time he founded SpaceX, Elon Musk could afford to wait for the start-up to earn profit from government contracts. Musk's team was given the time needed to challenge the chokehold that the major defense firms had on the Pentagon. SpaceX could endure cutting through that painful bureaucracy and to check the unfair privileges that the major defense contractors enjoyed. What's more, because SpaceX was Musk's side hustle, he did not need to sell his rising start-up to the bigger defense contractors, as so many start-ups are inevitably sold to one of the big contractors. So, Musk could focus on disrupting the military space industry over the long term. Due to the disruption that SpaceX caused, costs for taxpayers in this area have been drastically reduced. Thanks to SpaceX's incredible reusable rockets, cost cutting is allowing for more innovations to occur in the defense space sector—all because costs are being lowered and efficiency increased. Without competition, necessary capabilities are

---

15 Rodrigo Carril and Mark Duggan, "The Impact of Industry on Government Procurement: Evidence from Department of Defense Contracting," *The National Bureau of Economic Research,* October 2018. https://www.nber.org/papers/w25160

rarely created because monopolies tend to be averse to change. America needs more disruption in the space launch services sector, as that is the costliest part of national security space policy.

Toward that end, SpaceX's reusable rocket is the key to lowering launch costs. Not only is it much cheaper than its competitors in the United Launch Alliance, but the Falcon 9 rockets are entirely American-produced. Unlike some of the ULA's rockets that are used to place sensitive military equipment in orbit, SpaceX is not using the Russian RD-180 rocket engines in their Falcon 9 rockets. So, just like that, the South African–born Elon Musk has done more to secure the American space supply chain than the Department of Defense has been able to do! And for all of those who argue that Musk and those like him are nothing more than "crony capitalists," I say Musk and his government-subsidized ilk would have never needed to come into existence had the US government not spent the last several decades slashing its federal research and development budget.[16] Musk and other start-ups receiving federal monies are at least helping to fill the gap left by the unwise gutting of America's once-mighty federal research and development budget. Besides, the competition will force the larger defense contractors to lower their costs and focus on what they're usually very good at: scaling up advanced projects.

Once the Pentagon embraces the cost-saving techniques of Thomas D. Taverney, coupled with opening up its space launches and future military space system designs, costs for US military space operations will fall to affordable levels.[17] Suddenly, space won't be "so hard," even for the most harebrained swamp dwellers. If President Donald J. Trump's calls for an independent space force are to be fully realized, Washington

---

16 Russell Brandom, "Trump Is Turning Elon Musk into a Crony Capitalist," *The Verge,* 6 February 2017. https://www.theverge.com/2017/2/6/14520608/elon-musk-trump-tesla-business-deal-economic-advisory-team

17 "Reusability: The Key to Making Human Life Multi-Planetary," *SpaceX,* 10 June 2015. https://www.spacex.com/news/2013/03/31/reusability-key-making-human-life-multi-planetary

must defend its existing satellite constellations. To achieve this simple task, the Pentagon must first drastically cut costs. Only by opening up the space launch services sector and by making military satellites smaller, less complex, and more easily replaceable can America's vaunted satellite capabilities survive in the face of increased threats to those satellites. From there, America can address greater strategic challenges, such as those posed by nuclear-armed rogue states, like North Korea or Iran in the form of a space-based missile defense system. Expanding out from that point, over the longer term, then, the Pentagon can maintain a permanent presence deeper in the solar system to better counteract whatever malign intentions challengers in the new space race may have toward America's dominance in space.

Thankfully, President Trump signed the space force into law on December 20, 2019. This move is but the start of what I hope to be a much larger, longer-term reorganization of the national security apparatus to better defend against threats in space, and more importantly, to better dominate space. Part of the law that the president signed on December 20, 2019, gave explicit directions to the Pentagon's acquisition teams to begin embracing the cost-cutting measures for space launches and satellite design that I and others have been advocating for years. This will help to ensure that Washington does not abandon space as a strategic warfighting domain once President Trump is no longer in the White House.

Yes, "space is hard." At one time, cyberspace was hard, air was really hard, and the ocean was very hard to master. So, too, was land "hard" to control—particularly for a country, like the United States, which has been far removed from conflict zones. *All of the strategic domains are "hard" to operate in.* Listening to purported national security "experts" talk themselves—and policymakers—out of enacting an ambitious space policy is akin to watching a doctor opt to euthanize a patient because it's easier to do that rather than implement treatment for a complex disease. Ultimately, the doctor's job, to ameliorate the damage caused by disease, is not accomplished. Similarly, the Pentagon's job is to defend the United States against all manner of foreign threats. It has habitually failed to achieve this in space, so much so that the United States now faces the prospects of a Space Pearl Harbor.

STAR WARS AND OTHER COOL STUFF WE COULDN'T AFFORD

When Ronald Reagan took over as president in 1980, the country was in shambles. A turgid economy had produced "stagflation," inept foreign policy had encouraged America's greatest enemies to act bolder than they ordinarily would have on the world stage, and many people around the world had believed that Soviet Communism, not American capitalism, was on the rise. To much of the world, and some Americans, former president Jimmy Carter's claims that the United States was in a declinist "malaise" was true. But as the preeminent historian Niall Ferguson has argued, decline is a choice; it's a state of mind. Rather than give in to the prognostications of the elite and America's Communist foes, most voters turned out in support of Ronald Reagan and his "sunny" optimism about America's future, both economically and regarding his belief that the Soviet Union would be sent to history's "dustbin."

It was "morning in America" again under Reagan. Part of his national renewal program relied heavily on enhancing America's flailing space program. Reagan rightly recognized that the Soviet system and its foreign policy represented the "evil empire." He was clear-eyed in his assessment that the USSR had to die. What's more, Reagan believed it would die by his hand. Many believed that Reagan intended to start a nuclear war— including the Soviet leadership. The Left was so frightened of Reagan's tough stance toward the Soviet Union that they started a nuclear freeze movement (it would later be discovered that many of these anti–nuclear weapons movements were funded by Soviet intelligence).[18] As it would turn out, though, Reagan was not interested in starting a nuclear war to settle the Cold War with finality. President Reagan hated nuclear weapons.

---

18 Daniel Southerland, "Communist Influence in Peace Movement: Threat or Red Herring?" *Christian Science Monitor*, 29 November 1982, https://www.csmonitor.com/1982/1129/112936.html

His loathing of both the Soviet system and of nuclear weapons prompted him to announce the creation of the Strategic Defense Initiative (SDI). This became known as "Star Wars" in the press.

Speaking from the Oval Office in 1983, Reagan asked:

> What if free people could live secure in the knowledge that their security did not rest upon the threat of instant US [nuclear] retaliation to deter a Soviet [nuclear first-strike] attack? That we could intercept and destroy strategic ballistic missiles before they reached our own soil or that of our allies? I know this is a formidable, technical task, one that may not be accomplished before the end of the century. I call upon the scientific community in our country—those who gave us nuclear weapons—to turn their great talents now to the cause of mankind and world peace. To give us the means of rendering these nuclear weapons impotent and obsolete.[19]

Overnight, President Reagan and his supporters dreamed of a world not governed by the insanity of mutual assured destruction. Instead, Reagan wanted an international order defined by mutually assured survival.[20] He did not want to be held hostage to the Soviet Union's nuclear threat. For Reagan to move the world beyond MAD, though, he needed a defensive system that could utterly defeat the Soviet nuclear weapons threat. The Strategic Defense Initiative was that defensive weapons system. Utilizing the most advanced lasers and space technology, the SDI would be deployed to Earth's orbit, where it would be linked up to the most sophisticated early warning missile systems. If any Soviet launch were detected, it would attack the incoming missiles, all of which passed through space on their way down to targets in the United States,

---

19 ReaganFoundation, "National Security: President Reagans Address on Defense and National Security 3/23/83," YouTube, 16 April 2019. https://www.youtube.com/watch?v=ApTnYwh5KvE

20 Keven Bankston, "How Sci-Fi Like *WarGames* Led to Real Policy During the Reagan Administration," *Slate*, 8 October 2018. https://slate.com/technology/2018/10/reagan-wargames-star-wars-science-fiction-policy.html

and the lasers would destroy those incoming nuclear missiles before they could reach their target. The idea of intercepting nuclear weapons in the distant and isolated environment of space, as opposed to waiting for those missiles to reenter Earth's atmosphere, and rely on rudimentary ballistic missile interceptors, using missiles fired from defense batteries on the ground to track and destroy incoming nuclear weapons, was a far better concept of protecting against the Soviet nuclear threat—especially in the age of the multiple independent targeted reentry vehicle (MIRV).

It was a simple idea, a concept that maximized America's dominance in the high-technology arena. Many people opposed Reagan's policy. Some argued that it would upset the delicate balance of the Cold War; the mere announcement of such a system would prompt Moscow to take dangerous and aggressive actions to preserve their nuclear threat to the United States. In a use-it-or-lose-it approach, some worried that Moscow might opt to initiate a preemptive nuclear strike on NATO before the SDI system could be fully developed and deployed in orbit. Others feared that that SDI would precipitate another arms race within the Cold War. In typical fashion, the space policy community insisted that the United States should not invest in the system. Instead, the utopians decried the "weaponization" of space while the naysaying engineers focused their efforts on "proving" that such a system could not ever work (because, science).

Regardless, the idea scared the bejesus out of Moscow. Ultimately, the Soviets spent themselves out of existence, trying to better compete with Reagan's high-tech military buildup. The Soviets were interested in their own "Star Wars" and developed technology to support such a program. But by the 1980s, the Soviet high-tech industry was being gutted in the face of constrained budgets and increasing political instability. Had the Soviet tech and defense sector in the 1980s possessed the resources it did at the height of the arms buildup, things might have gone very differently. The Soviet version of Reagan's "Star Wars" would have been a formidable system. Yet, the Soviets were in no position to deploy such a system—let alone build it. The Soviet Union whimpered into collapse a few years after Ronald Reagan had left office. In many ways, Reagan's idea of a space-based missile defense shield broke the strategic back of the Soviet Union—where the primary threat Moscow posed to the West came from its nuclear weapons. Reagan essentially challenged the USSR in the one area where it believed the Americans could never defeat them: nuclear war. And Reagan aspired to accomplish this goal without ever firing a shot at the Soviets. Reagan's proposal forced the Soviets to play catch-up at a time when neither their scientific base nor the economy could handle it, and the psychological impact of SDI's potential was enough to put a final nail in the Soviet coffin.

Today, the threat posed by nuclear-armed rogue states to both the United States and its allies is high. The nuclear weapons programs of both North Korea and Iran are farther along than many are willing to acknowledge. At some point, given the tension level between Washington, Tehran, and Pyongyang, either North Korea or Iran (or both) will destabilize their regional political orders by using—or explicitly threatening to use or proliferate—nuclear weapons against the United States and its allies. The United States has spent decades worrying about this particular threat. Yet, neither North Korea nor Iran is in any way America's equal in terms of military capabilities or technology. Nuclear weapons are now almost a century old. How is it that no viable defense or countermeasure against them has been crafted?

Fundamentally, the technology that Reagan's SDI relied upon was sound.[21] While some of the detractors of Reagan's "Star Wars" program were correct—the technology was probably not yet ready for showtime in the 1980s—today, the technology is more than ready.[22] What is needed is someone at the leadership level to push for the creation of a

---

21 Sean Gallagher, "Trump's Missile Defense Strategy: Build Wall in Space, Make Allies Pay for It," *Ars Technica*, 29 January 2019. https://arstechnica.com/tech-policy/2019/01/new-dod-missile-defense-strategy-star-wars-ii-the-wrath-of-trump/

22 Harold Brown, "Is SDI Technically Feasible?" *Foreign Affairs,* 1985. https://www.foreignaffairs.com/articles/1986–02–01/sdi-technically-feasible

reliable space-based missile defense system. As Angelo Codevilla wrote in 2018, "America's need for serious capabilities to defend and attack satellites, as well as for a missile defense worthy of the name, has been debated for decades. Only under the Trump administration, however, have persons occupied senior positions for whom these needs override other considerations."[23] Now, the US military is testing weapons-grade lasers on its warships in anticipation of new and extraordinary forms of attacks that many American rivals can now employ against the US Navy.[24] The Air Force is developing similar systems for its warplanes, anticipating their need in the next round of great power conflict, which is coming.[25]

Meanwhile, as the Pentagon slowly embraces the reforms needed to reduce launch costs, the possibility that the United States could field a reliable missile defense system in orbit becomes probable. With North Korea's nuclear program already in possession of nuclear arms, with the last step being for them to develop an ICBM capable of hitting the United States, time is of the essence for President Trump's space force to not only be constituted but to make the deployment of such a big, beautiful defensive weapon system in orbit a priority. Defending satellites from attack is great. But that's irrelevant if North Korea or Iran manages to hit the United States with a surprise nuclear attack. A space-based weapons system would drastically reduce such a threat and make the leadership of even the most intransigent rogue state more pliable at the diplomatic level, as they would have little choice but to negotiate with the United States, since their nuclear threats would no longer hold sway. Also, such a system would place the Chinese and Russians

23 Angelo M. Codevilla, "The Space Force's Value," *Hoover Institution,* 15 October 2018. https://www.hoover.org/research/space-forces-value

24 Megan Eckstein, "Navy to Field High-Energy Laser Weapon, Laser Dazzler on Ships This Year as Development Continues," *USNI News,* 30 May 2019. https://news.usni.org/2019/05/30/navy-to-field-high-energy-laser-weapon-laser-dazzler-on-ships-this-year-as-development-continues

25 Kris Osborn, "Stealth Strike! U.S. Air Force Wants to Test F-35 Laser Weapons by 2021," *The National Interest,* 16 October 2019. https://nationalinterest.org/blog/buzz/stealth-strike-us-air-force-wants-test-f-35-laser-weapons-2021–88716

on the defensive—which is precisely where Washington should want them. Short of engaging in what most believe would be a bloodbath on the Korean peninsula, Washington needs a technological ace up its strategic sleeve. A space-based missile defense network would provide such a silver bullet.

What's more, the lowering of costs and streamlining of military space launches—along with the embrace of simpler satellite systems—would allow for the military to deploy smaller satellites (cubesats) to tag along with their existing expensive satellites in orbit. These smaller satellites would be "bodyguard satellites" in the words of my colleague Brian G. Chow. They would "operate near the satellites they are protecting. These bodyguards would have a range and lethality like those of an adversary's [space stalkers] attacking robotic spacecraft in order to generate a proportional response and minimize escalation during a crisis or conflict." Chow states that "not only can these bodyguard spacecraft block an adversary's robotic spacecraft from reaching and hurting US satellites, they can also use passive or reversible defenses that do not even harm these threatening invaders." Further, "these bodyguards can also protect vulnerable satellites against the approaching ground-based direct-ascent ASAT missiles by releasing decoys to confuse them, jamming their command, control, and communications links, or blinding their sensors."[26]

Chow believes that these bodyguard satellites would be key to "minimizing a space arms race."[27] In this, I respectfully disagree with my colleague. The arms race in space *is here*. What's needed now is for the United States to speed its innovation and deployment times, by making space easier and cheaper to access to maintain a dominant lead over any and all rivals. Where Chow is correct is his conclusion that "using bodyguards is a practical application of [the Mitchell/MITRE report]

---

26 Brian G. Chow, "Op-Ed: Nuclear Vulnerability: In-Orbit Bodyguards Would Help Protect NC3 Satellites from Attacks," *Space News*, 1 April 2019. https://spacenews.com/op-ed-nuclear-vulnerability-in-orbit-bodyguards-would-help-protect-nc3-satellites-from-attacks/

27 Chow.

to protecting [key early warning missile satellites] and other vulnerable satellites, especially during the 2020s while their resilient follow-ons are still unavailable."[28] American leaders have allowed critical space capabilities to atrophy. Therefore, things like bodyguard satellites to protect larger vital American military satellites in orbit, as well as the changes to the way the US military operates in space as discussed above, will be critical to ensuring America's competitive edge in the near term in space.

Eventually, conflict will migrate to the strategic domain of space. In the 1960s, the air force had the MOL Program, which was meant to place manned surveillance satellites in orbit. These tiny space stations, according to the recently declassified documents on the matter, were to be the nucleus of command-and-control facilities placed in geosynchronous orbit. Writing in 2009, George Friedman predicted that a space force would reconstitute these plans. The reason is not so much to have manned satellites monitoring from orbit (as noted, unmanned systems are good, if not better, than manned systems). The object would be to move the command-and-control (C2) functions of satellites from ground-based nodes to manned orbital ones. Remember the previous sections on how foreign adversaries, like Russia, can jam the links between satellites in orbit and units on the ground.

Specifically, countries seeking to deny space to another country usually target transmissions going from the ground up to the satellite in orbit with the exception of GPS, which is known as "downlink jamming," disabling the communication between the satellite overhead and receivers on the ground. Moving the C2 functions off Earth and into orbit would deprive an attacker of the relatively cheap and preferred method of jamming the signal linking a ground station with a satellite. Therefore, an attacker in space would have to risk using a more damaging form of attack like an ASAT weapon that could cause massive amounts of debris in orbit, potentially risking the national security of the attacking state. Plus, if an attacker were to risk the lives of American personnel in space, there would

---

28 Chow.

be international outcry—far more than if an attacker just disrupted the functioning of an unmanned satellite in orbit.

Besides, moving C2 functions into geosynchronous orbit would increase the functionality of the satellites of a US space force, because "space provides additional speed as well as natural line of sight and secure communications." By removing the C2 function from Earth, similar to what the MOL Program wanted to do in the 1960s, the United States could assert further dominance in orbit because these space-based C2 stations would "control swarms of satellites in orbit organized into satellite battle groups, where smaller defensive satellites protect more expensive military satellites that provide communications, navigation, and even strike platforms able to engage targets on Earth or in space." Friedman continues, "In addition to controlling America's space armada, the Battle Stars [would] provide targeting information and C2 to ground-based unmanned hypersonic aircraft, which act as America's primary strike capability." In the words of Brent Ziarnick, "Analysis of current trends in military doctrine and in the space industry reveals that many of Friedman's predictions may *sound* fanciful but are actually eminently *possible* and perhaps even *likely*."[29]

Another long-range concept is one that was proposed by a band of outside-of-the-box-thinking Marines in 2001. At the time, the US military was called to wage war in Afghanistan, in order to avenge the 9/11 attacks. The US Marine Corps was on the front line. Yet, the Marine Corps was designed as a maritime-based expeditionary force. Afghanistan was a landlocked country. Americans had to fly more than four hundred miles over Pakistan in unstable heavy-lift helicopters to get into the country. A Marine Corps officer, Roosevelt Lafontant, thought that the Marines could do better than that. Lafontant's idea was to create a spaceplane to transport a squadron of Marines—at hypersonic speeds—through space to their distant targets.[30]

---

29 Brent Ziarnick, "The Age of the Great Battlestars," *The Space Review,* 10 August 2009. http://www.thespacereview.com/article/1438/1

30 Sharon Weinberger, "The Very Real Plans to Put Marines in Space," *Popular Mechanics,* 14 April 2010. https://www.popularmechanics.com/military/a5539/plans-for-marines-in-space/

It was called the Small Unit Space Transport and Insertion plane (or, SUSTAIN). "In theory," writes David Axe, "SUSTAIN could have deployed forces from the U.S. to anywhere in the world in only two hours. By flying at suborbital altitudes, SUSTAIN would have been invulnerable to enemy air defenses and would have avoided violating the national airspace of countries bordering the war zone."[31] The program was kept in Development Hell at the Pentagon for about a decade. Despite its relative unpopularity among many of the military planners, Lafontant was passionate about his cause, and he had enlisted the help of several well-connected officers to back his play. Ultimately, the program never took off, but the Pentagon did move it to the planning phase, which is more serious attention than anyone would have believed it would get in 2001, when it was first proposed.

The SUSTAIN idea was modeled on Richard Branson's Virgin Galactic spaceplane, *SpaceShipOne.* Branson's spaceplane was to be the basis of a fleet of suborbital hypersonic fliers that take high-paying customers on incredible adventures for the budding space tourism industry. The Marine plan called for using this spacecraft to deliver a squadron of raging Marines on top of any war zone in the world rather than simply blasting wealthy tourists into space. It was yet another example of how the Department of Defense needs to work with all private-sector entities, not just the big defense contractors. Here again is how an innovative start-up, Virgin Galactic, designed to ferry wealthy lookie-loos into space, might be a key technology to allow for the US military to bust through another country's air defenses—without breaking any international laws. Usually, the Pentagon must request what's known as "overflight rights" from countries that are between the United States and the territory of an enemy state. During the invasion of Iraq in 2003, Turkey refused to allow US forces to use its territory to attack Iraq because Turkey disagreed with the American invasion—despite being a NATO member.

---

31 David Axe, "The Pentagon's Plan to Put Robot Marines in Space," *The Week,* 9 July 2014. https://theweek.com/articles/445664/pentagons-plan-robot-marines-space

Such a spacecraft would be essential for an expeditionary force, like the Marines or the Special Forces (who had also expressed interest in the SUSTAIN project). Engaged in a long-term war against terrorists who are mostly located in the distant and turbulent lands of Eurasia, suborbital spaceplanes could help keep the diplomatic, legal, and human costs of such operations relatively low for the United States. Or, imagine using these craft to resupply US forces engaged in tough combat. Eventually, thanks to the lowered cost of launches (should the recommendations of this book be taken to heart), the United States could place supply depots in orbit. These orbital supply stations could be used to launch supplies down to forces deployed throughout the world—at cost. Maintaining equipment in orbit would also cut down the future costs of launches, since the military would have to take less stuff into orbit as they deployed. The Defense Innovation Unit (DIU), one of the few truly cutting-edge programs coming out of the Pentagon these days, has placed calls for proposals to construct a military space station in orbit. The station would start small and be automated. Yet, DIU wants the station to have a "human rating"—in other words it needs to be legally safe for astronauts to live there. This could be the basis of an entirely new US military infrastructure in orbit that could further enhance the reach and scope of the space force.[32]

Also, the United States could eventually station marines or members of the Special Forces in orbit. These forces would be orbital shock troops, who stood ready to deploy down to the Earth, in order to help win whatever fight was occurring around the world. While current Pentagon plans for the US Space Force do not include elements from the Marine Corps, given the evolving nature of the space force, it is likely that a need for a space-based marine force will arise—especially if both the Marine Corps and Special Forces invest in a spaceplane

---

32 Theresa Hitchens, "Pentagon Eyes Military Space Station," *Breaking Defense,* 2 July 2019. https://breakingdefense.com/2019/07/pentagon-eyes-military-space-station/

capability.[33] Plus, in the age of anti-access/area-denial (A2/AD) strategies employed by countries like China, Russia, North Korea, or Iran, having a hypersonic spaceplane able to deliver forces or supplies to other parts of the world that may be blockaded by advanced air defenses would be be helpful in breaking through those defenses.

All of these scenarios become probable progressions for the United States in space *if* the cost for current space operations is lowered significantly. This requires fundamental reform at the institutional level. The recent creation of the US Space Force is a necessary first step. But that is not the last step. The Pentagon and civilian bureaucracy must fully embrace the implications of the new Space Force. America's elephantine national security bureaucracy will need to encourage sweeping reform within itself to ensure that the United States remains competitive and dominant in the strategic high ground of space. Such changes would mean altering the way that satellite systems are presently procured and then launched. It would also mean giving start-ups, like SpaceX, a real chance to prove themselves. The military space program must work in tandem with the private sector—and the private sector groups that the military space program partners with, cannot just be the legacy defense contractors. Innovation and cost cutting require competition, and new players must be welcomed on the field. Sure, the Pentagon is a closed system because it deals in secret projects. Somehow, though, Elon Musk has been involved with secret Pentagon launches for a few years and, despite publicly smoking marijuana on the set of Joe Rogan's podcast in California, where pot smoking is legal, Musk has never compromised US national security secrets.[34] So, the old argument that only the big

---

33 Alex Hollings, "No Space Marines: Pentagon Document Shows Space Force Plans to Draw Troops from Army, Navy, and Air Force," *SOFREP,* 23 October 2018. https://sofrep.com/news/no-space-marines-pentagon-document-shows-space-force-plans-to-draw-troops-from-army-navy-and-air-force/

34 Kyle Rempfer, "Air Force May Investigate SpaceX CEO for Smoking Pot, While Canada Greenlights Its Use for Troops," *Air Force Times,* 11 September 2018. https://www.airforcetimes.com/news/your-air-force/2018/09/11/air-force-may-investigate-spacex-ceo-for-smoking-pot-while-canada-greenlights-its-use-for-troops/

defense contractors can be trusted with the air force's "crown jewels" is an absurdity for the ages. One can still have free market competition and protect state secrets.

For the United States to leverage its immense though declining technological advantages to remain the dominant space power on Earth, the United States must be willing to bring all of its institutional and private-sector heft to bear. In order to accomplish such a herculean task, the entire worldview of American leaders regarding space must be shifted from "space is hard" to "space is essential." Meanwhile, the minds, hearts, and souls of the American voters must be stirred in ways they have not been since the glory days of NASA in the 1960s, when then president John F. Kennedy told the world that the United States would make it to the Moon within the decade. And Americans did make it to the Moon by the close of that tumultuous decade. In the next chapter, you will see why NASA has declined, despite enjoying a budget that is larger than most other national space programs, how we can fix it, and what happens if we don't appeal to the American people to "Make Space Great Again."

21

NASA'S CATWALK IN SPACE

"WE CHOOSE TO GO TO THE MOON," President John F. Kennedy proudly declared to a United States at the height of its power and national confidence. "We choose to go to the Moon in this decade and do the other things, not because they are easy, but because they are hard, because that goal will serve to organize and measure the best of our energies and skills, because that challenge is one that we are willing to accept, one we are unwilling to postpone, and one which we intend to win, and the others, too." Standing in a football stadium at Rice University on September 12, 1962, John F. Kennedy understood the significance of space better than any other US president after him—until Donald J. Trump.

Kennedy's space quest is all the more fascinating because, as one of Massachusetts' senators, Kennedy was skeptical of developing America's spaceflight capabilities. Things changed, however, when JFK was elected president and he saw not only the threat that the Soviet space program posed to the United States but also the immense opportunities that existed in space. By 1962, US satellites were already saving the private shipping industry money by, as Kennedy explained in his famous Moonshot speech, "helping our ships at sea [steer] a safer course," while early US environmental satellites gave the American people "unprecedented warnings of hurricanes and storms," and satellites at that time started to "do the same for forest fires and icebergs." At the time of JFK's Rice University speech, the United States was "behind" the Soviet Union in space. But Kennedy assured the American people that he did not "intend to stay behind, and [by the end of the 1960s the United States] shall make up and move ahead." Lastly, President Kennedy explained, "The growth of our science and education will be enriched by new knowledge of our universe and environment, by new techniques of learning and mapping and observation, by new tools and computers for industry, medicine, the home, as well as the school."[1]

NASA would be the primary element of Kennedy's plan to leapfrog the Soviet Union in space by the end of the 1960s. Some readers will rightly point out that JFK's Moonshot speech was as much of an appeal to globalism as it was to national pride. Yet, as I've elaborated in the previous chapter on space nationalism, a nation's embrace of nationalism and a focus on its national interests does not inherently preclude that country from working alongside other countries. Far from it. When national interests coincide, whether it is on space policy or another policy area, countries should work together. Reading the transcript of

1 John F. Kennedy, "John F. Kennedy Moon Speech—Rice Stadium," NASA, 12 September 1962. https://er.jsc.nasa.gov/seh/ricetalk.htm

JFK's speech, though, one can see that the bulk of his comments and his overall actions in space policy were most assuredly meant to enhance the strategic interests of the United States in space—even if it was, at times, couched in globalist rhetoric.

BEING MISSION-ORIENTED RATHER THAN CAREER-MINDED

When NASA had strong leadership, consistent funding, and a mission, such as landing Americans on the Moon, there was no hurdle that the organization could not overcome. NASA, like most of America's present institutions, was birthed in response to the pervasive threat that the Soviet Union posed the country. A retired CIA officer once explained to me that "after the Soviet threat was gone, the mission-oriented professionals [of that first generation of CIA officers] were replaced by career-minded bureaucrats." This explanation applies as much to NASA today as it does to the CIA. The threats today are more diffuse and complex than the centralized, relatively isolated, Soviet threat was. Competition is fiercer, and American rivals are closer to beating the United States than the Soviets were. Yet, America's byzantine bureaucracy is incapable of seeing these trends, despite the fact that these changes have been underway for decades.

Once the United States made it to the Moon, interest in space waned. This loss of interest correlated closely with the cultural malaise the United States was experiencing in the terrible wake of the turbulent 1960s and the cultural shifts that forever changed the country. Soon, America's space policy lost focus until, as you've seen today, NASA has done little to further the big-ticket space missions. Yes, they've made speeches indicating a desire to return to the Moon and go beyond to Mars, although, little progress has been made. And in the key area of lowering costs, NASA is as bad and lacking innovation as their partners at the Pentagon have been. American leaders soon embraced a schizophrenic view of space: most wanted to exploit space the way that JFK had dreamed of doing. Yet, few from either political party wanted to spend the money to do it.

NIXON'S TERRIBLE SPACE LEGACY

This policy schizophrenia is encapsulated in former president Richard Nixon's simultaneous commitment to space and his opposition to investing in that strategic domain. As John M. Logsdon analyzed in his book *After Apollo? Richard Nixon and the American Space Program*, the Nixon administration's space doctrine of incremental, evolutionary change, as opposed to the redirecting of a "'massive concentration of energy and will,' not to mention adequate financial resources, was a dead-end for NASA."[2] In fact, Nixon's official space doctrine was composed of two components, according to Logsdon. First, "to change the status of the space program from an effort formally assigned the highest national priority." Second, was to "declare that the space program from 1970 forward would have to compete with other discretionary government activities for priority and corresponding budgetary support." Space minimalism thus became the basis of US space policy from Nixon onward. Even President Jimmy Carter, in many ways the anti-Nixon, had a space doctrine that parroted Nixon's space policy.[3] Thanks to the Nixonian model of space minimalism, NASA existed in an aimless and listless state. NASA allowed critical capabilities to wither on the vine as American rivals slowly but surely caught up to the United States.

Logsdon provides another hilarious, though disturbing, look into the mindset of every US president since Richard Nixon and until Donald J. Trump. On March 9, 1971, President Nixon made an artful case to a group of astronauts who had visited him in the Oval Office. He said:

> I know what people say, we are being jingoistic. America stays number one and so forth. In the history of great nations, once a nation gives up in the competition to explore the unknown, or once it accepts a

---

2 John M. Logsdon, *After Apollo? Richard Nixon and the American Space Program* (New York: Palgrave Macmillan, 2015), 279.

3 Logsdon, 279.

position of inferiority, it ceases to be a great nation. It happened to Spain. It happened in the twentieth century to the French and then to the British. And it could happen to the United States. That is what it's all about, and so when we look at . . . the space program, whether it's Mars or whether it's the shuttle or who knows what it is. I don't care what it is, but the main thing is we have to go, we have to go, and we've got to find out.

The majority of people in all of the polls show that they are against the SST [supersonic transport], they are against the space program. They just want to sort of settle down . . . If the United States just didn't . . . have the problems of going to space, then what a wonderful country that would be. And the answer is it wouldn't be at all. It would be a terrible country. It would be a country big, fat, rich, but with no sense of spirit . . . If an individual does not want to do something bigger than himself, he is selfish. That's what space is about.[4]

As Logsdon observed, though, Nixon held a completely different point of view when speaking with an aide not associated with the space program. A few weeks after his meeting in the Oval Office with the astronauts, Nixon attempted to secure the funding for NASA's supersonic transport from the United States Senate. He was unsuccessful. Afterward, Nixon told one of his congressional liaisons that "the 'United States should not drop out of any competition in a breakthrough in knowledge—exploring the unknown. That's one of the reasons I support the space program.' Without pausing, he added, 'I don't give a damn about space. I am not one of those space cadets.'"[5] Herein lay the basis for America's space policy going forward: rhetorical ruminations about the possibilities of tomorrow coupled with serious inaction in the real world. Over time, such dithering has allowed American rivals to make stunning and rapid technological progress, which directly threatens the

---

4 Richard Nixon, "Conversation 7, Tape 464," Richard Nixon Presidential Library, 9 March 1971.

5 Logsdon, 179–80.

safety of every American living today and could permanently change the historical trajectory of the world in America's disfavor. America today, as Nixon warned, is a "country big, fat, rich" but, quite possibly, it is losing its "sense of spirit." The decline of the United States as the dominant space power is a perfect example of this unfortunate trend.

Since NASA's embrace of the space shuttle program, things have only deteriorated for the once-august space agency. Every so often, NASA proliferates grand plans for returning astronauts to the Moon and going to Mars. Many taxpayer-funded conferences are held in which scientific luminaries gather to discuss the endless possibilities. Yet, no real movement is taken. The Trump administration, for its part, announced a necessarily ambitious civilian space policy agenda: it planned to return astronauts to the Moon by 2024. Of course, since President Donald Trump is behind this movement, the congressional Democrats—the party of JFK—have not only resisted Trump's space force proposals, but they are also now insisting that NASA not return to the Moon by 2024. The Trump administration's NASA director Jim Bridenstine had requested "an additional $1.6 billion [in congressional funding]." Bridenstine referred to this as a "down payment" to help NASA achieve its "two-phased approach: the first is focused on speed—landing astronauts on the Moon in five years—while the second will establish a sustained human presence on and around the Moon by 2028."[6]

Alas, Representative José Serrano (D-NY), who serves as the chairman of the House Appropriations Committee's Commerce, Justice, and Science Subcommittee, argued in October 2019 that "it's hard to justify any extra spending on this effort in the current fiscal year when we don't know the costs down the road." Serrano's comments may seem reasonable, but then his political bias poured out when he followed those comments with "To a lot of members,

6 David Grossman, "Actually, About That Whole 'We're Going to the Moon in 2024,'Thing . . ." *Popular Mechanics,* 17 October 2019. https://www.popularmechanics.com/space/moon-mars/a29492598/congress-nasa-2024-moon-return/

the motivation [for going to the Moon by 2024] appears to be just a political one—giving President Trump a Moon landing in a possible second term, should he be reelected."[7] Of course, Bridenstine rebuffed this assessment by the politically motivated Serrano. He argued that the additional, up-front cost would help to solidify NASA's Artemis Moon mission plans "without cannibalizing" other essential, scientific missions that NASA maintains.[8]

Yet, the domestic political ramifications of sending US astronauts back to the Moon in what will likely be a second Trump term should not be overlooked. For decades, the Democrats have insisted that theirs was the party of science and progress. Very often, they would point to the various successes in manned spaceflight that occurred under both JFK and LBJ presidencies. For far too long, the Democratic Party has taken solace in the history that previous Democratic Party leaders supported NASA. Those days, however, are long gone. A succession of Democratic Party leaders, starting with Jimmy Carter, then Bill Clinton, and Barack Obama, have all presided over the systematic dismantling of America's once-enviable civilian space program.

Yes, Richard Nixon began the slow deterioration of NASA with his advocacy for less intensive space development. But the Democrats hopped onboard that policy train to nowhere with great élan. There were few differences over the decades between the Republicans and Democrats when it came to space—save for rhetoric. Oddly enough, Republicans tended to talk a much bigger game in space than Democrats did. Reagan championed the space shuttle as the beginning of a new era of sustained human spaceflight as well as his much-ballyhooed "Star Wars" initiative. His successor, President George H. W. Bush, supported the Space Exploration Initiative, which was designed to establish a long-term American presence on the Moon. This was, of course, canceled by

7 Grossman, Ibid.

8 Jeff Foust, "Key House Appropriator Remains Skeptical About Artemis," *Space News,* 16 October 2019. https://spacenews.com/key-house-appropriator-remains-skeptical-about-artemis/

Bush's successor, Bill Clinton, in 1996, due to cost overruns. Clinton's Republican successor, George W. Bush, crafted a magnificent-sounding space policy agenda in 2004, only to have his signature Constellation Program terminated by his Democratic successor, Barack Obama. And of course, Obama quickly aborted the space shuttle program with no follow-up, meaning that the United States came to rely on Russia for launching its astronauts into orbit.

President Donald J. Trump has both talked a big game *and* started the process of rehabilitating the civilian and military space sectors. The Democrats in Congress and the sclerotic bureaucracy, unfortunately, have worked to undermine Trump's renewal of US space policy. Meanwhile, the stifling utopianism and naysaying of many space policy leaders has filtered into the very institutional bones of NASA. Anything that smacks of a new NASA mission to enhance American prestige and power in space is very often met with internal resistance. As much as there is a deep state in the national security bureaucracies opposed to Trump, there exists a parallel one in the halls of NASA.

CULTURAL MARXISM BLASTS INTO ORBIT

Why would NASA's bureaucracy attempt to prevent its own spaceflight operations from being furthered? Well, as Columba Peoples argued in 2006: America's space program is fascist. Another space cadet, Penny Griffin, said, US space policy is predicated on "heteronormative tropes of masculinization and feminization." As I noted in 2019, "Griffin takes particular offense at the language undergirding the American space program."[9] She writes, "The US techno-strategic discourse reconfigures all other space-able nations as subordinate, constructing a binary, heterosexual relationship of masculine hegemony/feminine subordination." This type of thinking has become endemic in the NASA bureaucracy over the last several decades. Gone are the days when America's space

---

9 Brandon J. Weichert, "A Catwalk in Space," *American Greatness,* 20 October 2019. https://amgreatness.com/2019/10/20/a-catwalk-in-space/

policymakers would "pay any price and bear any burden" in order to ensure that America not only led the world in space exploration, but the United States also gained the most economically and technologically from space all to maintain a strategic dominance over its rivals in space. No, that's too sexist—or something.

Things have gotten so ridiculous that NASA has toyed with the notion of sending an all-female crew to the Moon in 2024, in order to win over recalcitrant Democrats in Congress. Identity politics is now blasting off to the cosmos! Well, probably not, because it won't matter if Trump is in office in 2024. Democrats will insist that NASA hold off until they can figure out how to get enough people to vote for the non-gendered socialist candidate of their choice in the 2024 presidential election. Meanwhile, the threats America faces from China, Russia, North Korea, and Iran in space will have to wait. The grand innovations that the United States could pioneer from a greater and more robust commitment to its manned spaceflight capabilities will be inconsequential. The Orange Man in the Oval Office is bad. So, everything must be resisted, because space dominance is, like, totally fascist.

Even under the Trump administration, NASA's leadership has fallen prey to the idea that gender and identity politics should take priority over sensible policy. In an effort to attract attention from the "Never Trump" media, NASA ordered that an all-female spacewalk occur outside of the International Space Station. The media was irrationally exuberant over the all-female spacewalk. Seeing the media excitement, President Donald Trump called the two female astronauts, Jessica Meir and Christina Koch, to congratulate their success as being the "first women in space."[10] To which, Meir chided her commander-in-chief that many women had gone to space, some of them Americans, for

---

10 Courtney Subramanian and John Fritze, "Trump Calls NASA Astronauts Jessica Meir and Christina Koch to Congratulate Them on First All-Female Spacewalk," *USA Today,* 18 October 2019. https://www.usatoday.com/story/news/politics/2019/10/18/trump-calls-nasa-female-astro-jeissca-meir-and-christina-koch-congratulate-them-first-all-female-spa/4022523002/

decades.[11] In fact, the first American woman in space was Sally Ride, who blasted off to the cosmos aboard a space shuttle in 1983 (she joined NASA in 1978). What was important about the 2019 all-female spacewalk was that it was the first time *more than one woman partook in such an activity*.

Well, whoop-de-doo, NASA.

Rather than inspire Americans of all backgrounds, the NASA catwalk exacerbated preexisting political divides. It was identity politics on rocket fuel. Does it really matter if more than one woman partakes in a spacewalk at once? Was it that big of a deal when more than one man conducted a spacewalk? And if it was a momentous ordeal, was the focus on their gender or the simple fact that more than one human being was doing such a difficult task for the first time? Is it that important that an all-female crew lands on the Moon by what is unlikely to be NASA's target date of 2024? Is this PR stunt intended to win over the support of the "Never Trump" elected leaders in Congress and the media the most judicious use of what are clearly limited resources?

During the Obama administration, NASA conducted an intensive study that found there were stark differences between men and women who lived in space over prolonged periods of time such as would be required on trips to either the Moon or, eventually, Mars. While both men and women experienced physical deterioration, the study determined that female astronauts were more likely to suffer grave physical degradation from prolonged exposure to the radiation and zero-gravity environment of space than men.[12] Radiation and zero-gravity are two of the most ubiquitous elements of space. Long-term colonies on the Moon or Mars, staffed by a majority-female crew, could suffer from significant losses of capabilities over time.

---

11 Ephrat Livni, "Female Astronauts Schooled Trump from Outer Space," *Quartz,* 20 October 2019. https://qz.com/1731692/female-astronauts-school-trump-from-outer-space/

12 "Study Investigates How Men and Women Adapt Differently to Spaceflight," NASA, 17 November 2014. https://www.nasa.gov/content/men-women-spaceflight-adaptation

A bevy of medical studies have been conducted on female astronauts over the last several years to determine how space affected their bodies. The general consensus among the scientific community is that women are only slightly better adapted to space travel than men.[13] Yet, the fact that women's bodies are more sensitive to certain aspects of radiation exposure and microgravity than men begs the question how these scientists came to their conclusion. Regardless, there have not been enough women astronauts to reliably answer these scientific inquiries, which is one reason that the Obama-era study urged NASA to send more women into space. Thus far, the differences are so insignificant that they hardly warrant the fanfare, to say nothing of the resource expenditure, that has surrounded the topic of women in space. What should be clear is that NASA needs mixed crews of both men and women for space exploration, since men also suffer different forms of physical degradation during long-term space travel, according to reports. The women can, therefore, augment the weaknesses of their male counterparts and vice versa. Plus, all of the time that NASA has spent in advertising for all-female crews or championing their level of "wokeness" has taken away from the larger NASA goal of enhancing American prestige and capabilities. Virtue-signaling to the Left has clearly replaced practical mission sets. This is evident by the continued failure to send any American astronauts into space aboard an American-built spacecraft.

NASA'S BUREAUCRACY MAKES SPACE TRAVEL EXPENSIVE

When it comes to the matter of affordability, just like their colleagues at the Pentagon, the NASA bureaucracy has consistently failed to bring down costs. Building the capability to launch astronauts to the Moon is far costlier than the launches of the space shuttle program were. This is because greater amounts of thrust—fuel—is required to send astronauts

---

13 Belinda Smith, "Women Astronauts May Be Better Suited to Space Than Men—but not by Much," *AU News,* 7 April 2019. https://www.abc.net.au/news/science/2019–04–08/why-women-may-be-slightly-better-suited-to-space-living-than-men/10941616

to the Moon and back. Going back to the matter of start-ups in space, Jeff Bezos's firm, Blue Origins, recently said they could help NASA develop a second-stage rocket for their Space Launch System (SLS). What's more, Blue Origins said that their system would be cheaper than the one that NASA is currently considering. Rather than hear Blue Origins out, NASA said "no, thanks."

At present, the cost for the SLS rocket is one of the major stumbling blocks for getting continued congressional support to send astronauts to the Moon (beyond the whole Orange Man Bad thing). In typical fashion, NASA had signed on with Boeing to develop the new system. But after years of cost overruns, in 2017, NASA requested for the space launch industry to offer a "low-cost replacement" for the upper stage rocket or "perhaps an entirely new stage itself." After issuing the request to the wider launch industry for replacement proposals, NASA "decided to stick with Boeing's version of the Exploration Upper Stage." Blue Origin, however, submitted an alternative concept "based upon Blue Origin's BE-3U rocket engine, a modified version of the motor that powers the New Shepard launch system, which will also fly in the upper stage of the company's New Glenn rocket. A single BE-3U engine has more thrust than four RL-10 [the rockets Boeing is using on the second-stage SLS rocket] engines combined." Needless to say, NASA rejected "Blue Origins' less-costly, and potentially more efficient, alternative."[14]

While it is true that the incorporation of the Blue Origins rocket would have forced changes to other parts of the overall SLS, it is likely that, had NASA opened the bidding process up to *all* launch services, some cheaper variation would have been found through the competitive process. Yet, the government's well-funded and highly favored, far costlier, contractors won the day—only empowering NASA's detractors in Congress, who are looking for excuses to squelch any real progress

---

14 Eric Berger, "NASA Rejects Blue Origin's Offer of a Cheaper Upper Stage for the SLS Rocket," *Ars Technica*, 5 November 2019. https://arstechnica.com/science/2019/11/nasa-rejects-blue-origins-offer-of-a-cheaper-upper-stage-for-the-sls-rocket/

for NASA because of their hatred for the current president. The SLS is another example of how the Nixonian doctrine of incremental evolution in the space program is insufficient for winning the new space race.

For all of the rhetoric mocking China's aforementioned Long March V heavy-lift rocket woes, NASA's new heavy-lift rocket is experiencing huge development delays itself. For all of the talk from those on the right about how America's private space sector will make up for any delays in its public space program, few seem to understand that, under existing limitations of both international law and lack of federal research and development funding—let alone the political hurdles these private space companies face in America today—American private space companies will struggle to thrive without consistent support from government organizations, like NASA. The cost of space operations will not decrease so long as the US government continues favoring the monopolistic legacy contractors, like Boeing. Certainly, these larger companies have important roles to play. But these larger corporations must get far more competitive and provide cheaper services. Such changes will never happen without considerable competition from smaller start-up firms. This is not about getting rid of the bigger firms and totally replacing them with smaller start-ups. This is instead about making those larger firms, like Boeing and Lockheed Martin, more competitive and less costly so that everyone in the space sector can do more business which, in turn, will benefit the country.

NASA'S CURRENT PLANS: RETREAT TO MOVE FORWARD?
All of the grand plans NASA has proposed since it mothballed its Saturn V heavy-lift rocket systems have been terminated or seriously curtailed, in large part, out of economic reasons. Usually, these programs experience massive cost-overruns. Of course, NASA and the major contractors insist that this is the nature of high technology development: there will be bumps along the way to greatness. This is certainly a fair point. Yet, one cannot help but to wonder how much cheaper things would be if NASA had a diversified domestic supply chain that could provide all

NASA projects with the most efficient, cheapest systems imaginable. For all of the talk about lowering costs and for all of the complaining about short-term-minded politicians, NASA has the ability to force competition on the vendors who serve NASA's needs *without* congressional approval. This was why so many people applauded the initial NASA request for lower cost proposals on the second-stage SLS rocket in 2017—and why so many more people were upset when NASA ultimately chose to stay with Boeing, even though NASA never officially opened up the competitive bidding process to firms, like Blue Origins.

Also, what does it say about NASA as an institution when it struggles to reconstitute a capability that it had in the 1970s? NASA is the embodiment of what happens when bad leadership meets bureaucracy. Today, NASA should be able to rest on a record of continuous success in order to meet reduced congressional resistance when it launches new proposals. Instead, NASA struggles even to maintain its already reduced human spaceflight capabilities, while touting its robotic exploration missions, despite the fact that now, countries like Japan and India can do similar missions far more cheaply.

The Nixonian concept of space minimalism is wrong. NASA today finds itself in much the same position as NASA found itself in 1962, when JFK made his Moonshot speech at Rice University. Today, the United States, if you'll pardon the expression, needs a crash program in human spaceflight. There are arguments from NASA Director Bridenstine warning of creating serious friction between institutional stakeholders at NASA, should Congress not give NASA's Artemis Program the $1.6 billion down payment it claims to need. After all, Bridenstine is keenly aware that President Trump has issued directives aimed at putting astronauts on the Moon during his tenure in office—as well as sending Americans to Mars. Ultimately, Bridenstine and the NASA bureaucracy will have to start cannibalizing some scientific missions like their obsessive quest for global warming in order to complete the manned spaceflight missions that generate public interest and further empower America at the expense of its rivals.

What's needed is a Kennedy-like Moonshot program. The United States cannot do incrementalism any longer. It needs rapid, fundamental change across the entirety of its space and technology sectors. Since the Nixon administration's decision to terminate the Apollo Moon missions and replace them with a multipurpose space shuttlecraft, as well as the Nixon administration's short-sighted decision to terminate the Air Force's MOL Program in favor of automated satellite constellations, US space policy has taken an evolutionary dead end. In 2012, when the Obama administration ordered the termination of the space shuttle program, this assessment was proffered:

> The National Aeronautics and Space Administration (NASA) is at a transitional point in its history . . . The agency's budget . . . is under considerable stress, servicing increasingly expensive missions and a large, aging infrastructure established at the height of the Apollo program. Other than the long-range goal of sending humans to Mars, there is no strong, compelling national vision for the human space-flight program, which is arguably the centerpiece of NASA's spectrum of mission areas. The lack of national consensus on NASA's most publicly visible mission, along with out-year budget uncertainty, has resulted in the lack of strategic focus necessary for national agencies operating in today's budgetary reality. As a result, NASA's distribution of resources may be out of sync with what it can achieve relative to what it has been asked to do.[15]

Yet, what few bureaucrats and intellectuals appear capable of understanding is that consensuses are forged—even radical ones. It takes leadership to forge consensus; very often one person breaks through the stifling miasma of the status quo to reorient the culture away from its current course in a given area and toward a new path breaking one. Donald Trump has already indicated his desire to get to Mars rather

---

15 "NASA's Strategic Direction and the Need for a National Consensus," *The National Academies of Sciences, Engineering, and Medicine,* 2012, p. 1. https://www.nap.edu/read/18248/chapter/2

than returning to the Moon. While I remain hopeful that the Moon is front and center for the US military's space efforts (it is a strategic and economic vantage point that rivals, like China, are planning to take for themselves), perhaps NASA's focus should be on putting some astronauts on Mars first. Just as with the Moon program, this would drive America's fascination with discovering a new world as opposed to visiting one we've already been to.[16]

NASA'S FOCUS SHOULD BE MARS;
LEAVE THE MOON TO THE MILITARY

In fact, there's no reason why NASA could not return astronauts to the Moon on its preferred timeline (2024) while planning to send other astronauts beyond the Moon to Mars. In any event, telling the public that NASA was planning to land a person on Mars by the end of the decade, and focusing time and effort on generating the funding and interest for such a manned spaceflight endeavor, as JFK did with the Moonshot speech in 1962, would likely grant President Trump his desire. Let the military space program deal more completely with securing the Earth-Moon system, while the private sector develops the resources on the Moon, as NASA pushes the boundaries of human spaceflight abilities by getting to Mars. Perhaps such a mission would give NASA the added leverage to develop an economical engine for getting astronauts to and from Mars quicker than is possible with current chemically propelled rockets. The EmDrive or a larger ion drive might do the trick. None of these systems have been fully experimented with, because NASA has never had the leadership or the consistent commitment to achieving such a profoundly important mission.

The Trump administration, particularly by the second term, must *push* NASA toward this, while encouraging the cost-saving development measures that I wrote on earlier in this chapter. What's needed is more

---

16 Donald J. Trump, *Twitter*, 7 June 2019, 1:38 PM. https://twitter.com/realdonaldtrump/status/1137 051097955102720?lang=en

Kennedy boldness and less Nixonian cynicism. Space remains the final frontier, and whoever gets there first—and stays there—wins more than a footnote in the history books. America's civilian space agency has a vital role to play. If they do not fully embrace it but instead focus on the smaller, esoteric missions that gain a few minutes at the end of a news program rather than dominate the headlines as they did during the Apollo era, then a more draconian decision on NASA's future will have to be made.

The American people do not support mindless bureaucracy with cool logos. They want results. The manned spaceflight program is NASA's most important contribution to both science and the country. It's a way to enhance our national prestige over that of China or Russia. The continual failure to make new strides in the manned spaceflight program is NASA's enduring failure to all Americans. And if NASA is no longer serious about pursuing human spaceflight *beyond* Earth's orbit; if they will continue insisting on receiving tax dollars while not bringing anything new to the table, then it will be time to cut this program, as its existence will be an affront to the once-great memory of NASA.

Spending $21.5 billion a year just for NASA to repeal its former claim that Pluto is not, in fact, a planet is insufficient for most taxpayers.[17] NASA must go big or go home on their human spaceflight mission, even if that means cannibalizing other scientific missions or abandoning their delusions of all-female crews to distant worlds. Let's just get people to Mars first and colonize the Moon before the Chinese do. After that, NASA can worry about taking those pretty pictures of the universe they've come to specialize in and celebrating the gender-bending roles that future female astronauts will play in future NASA missions.

---

17 Stacy Liberatore, "NASA Chief Jim Bridenstine Doubles Down on Claim Pluto Is a Planet, Citing Its Moons, Oceans and Organic Compounds," *Daily Mail*, 6 November 2019. https://www. dailymail.co.uk/sciencetech/article-7653569/NASA-chief-Jim-Bridenstine-says-Pluto-planet-AGAIN. html?ito=social-facebook&__twitter_impression=true&fbclid=IwAR3–0Alq5Q2UPfdMatPdFAA01I FeDw3yALTpneK_UyCOaNPMWUJ6fSLEQvU

22

STAYING TO THE LEFT OF BOOM

BY GOING HIGH

IN THE MIDNIGHT HOUR of a cool New England night of April in the year 1775, Paul Revere, Samuel Prescott, and William Dawes—Revere's Midnight Riders—rode out to warn the American colonists of an impending attack on colonial ammunitions stores by the British Army. As the apocryphal tale goes, published decades later in 1863 by the great American writer Henry Wadsworth Longfellow, Revere said to his friend:

> If the British march by land or sea from the town to-night,
>
> Hang a lantern aloft in the belfry arch of the North Church tower as a signal light,

—One if by land, and two if by sea;

And I on the opposite shore will be,

Ready to ride and spread the alarm

Through every Middlesex village and farm,

For the country folk to be up and to arm.' [. . .]

So through the night rode Paul Revere;

And so through the night went his cry of alarm

To every Middlesex village and farm,

—A cry of defiance, and not fear,

A voice in the darkness, a knock at the door,

And a word that shall echo for evermore! For, borne on the night-wind of the Past,

Through all our history, to the last,

In the hour of darkness and peril and need,

The people will waken and listen to hear

The hurrying hoof-beats of that steed,

And the midnight message of Paul Revere.[1]

Of course, as the real history would go, Paul Revere didn't ride through the streets of Concord, Massachusetts, hollering a warning about the impending British attack. According to some sources, Revere never even made it to Concord (he did, however, warn the denizens of Lexington about the British attack). As Laurie L. Dove details, it was

---

1 Henry Wadsworth Longfellow, "The Midnight Ride of Paul Revere," National Center, 19 April 1860. https://nationalcenter.org/PaulRevere%27sRide.html

actually Samuel Prescott who made it to Concord, where "he warned its residents to protect the ammunition and weapons stored in a hidden depot near the town." Yet, the effect was the same: Revere and his fellow Midnight Riders, using intelligence and common sense, understood that the situation between the British forces that had occupied the colonies and American colonists was breaking down. The British leadership wanted the ongoing rebellion in the Thirteen Colonies resolved. Revere and his compatriots also knew that the presence of the colonists' ammunition stores was a strategic threat to the ongoing British attempt to pacify the unruly colonies. Eventually, the British would have to remove those ammunition stores to take control of the deteriorating situation there.

Rather than wait for impending doom, Revere, Prescott, and Dawes struck out in the cold night to warn their fellow colonists of what the British were secretly planning. They got ahead of the advancing British force by several hours. Although the men were detained by the British at one point, they were able to escape in the confusion of the British advance on Concord and Lexington after which Prescott got to Concord and successfully warned them of the British surprise attack. The Midnight Riders in 1775 anticipated a grey rhino event and were able to help their countrymen avoid the stampede that night. Had the colonists not heeded Revere's warnings, the American War for Independence would have gone very differently. For without those ammunition stockpiles, the colonists would have little with which to defend themselves. A similar situation exists for the United States in space today.

The military has a term, "left of boom," which is used to describe that moment just before a bomb detonates. This is when "you still have time to prepare and avert a crisis. Right of boom, by contrast, includes the chaotic and deadly moments after the explosion or attack."[2] In the war game scenario presented in the first chapter of this book, you saw

---

2 Benji Hutchinson, "Left of Boom—Defeating the Threat Among Us," *NEC Today*, 11 July 2017. https://nectoday.com/left-of-boom-defeating-the-threat-among-us/#

the moments of what will likely happen under the present conditions of US space policy when the Russians or another rival, like China, no longer believe their strategic goals can be achieved through all measures short of war. When Donald J. Trump campaigned for president in 2016, he routinely told audiences that America's enemies were smart and highly dangerous.[3] Countries like autocratic Russia, seeking to gain territorial advantage over the United States in the European plain, know that they cannot engage in a fair fight against the United States, and hope to win. What's more, despite Russia's immense nuclear arsenal, the Kremlin is not foolish enough to risk a wider nuclear exchange with the United States without first having debilitated America's ability to respond to their next large-scale attack on Europe. The moments presented in the initial war game, then, are about those terrible moments to the "right of boom," after Russia has successfully crippled America's satellite constellation. The results are not pretty.

Not only would the US, NATO, and other allied European forces inevitably be overrun under such conditions as presented in the war game scenario in the first chapter, but the strategic position of the United States in the world would be destroyed. Think about it: US foreign policy is predicated upon a web of globe-spanning, interlocking security alliances. The fact that NATO has for years been under increasing pressure from Russia, and that the United States and its allies have failed to reinforce themselves under the pressure imposed by Moscow, has had a chilling effect on America's global military alliances. US allies, such as Saudi Arabia and Japan, are looking to see what US security guarantees are worth. This book is not about how the United States should change its overall foreign policy outlook. What I am endeavoring to do in this book is to show you how, under present

---

3 Maya Oppenheim, "Donald Trump Shaken by 'Scary' Intelligence Briefings: 'We Have Some Big Enemies Out There,'" *Independent,* 18 January 2017. https://www.independent.co.uk/news/world/americas/donald-trump-scary-intelligence-briefings-interview-big-enemies-out-there-a7534031.html

conditions, the United States is slated to lose an engagement with a rival state because of its weakness in space. The goal of this book, then, is to show how a reorientation of America's overall space policy—at the national security, economic, and scientific levels—will keep the United States to the "left of boom." Basically, the United States does not want to be to the "right of boom." However, it must do a better job of planning to survive a "right of boom" moment in the near term.

America's institutions should be in the position of Paul Revere and his Midnight Riders in 1775, warning their countrymen of incoming danger, giving them time to prepare. American intelligence analysts instead are in a position similar to that of Army Air Corps Lieutenant Kermit Tyler. Tyler was a radar officer in training who had only been on duty for two weeks and who had found himself in the unenviable position of being the only officer on duty when the Opana radar station in Hawaii detected "something completely out of the ordinary" at 7:02 a.m. (Pacific time) on December 7, 1941. Lieutenant Tyler, unlike Paul Revere and his Midnight Riders 166 years before, "told [the two privates at Opana station] to forget about" the strange radar tracks they were monitoring. Despite having *no experience in his position*, Lieutenant Tyler, according to Roberta Wohlstetter, told the enlisted men at Opana that "what they were seeing were friendly craft, probably the flight of B-17's from the mainland that was due that morning and in fact arrived in the middle of Japanese attack [on Pearl Harbor]." Wohlstetter concluded that had Hawaii's various early warning radar stations been operating as they were supposed to, staffed by at least one experienced officer during every rotation, the Opana station's warning would have "provided a forty-five-minute warning to the Army, and perhaps a thirty-minute warning to the Navy."[4] Revere gave American forces just a few hours heads-up. Yet, in 1775 that was just enough time for American colonists to stand their ground against the British. In 1941, a forty-five-minute or thirty-minute warning about an impending attack

---

4 Wohlstetter, *Pearl Harbor*, 12 (see chap. 1, n. 39).

could have potentially allowed for some defense to be taken against the Japanese attackers. Such a defensive posture would have then reduced the nearly catastrophic damage that was ultimately incurred by the United States military at the hands of the Japanese raiders.

Both Pearl Harbor and 9/11 were instances where the United States found itself to the "right of boom." In the former case, America ultimately prevailed—but at great cost. In the case of the latter example, the United States has waged an otherwise ineffectual "Global War on Terror" that has entirely destabilized the "Greater Middle East," while it has unduly distorted the political situation in the United States. In each instance, had the United States simply paid closer attention to intelligence signals it had already gleaned from its enemies, both Pearl Harbor and 9/11 could have been avoided or, at least, mitigated. Contrary to the popular opinion of how the United States manages to "dust itself off" after enduring a grey rhino event, the United States is *not* in a position today anywhere near what it was in during the Second World War.

After the Pearl Harbor attacks, the United States had the infrastructure in place to respond to the Axis powers across the world. During the 1930s, the decade that preceded the breakout of the Second World War, the first wave of massive modern infrastructure was built throughout the country. Massive industrial centers in America could churn out the planes and weapons that were needed to fight the war. The United States was not in the massive level of debt that it currently finds itself in, meaning it was well positioned to benefit from the wartime and postwar boom in productivity. Plus, the entire culture was fundamentally different. American companies happily gave themselves over to patriotic pursuits to assist the war effort, rather than the craven pursuit of global, quarterly profits, as many major corporations do today under the shareholder model of capitalism. As I wrote earlier, 12 percent of the country's population joined the armed forces with intentions to fight the Axis powers to the finish. Today, less than 1 percent of the nation's population serves. Plus, unlike the culture during the Second World War, today's culture is not built on unity and a shared love of Western culture. It is the shallow product of decades of cultural Marxism meant to hollow out and replace Western culture with Marxist thought and dialectic. Decades of deindustrialization have ensured that the United States will be hard-pressed to withstand the kind of assault that large countries, like Russia or China, would force America to endure in war. The myth of America's "come-from-behind" victory over the Axis powers had to do with latent industrial production becoming realized over the course of the war. It also happened thanks to distance from the war zones allowing for America's industrial and economic base to be unaffected by the fighting. What's more, Americans of yesteryear were a much sterner and united people, taking the fight to those who had attacked the United States. Those Americans were also willing to bear much heavier costs and burdens than succeeding generations likely were.

Presently, the Pentagon fears that it will not be able to indigenously produce enough weapons and materiel in the event of an outbreak of a prolonged struggle with either Russia or China.[5] Meanwhile, the world has been made much smaller due to globalization, high technology, and open borders. We have been linked together by a web of advanced technologies that have not only made the United States dependent on sensitive technologies but have also restricted America's ability to protect itself. And because the United States has sent many capabilities overseas, the US military has to rely on a variety of foreign vendors—including some in China—to supply it with much-needed equipment. Should disaster strike, such as a Space Pearl Harbor, being on the "right of boom" could be the end of the United States' role as a global superpower. Perhaps that is something you support. I certainly have issues with the way the United States conducts itself internationally. But such changes to US foreign and national security policies should be in accordance with the will and desires of the American people. Those

---

5 Aaron Mehta, "Here Are the Biggest Weaknesses in America's Defense Sector," *Defense One*, 27 June 2019. https://www.defensenews.com/pentagon/2019/06/27/here-are-the-biggest-weaknesses-in-americas-defense-sector/

changes should *not* occur because the nation is being held at proverbial gunpoint by autocracies, such as Russia or China.

Space is a domain where all manner of advanced technologies are taken to their limits and where new innovations are crafted. These innovations have grave strategic and economic implications. Other countries, as you've seen, are happily developing their own space capabilities in order to disproportionately benefit from those newfound abilities. Meanwhile, America's position continues to falter. And its weakening position invites challenge. The objective for US policymakers regarding space should be threefold: first, to ensure that the United States suffers no major disruption in satellite functions due to foreign attack or meddling. Second, that the United States develops the technology and organization needed to protect itself against the continuing threat posed to the homeland by nuclear-arming rogue states, like North Korea and Iran. Third, the United States must create policies that will encourage the private economic development of space—all before an American rival can exploit the precious assets of space. Whoever gets to space first not only wins, but they also get to write the *real* rules and codes for conduct in this amazing new domain.

Some theorists like to study the strategic implications of America's flailing space policy in stilted labels. As has been noted, there are four major military theories regarding the strategic use of space. The first is the sanctuary view, which sees space as a weapons-free domain and, therefore, should also limit economic development of space. Much of the utopianism comes from this outlook. Then, there is the survivability school, which basically tries to figure out how to make current US satellites more survivable in a contested time, but also how to make American systems less dependent on these vulnerable satellites. From there, the space superiority or space control model bills itself as a defensive military space policy built off the concepts of deterrence and balance of power. Lastly, the space dominance theory is built on the controversial aspects of hegemony, preemption, and, when necessary, unilateralism.[6] I also tend to throw in the concept of compellence with space dominance.

Thomas C. Schelling coined the term *compellence*, and he differentiated it from the more commonly understood strategic term of "deterrence" by describing compellence as "a direct action that persuades an opponent to give up something that is desired." Deterrence, on the other hand, "is designed to discourage an opponent from action by threatening punishment." Right now, space dominance is impossible, if only because America's capabilities in space have deteriorated too far to allow for such a strategy to be believably employed. But should the reinvigoration occur over the next several years, space dominance will become a *fait accompli*. Hopefully by the mid-2020s, and assuredly by the dawn of the 2030s, America would be able to employ the space dominance model with vigor. Conceptually, a compellent model, as opposed to a deterrent model, in space would allow for the United States to dictate the terms of each engagement in the cosmos. Such a model would also allow for US policymakers to force its rivals to abandon their own quests for regional (and, in the case of China, global) dominance in favor of the American-dominated world order that has persisted since the fall of the Soviet Union.

In my opinion, none of these strategic doctrines are mutually exclusive. I mentioned at the start of this book that, rather than existing as competing ideas, with the exception of the sanctuary model, these ideas—survivability, space control/superiority, and space dominance—should be viewed as inhabiting a spectrum. One feeds into, and undergirds, the other. Starting at survivability, the military should progress to space control/superiority.[7] Once its satellite constellations have been secured and it has

---

6 John Lewis Gaddis, *Surprise, Security, and the American Experience* (Cambridge: Harvard University Press, 2004), 13.

7 Travis C. Stalcup, "U.S. in Space: Superiority, Not Dominance: Trying for Dominance in Space Is Counterproductive: The U.S. Should Settle for a More Modest Goal," *Diplomat*, 2014. http://thediplomat.com/2014/01/u-s-in-space-superiority-not-dominance/?allpages=yes

kept American rivals in space at bay, the United States *must* then look at getting ahead of its rivals in space yet again. Remember, modern combat is predicated on being faster than your rivals. American forces require access to space to function properly—and operate quickly. American enemies are working hard to deny US forces this capability in the near term, should a major geopolitical crisis erupt. Given that the United States will remain far more dependent on space systems for its most basic military and civilian functions, there is no way that a space control/superiority concept will sustain the United States over the long run. Instead, US policymakers should plan to eventually shift from the deterrent and defensive-minded space control/superiority model to the more hegemonic outlook of space dominance. Embracing a space dominance strategy will allow for the United States to remain faster than its rivals.

THE SPACE FORCE: AN IDEA FOR WINNING THE FUTURE—*TODAY*

President Donald Trump and his supporters have advocated for an independent, sixth branch of the United States Armed Forces dedicated solely to space. Some in Congress, Republicans and Democrats alike, have challenged the need for an independent space force.[8] Yet, the track record is clear: under the current parameters, the Air Force has "gotten in its own way" when handling military space issues.[9] Plus, while we should encourage greater decentralized networks as a way of surviving future attacks and for streamlining efficiency in the modern threat environment, too much decentralization is just as ineffective as too much centralization. During budget crunches, the Air Force has been known to pilfer funds meant to maintain America's already-exposed

---

8 Valerie Insinna, "Rep. Mike Turner on Why He's Softened on Space Force, and the Importance of an East Coast Missile Defense Site," *Defense News,* 8 April 2019. https://www.defensenews.com/space/2019/04/08/rep-mike-turner-on-why-hes-softened-on-space-force-and-the-importance-of-an-east-coast-missile-defense-site/

9 Oriana Pawlyk, "Air Force to Congress: 'No Space Corps,' *Military.com,* 22 June 2017. https://www.military.com/daily-news/2017/06/22/air-force-congress-no-space-corps.html

satellite constellations to pay for other "mission priorities," like keeping the overpriced F-35 program going, according to Representative Mike Rodgers (R-AL).[10] While the United States could once afford to wait for a space force to come into fruition "naturally," as the vulnerability phase of the 2020s approaches, this is simply impossible. Thankfully, a small space force has officially been created under the Department of the Air Force as of December 2019.[11]

We should be grateful that the Air Force officers currently assembled by the Trump administration to help shepherd the US Space Force into existence are clearly dedicated to the cause of creating a new service with a unique identity. Of course, the bigger question must be: What happens once the pro–space force president, Donald J. Trump, is no longer in the White House? What happens to the space force once a president who is less supportive of its mission comes along and encourages elements opposed to the space force within the Department of the Air Force to descend upon the space force's resources like vultures descending upon the carcass of a lion in the African bush?

The false cries of terror about the purported wastefulness of an independent space force have resounded not only throughout the sacred halls of Congress, but also in the sterile offices of the Pentagon's leadership. When he was secretary of defense, James Mattis did not overtly support the creation of an independent space force.[12] Like Representative Dan Crenshaw, Mattis believed that the creation of a new military branch would be wasteful and, inexplicably, weaken America's defensive

---

10 Russell Berman, "Does the U.S. Military Need a Space Corps?" *Atlantic,* 8 August 2017. https://www.theatlantic.com/politics/archive/2017/08/military-space-corps/536124/

11 Merrit Kennedy, "Trump Created the Space Force. Here's What It Will Actually Do," NPR, 21 December 2019. https://www.npr.org/2019/12/21/790492010/trump-created-the-space-force-heres-what-it-will-do

12 "Old School Air Force Can't Handle 'Space Corps' Challenge?" *Military.com,* 9 October 2017. https://www.military.com/defensetech/2017/10/09/old-school-air-force-cant-handle-space-corps-challenge

readiness.[13] Of course, the claims of Representative Rodgers and other supporters of a space force have proven that, at times, the Air Force has not truly supported its space components. Also, with so many other hands in the space pie, crafting a concise and decisive national security space strategy has proven difficult. Think about it: not only does the United States Air Force have a say in America's global military space operations, but so too do the navy and army. Then, there is the Missile Defense Agency that was birthed from President Reagan's SDI and continues working on ways to protect the United States from ballistic missile attack. The National Geospatial Intelligence Agency (NGA) also has a role in space. And as the civilian space sector increases its operations in space, soon the Federal Aviation Administration (FAA) will have a role to play in space traffic management.[14] As will the Department of Commerce, under the Trump administration's secretary of commerce, Wilbur Ross, who wants to develop an office dedicated to the commercialization of space.[15] So, while opponents of a space force are right to worry about wasteful spending, creating a real branch of the military with the heft that comes along with being an independent force, would allow for US leaders to cut through the clutter.

Still, understanding and preparing for the days when space becomes a warfighting domain itself is critical for the United States going forward. Staying to the "left of boom" is the only thing that will prevent a strategic catastrophe from befalling the United States in the 2020s, which will undoubtedly be the most crucial decade for the United States in the first half of the twenty-first century. Opponents of the space force

13 Brandon J. Weichert, "The Case for an American Space Corps," *American Greatness*, 3 December 2017. https://amgreatness.com/2017/12/03/the-case-for-an-american-space-corps/

14 Jacqueline Feldscher and Bryan Bender, "FAA Put on Notice That It Needs to Catch Up with Spaceflight Boom," *Politico*, 8 November 2019. https://www.politico.com/newsletters/politico-space/2019/11/08/faa-put-on-notice-that-it-needs-to-catch-up-with-spaceflight-boom-487620

15 Jeff Foust, "Commerce Department Seeks to Increase American Space Industry's Global Competitiveness," *Space News*, 9 April 2019. https://spacenews.com/commerce-department-seeks-to-increase-american-space-industrys-global-competitiveness/

have successfully argued that any new force should be subordinated to the Department of the Air Force. This was the same tactic that opponents of an independent air force employed. Ultimately, though, as the technology progressed and the threat posed to the United States by the Soviet Union in the atomic age progressed, leaving the nation's air capabilities under the imprimatur of the army was simply an insufficient response to the threat the country faced.

Inevitably, the risks in space will become so profound that the space force will have to be cleaved away from the Department of the Air Force for it to be truly effective. In the meantime, the members of the new space force will have to be imbued with a culture of innovation and independence from the start. Just as the United States Marine Corps is technically part of the Department of the Navy, yet marine officers employ army ranks as opposed to naval ranks, so too will the space force have to utilize naval ranks rather than the air force's ranking system, which it imported from the army.[16] The reason is simple: it will help to foster a separate service culture, even if the space force is part of the Department of the Air Force. After all, the air force presently views space as a mere auxiliary to their wider mission of air dominance. While space certainly plays a role—a crucial one, at that—the air force has been slow to recognize this fact and will likely be unable to recognize this reality, given their institutional biases because space will always be ancillary to the strategic domain of air in the eyes of many air force leaders. Instead, the embrace of a naval ranking system will help to imbue the future space troopers of the space force with a strong sense of unique identity.[17] This, in turn, will allow space force strategists to envision strategies based on space being viewed as *more* than a mere auxiliary component to the other strategic domains. In fact, space is

---

16 Brent D. Ziarnick, "Why the Space Corps Needs to Use Naval Rank," *The Space Review*, 22 July 2019. http://www.thespacereview.com/article/3761/1

17 Dennis Wille, "What Should We Call Members of a Space Force?" *Slate*, 4 February 2019. https://slate.com/technology/2019/02/space-force-members-names-sentinels-troopers.html

very similar to the ocean environment that the US Navy operates in, so it will be key for future space troopers to comprehend the cultural similarities they share with their navy siblings even more so than they do with their air force parents.[18]

In the future, space force missions will include everything from surveillance and reconnaissance, to maintaining logistical supply lines, to conducting orbital strikes on rogue states and terrorists. And as space mining and tourism intensifies, a space force would inevitably be charged with providing protection for these civilian functions as they transit through the critical domain of space. After all, the flag and trade often march hand in hand into new virgin lands. All of this will be placed on one part of the air force, and it will be too much for what is set to be the smallest service of the United States military.[19] This is why, inevitably, the space force will morph into own department, just as the army, navy, and air force have today. Plus, the space force may need to incorporate America's cyber warriors, who have often been given the budgetary and policy short shrift, until recent days, at least.[20] Further, with the renewed threats that US forces face on the electromagnetic (EM) spectrum, it might be necessary to fuse an EM defense corps into the space force, since the space force will already be the most technologically sophisticated part of the US military.[21] By that point,

---

18 Jurica Dujmovic, "This Is What the U.S. Military's 'Space Force' Could Look Like," *Market Watch,* 16 June 2018. https://www.marketwatch.com/story/this-is-what-the-us-militarys-space-force-would-look-like-2018–04–18

19 Robert Burns, "Space Force Would Be by Far the Smallest Military Service," *Air Force Times,* 4 March 2019. https://www.airforcetimes.com/news/your-air-force/2019/03/05/space-force-would-be-by-far-the-smallest-military-service/

20 James Stavridis, "The U.S. Needs a Cyber Force More Than a Space Force," *Bloomberg,* 14 August 2018. https://www.bloomberg.com/opinion/articles/2018–08–14/u-s-needs-a-space-force-and-a-cyber-force

21 Lauren C. Williams, "Navy Declares EMS a Full-Fledged Warfighting Domain," *Defense Systems,* 23 October 2018. https://defensesystems.com/articles/2018/10/24/navy-electronic-warfare-williams.aspx

then, the Pentagon will have no choice but to genuinely reorganize in order to create that fully independent Department of the Space Force that President Trump has actually been calling for.

SPECIAL FORCES IN SPACE?

Another component of the US Armed Forces that is often overlooked for space operations is the Special Operations Command (SOCOM). Very often, when one thinks of special forces operators, invariably images of the Bin Laden raid in 2011 would come to mind. Unassuming, bearded men in camo is another image many Americans have. To America's terrorist enemies, these forces are the equivalent of the grim reaper. In fact, US Special Forces are more than modern-day Spartan warriors. They are technological marvels. Special forces operators are among the best-trained troops in the world. These fighters are sent into the toughest terrain and they are asked to either hunt down, capture, or kill elusive enemies such as Bin Laden or to help train troops in foreign countries. For these silent professionals, technology, like all of America's military (only more intensely), has become the critical force multiplier for these small teams of elite forces.

When 9/11 happened, the US Department of Defense was left scrambling for their maps of Afghanistan. The Pentagon wasn't ready for the kind of war that US leaders required them to fight. The DoD was effectively caught flat-footed. In the vacuum created by the Pentagon's lack of preparation for a potential campaign in landlocked Afghanistan, the director of central intelligence, George Tenet, stepped forward with a plan to deploy dozens of CIA paramilitary officers virtually overnight into Afghanistan. These small units would stir the shit-pot against al-Qaeda and the Taliban, which had offered bin Laden safe haven in that ancient land. This left former secretary of defense Donald Rumsfeld in a proverbial lurch: he wanted his people, the military, who were charged with both defending and avenging the United States against foreign attack to take the lead. Since the conventional forces were not prepared for an invasion of Afghanistan, Rumsfeld ordered US Special

Forces into Afghanistan to support CIA's efforts there.

One of the first major decisions that Rumsfeld made at the Department of Defense was to reform special forces. Despite his controversial tenure as secretary of defense, Rumsfeld got space and special forces right. Rumsfeld believed that special forces needed to be nimbler. Special forces needed to become a high-tech, though small, cadre of elite warriors to wage twenty-first-century warfare upon America's decentralized enemies spread throughout the world.[22] Rumsfeld was reportedly upset with the US Central Command's inability to "find and target top [al-Qaeda] leaders," according to a 2002 CNN report from Barbara Starr. Thanks to the institutional reforms that had occurred under the rubric of Rumsfeld's controversial "Revolution in Military Affairs" (RMA), the Pentagon's leaders crafted "a phased plan that [involved] the use of special operations forces for months in highly covert missions, all outside the regular scope of military or law enforcement operations." In fact, as history played out, the Pentagon has used special forces continuously across the world to combat perceived, developing, and unconventional threats. It all started, though, with Rumsfeld's historic decision to put "Special Operations, instead of the US Central Command, in charge of [counterterrorism] efforts."[23]

The famous images of American Special Forces operators riding on horseback, flanked by countless Afghan forces, as they rode into battle against al-Qaeda and Taliban forces, was the apotheosis of America's satellite dependency.[24] The United States essentially invaded Afghanistan following 9/11 with a handful of CIA and special forces operators. CIA

---

22 Greg Jaffe, "Rumsfeld Aims to Elevate Role of Special Forces," *Wall Street Journal,* 18 February 2006. https://www.wsj.com/articles/SB114020280689677176

23 Barbara Starr, "Sources: Rumsfeld Calls for Special Ops Covert Action," *CNN,* 2 August 2002. https://www.cnn.com/2002/US/08/02/rumsfeld.memo/index.html

24 Diana Stancy Correll, "How the 'Horse Soldiers' Helped Liberate Afghanistan from the Taliban 18 Years Ago," *Military Times,* 18 October 2019. https://www.militarytimes.com/news/your-military/2019/10/18/how-the-horse-soldiers-helped-liberate-afghanistan-from-the-taliban-18-years-ago/

operatives were deployed from Russian military helicopters flying in from neighboring central Asian states, with literal backpacks full of cash, to buy the loyalties of local Afghans in the push to punish al-Qaeda and the Taliban. Meanwhile, US Special Forces deployed from airplanes or helicopters and enmeshed themselves with local fighters. With small arms and satellite communications, these very small US forces were able to inflict maximum harm on their more numerous Taliban and al-Qaeda adversaries. The combination of American satellite interlinks, allowing for copious and precise air strikes, coupled with the will of the Afghan Northern Alliance forces, routed the Taliban and busted apart al-Qaeda in Afghanistan (at least for a time). From 2001 to the spring of 2002, these were high points of America's initial foray into Afghanistan. The outsized initial victories of these small American forces were entirely made possible by satellites overhead. If either the Taliban or al-Qaeda had the ability to interfere with US satellite constellations, then those early days in Afghanistan might well have been the quagmire that Afghanistan ultimately became after 2002.

Afghanistan taught US policymakers that special forces could be used when in a strategic tight spot. American leaders built out from this model, and eventually SOCOM became one of the central pieces in US national security policy. The persistent use of special forces globally as the tip of America's spear necessitated technological changes. Beginning in 2013, SOCOM invested in cubesats to enhance their capabilities. Special forces have been on hunter-killer/capture missions since the days when Rumsfeld commissioned SOCOM to track down those high-ranking leaders of al-Qaeda whom CIA and other intelligence agencies had not been able to find. To assist US Special Forces operators, SOCOM invested in "tags," which are pieces of tracking technology used to "clandestinely mark [special forces'] prey." These tiny beacons are placed throughout the world by special forces. The beacons have an array of capabilities, like employing infrared flashes to signal their location, while still others are covertly hidden within commercial electronics, such as cell phones or key fobs. Some beacons are affixed to cars or people—all of which transmit the whereabouts of the unsuspecting target to US Special Forces by relaying the data via satellites. To maintain—and enhance—this critical capability, US Special Forces decided to invest in a cluster of eight cubesats to cover "areas of the world where the satellite coverage is thin, and there aren't enough cell towers to provide an alternative." The eight cubesats were launched atop one of SpaceX's Falcon 9 rockets. Those eight cubesats remained in orbit for three years.[25]

Similarly, SOCOM relies more heavily on satellite communications than most of the other branches. With its forces deployed into the most dangerous parts of the world on top-secret missions that could decide the fate of their countrymen, SOCOM needs real-time communications. The capabilities of America's small and nimble special forces are greatly amplified by satellites. Should a Russian or Chinese counterspace attack occur on the satellite networks that special forces use, catastrophe would ensue as US Special Forces would be unable to effectively operate.

Given their use of space assets, SOCOM will invariably have a role to play in the strategic domain of space, regardless of whatever the future of the space force may be. The objectives of special forces will not involve either space control/superiority or the space dominance theories of military space power. For SOCOM to be effective, they must bear in mind that their needs are far more tactical than those of other parts of the military that use space. Constant contact with distant enemies is what special forces is involved with. The school of space power theory that SOCOM must embrace is the survivability school. SOCOM appears to have gotten this concept right off the bat. In keeping with Thomas D. Taverney's ideas of mixed constellations, "the Pentagon is looking to make its communications systems more resilient against interference or hostile jamming. One way to do that is to allow users of satellite communications to roam across commercial and military

---

25 Noah Shachtman, "With New Mini-Satellites, Special Ops Takes Its Manhunts into Space," *Wired*, 21 May 2013. https://www.wired.com/2013/05/special-ops-mini-sats-manhunts/

networks so if one system is disrupted, they can switch to a different provider." Along with Kratos Defense & Security Solutions, the air force conducted a test wherein they used a Special Operations Forces Deployable Node and made it compatible with an SES commercial satellite network. The results of the successful experiment "is that the military does not need to buy brand new terminals to get roaming services. With minor updates to legacy terminals, the military can get access to multiple satellites and services, including the Time Division Multiple Access (TDMA) and Frequency Multiple Access (FDMA) networks."[26]

SOCOM should look to firms developing more advanced and cheap cubesats, such as Israel's NSLComms, which was mentioned in the chapter on Israel's space program. With their successful tests of an expandable antenna aboard a small satellite, NSLComm's technology could provide SOCOM with greater capabilities in a crisis. To avoid having to rely on costlier, larger communications satellites placed in geosynchronous orbit that will be required to last for eight, perhaps even fifteen years—increasing the costs significantly—NSLComm's "solution was to develop a new kind of antenna that can deploy on its own, without the help of any additional heavy machinery, and that can extend to the sizes needed to provide truly high-throughput connectivity on a satellite that's small and much easier to launch, providing about 100 times faster connectivity than the fastest nano-satellites in the same size class today at about one-tenth the launch cost." The company plans on launching thirty satellites by 2021 and hundreds by 2023. These tiny marvels will transform the civilian communications business. NSLComm ultimately intends "to offer a 'private constellation' offering, where, for example, a cruise ship operator could build, launch, and operate its own network constellation for its customers at

---

26 Sandra Erwin, "Commercial Satellite Roaming Possible with Existing Military Terminals, Experiment Shows," *Space News,* 4 February 2019. https://spacenews.com/commercial-satellite-roaming-possible-with-existing-military-terminals-experiment shows/

minimal cost."[27] Just imagine SOCOM embracing this model and this technology. Not only could they make their vital communications and command capabilities more survivable in the inevitable satellite war, but these tiny satellites might allow for SOCOM to offer last-ditch augmentation to other US military forces harmed by any attack on America's satellite constellations.

In keeping with the covert nature of special forces, and their interest in developing cube satellites, it might be time for this clandestine force to develop "space stalkers" of their own. Presently, the United States operates in a legal grey zone when it comes to military operations in space. We are bound by the principles found within the Outer Space Treaty of 1967, yet as you've seen time and again in this book, American rivals don't play by the rules. As the United States refuses to decide whether weaponization of space should occur or not (of course, it already is occurring), American rivals are placing dual-use systems in orbit. The Chinese are developing their space laser to supposedly help in clearing orbital debris, and the Russians have already placed co-orbital satellites meant to supposedly repair their communications satellites should anything happen to these systems while in orbit. Of course, these space stalkers could just as easily be used to attack American satellites, as was the scenario presented in the first chapter of this work. Until an actual space warfare doctrine can be developed by a space force, then SOCOM might be able to fill in the gap during the next decade of vulnerability. SOCOM could easily spend a small amount of money to develop and deploy rudimentary space stalkers. These covert devices, controlled from MacDill Air Force Base in Florida, could be maneuvered to tailgate behind sensitive satellites belonging to American rivals and could be used in the same manner that Russia and China plan to use space stalkers. They could be used to either collect surveillance or to destroy the systems of our enemies.

---

27 Darrell Etherington, "NSLComm's First Spring-Loaded Expanding Antenna Satellite Is Headed to Space," *Tech Crunch,* 3 July 2019. https://techcrunch.com/2019/07/03/nslcomms-first-spring-loaded-expanding-antenna-satellite-is-headed-to-space/

Looking beyond the next decade, as space becomes more accessible, the aforementioned SUSTAIN concept will have to be supported—if not by the marines as the original concept was, then definitely by SOCOM. In 2017, US Air Force General David Goldfein declared that "special operators" would "be able to strike anywhere in the globe within minutes" from military space stations in low-Earth orbit. SOCOM should immediately begin quietly partnering with Richard Branson's Virgin Galactic to gain access to a military version of the company's suborbital plane. The reason behind Richard Branson's obsession with suborbital spaceflight has, as always, to do with the costs of rocket fuel. Branson has correctly observed that "the most rocket fuel costs are from the surface to 50,000 feet because that is where the air is the thickest." Yet, if Virgin Galactic could take people to 50,000 feet and launch into space from there, the costs drop precipitously.[28] As Goldfein speculated, "The question [for the Pentagon] is, what does it mean if I take seven special operators, put them on [a sub-orbital spaceplane] and then can get to any place on the planet in less than an hour?"[29]

SOCOM will play a crucial role for supporting overall US military capabilities. What's more, given SOCOM's small and nimble nature, as the bureaucratic battle over space force rages throughout the 2020s, when American satellite constellations will be their most vulnerable, a special forces satellite capability might prove decisive in helping to mitigate the damage of a Space Pearl Harbor. Special forces embody what Winston Churchill called the "guerilla idea: guile combined with courage and imagination."[30] Basically, "leverage is the essence of special

28 Marco della Cava, " 'The Ultimate Space Adventure': Virgin Galactic's Space Tourism Plan Demands Courage and $250K," *USA Today,* 13 December 2018. https://www.usatoday.com/story/news/2018/12/07/want-travel-space-virgin-galactic-pack-courage-and-250-000/2140284002/

29 David Willetts, "Forces' Star Wars: Special Forces Soldiers Could Soon Be Deployed in SPACE to Strike Anywhere on Earth in Minutes," *The Sun,* 14 July 2017. https://www.thesun.co.uk/news/4015794/special-forces-soldiers-could-soon-be-deployed-in-space-to-strike-anywhere-on-earth-in-minutes/

30 Derek Leebaert, *To Dare and to Conquer: Special Operations and the Destiny of Nations, from Achilles to Al-Qaeda* (New York: Little, Brown, 2006), 24.

warfare—leverage usually obtained by careful planning, extraordinary risks, and exceptional temperaments."[31] This is the sort of mentality that would prove decisive for SOCOM being a pioneer in limited, though essential, space warfare concepts. Yet, for all of the help that SOCOM may provide the wider military in space, as Derek Leebaert assesses in his work on special forces, no amount of guile and imagination can make up for policy failures. For twenty years, the United States has engaged in a series of policy failures in space that weakened our presence there and inspired rivals to challenge us in this dangerous domain. A total reassessment of American policy in space will be required to stave off the disaster that is coming.

TAKING THE DEEP BLACK

When the Portuguese liberated themselves from Muslim rule in 1249, the tiny country on the Iberian Peninsula of Europe was able to establish itself as the maritime power of Europe. By 1415, Portugal ruled the waves; its innovative people, notably Henry the Navigator, had built the powerful Caravels that brought Portuguese sailors around the world. The tiny European state charted the first course from Europe around the Horn of Cape Hope in Africa, all of the way to the rich trading outposts of India. Portugal cornered the market when most European states were still mired in what has loosely been referred to as, the "Dark Ages." Portugal was the envy of Europe, and its future looked bright.

Its larger neighbor, Spain, wallowed in internal division during this time. That is, until 1469. Seemingly overnight, Spain's political situation stabilized with the marriage of Queen Isabella of Castile to Ferdinand of Aragon. By 1492, the newly united Spanish kingdom was able to break the chains that the Muslim Moorish invaders had kept Spain in. Afterward, Spain became the European powerhouse—at least

---

31 Dennis E. Showalter, "Derek Leebaert, to Dare and to Conquer: Special Operations and the Destiny of Nations, from Achilles to Al-Qaeda," *Michigan War Studies Review,* 1 December 2006. http://www.miwsr.com/2006/downloads/20061201.pdf

for a while. During that time, though, Portugal watched in horror as its larger, once divided Spanish neighbor became strong. The stronger Spain became, the weaker Portugal seemed. This was particularly true after Henry the Navigator died, and Portugal's innovation in the maritime sector apparently died with him, too.

Portugal's leaders, unlike Spain's, were no longer interested in funding the great maritime expeditions that had come to define Portuguese life with the passage of Henry the Navigator. Strategic gaps quickly formed that Portugal's upstart neighbor, Spain, was more than happy to exploit. As Portuguese investment in its immense maritime capabilities declined, Portugal's government removed the restrictions on exporting mapmaking and shipbuilding techniques to other states. Portugal made money off the removal of these restrictions in the near term. Yet, in the long term, the proliferation of their once-unique capabilities weakened Portugal. Spain was the most immediate beneficiary of Portuguese shortsightedness, although, France, Britain, and the Netherlands soon came to benefit from this knowledge during the Age of Sail.

The Genoese explorer, Christopher Columbus, first went to Portugal's king for patronage to find a better trading route to India. European vessels had to pass through Ottoman-controlled waters, subjecting those vessels to attack and enslavement by Islamic pirates. But the passing of Henry the Navigator killed the culture of exploration in Portugal. The government did not see any benefit to making a modest investment in the Genoese sailor's project. Not to worry, Columbus quickly found a patron in the Spanish crown. In 1492, Christopher Columbus sailed across the Atlantic Ocean in search of a new trading route to India. What he found instead was the New World. The Spanish, rather than the Portuguese, flag was raised high over this virginal territory. Thus, the eponymous Columbian epoch of human history was birthed. Spain became Portugal's primary nemesis on the high seas, having metamorphosed from what we might refer to as a "failed state" today into a global empire in a few short decades time. That Spain became the world power that it did after 1492 was not a *fait accompli.* Had the Portuguese simply not allowed for their monopoly on mapmaking and superior shipbuilding to be traded away to the highest bidders and had the Portuguese government simply invested in maintaining the immense naval capability that it had created under Henry the Navigator, it is unlikely that Spain would have displaced Portugal when it did.

To avoid a great power conflict, the Spanish crown appealed to the Vatican to mediate a truce between the two Iberian powers. Pope Alexander VI came up with a compromise: he split the New World between Spain and Portugal by drawing a north-south demarcation through the map of the New World. To the west of that line belonged to Spain. Everything to the east of that arbitrary line would fall to Portugal. This led to the 1494 Treaty of Tordesillas, which formalized this grand bargain mediated by Pope Alexander VI. Of course, it was a bad deal for the Portuguese, which explains why only one country in what we now refer to as Latin America speaks Portuguese while the rest speak Spanish.

Inevitably, Spain and Portugal were joined by other European powers seeking access to the gold and other booty found in the Americas (to say nothing of the prestige). Portugal's fateful decision to abandon its innovative edge by no longer investing in high-risk, high-reward maritime missions, such as the one that Columbus proposed, as well as Portugal's tolerance of its mapmakers and shipbuilders trading away their knowledge for trinkets, ensured that Portugal would be nothing more than a second-rate power as the Age of Sail progressed. Today, space is akin to the sea, and exploration of this vast and forbidding domain will be most like the Age of Sail.

At the start of that previous bygone age, Portugal and Spain blazed a new path that ultimately changed human history forever. Yet, both powers ultimately ceded their technological and geopolitical dominance to newer, rising powers. This is not because the Portuguese or Spanish became less innovative or powerful as a people. It is because bureaucracy overcame common sense and, at different times, both powers embraced staggering levels of ignorance and arrogance that eventuated in their downfall. By the end of that great age of exploration, the British Empire would stand victorious over all other European empires. However, even Britain's dominance would collapse in the great hurly-burly of evolving great power competition—where Britain's power yielded to that of the United States.

At the start of the Space Age, the United States and Soviet Union vied for power in the cosmos. The Soviets lost. The United States stood above all others until it faltered, and its leaders became irredeemably feckless. Just as the Portuguese did with their mapmaking and ship-building abilities and as the British did with their own industrial secrets, the United States allowed for its own people and economy to be pilfered under the false name of "free trade" by the Chinese. Now that America's once-unquestionable dominance in space has been challenged, a new wave of powers are rising to bring the United States down to size.

The United States must reinvigorate itself in the high-tech and space sectors. It must not allow its own internal contradictions and short-sightedness to fog its strategic view of the world. The United States must dominate the high ground of space and never let it go. If it does lose its perch in the heavens, its crash to Earth will be catastrophic for its overall power. An independent space force committed to a doctrine of space dominance and an unapologetic private sector development of space will be key if America is to avoid the fate of the Portuguese and so many other great powers that took their great power status for granted.

FORWARD, TOGETHER

What should be clear is that the status quo in space will no longer hold. For years, the United States has practically ignored the fact that it has been engaged in a long-term struggle for dominance of the Deep Black. Yet, as the second space race continues, as well as the arms race in space, it is becoming impossible to ignore. This is especially true in the face of both Russia and China's growing strategic threats to America's access to space. Over time, if left in their current configuration, America's greatest strategic assets—its satellite constellations—will become the country's greatest vulnerability. Once the United States loses access to space, it will quickly lose its global strategic dominance. And once it loses this, American security will forever be threatened as rival states take the strategic high ground in orbit and threaten the United States from above.

To avoid such a bleak future, the United States must engage in the arms race in space. It must best its competitors in the second space race as well. Further, the United States must now look to upstart nations, like India or Japan, for inspiration as it endeavors to efficiently compete and stay ahead of its rivals in space. American competitors have effectively synthesized their private and public sectors into a seamless entity dedicated to space exploitation and dominance, as the United States once did in the 1960s, to win the Moon Race during the Cold War. American competitors have also taken to the stars, not for airy globalist notions. They have done so to gain advantages over their fellow nation-states. To best their rivals, America should embrace the reforms I've advocated throughout this book; it should remember the reason nations go to space, and why it is important for the United States to remain the dominant power in space. It should learn, as the Indians have, to fuse its civilian space agency's development alongside its military space development and its private economic development of space—all to gain profound advantages over its global competition.

What's more, rather than simply following the Nixonian model of incremental space development, the United States Congress should heed the president's calls to engage in a grand program of national space exploration, exploitation, and defense. This should not be a political matter. It is not important who began this push and which political party may benefit most from whatever successes are had by implementing these reforms. Instead, this is about our nation's survival and its continued existence as the most dominant power in the world. Ceding the strategic high ground to rivals, as Washington has done for decades, or simply allowing for America's critical satellite constellations to exist over the next state in vulnerable states is unacceptable. Congress and the Pentagon must spend considerably more money and effort at protecting those systems and building the infrastructure necessary to ensure America's continued space dominance.

The United States risks a Space Pearl Harbor. In order to avoid this from happening—which should be Washington's primary goal, given the costs of having to repair and replace damaged space systems— Washington must take proactive measures today. By shoring up the defenses and survivability of America's vaunted satellite constellations, the United States can better deter Russia, China, and other actors from threatening the United States in space. By expanding America's strategic role in space, Washington might soon be able to compel nuclear-armed rogue states to entirely abandon their hostility to the United States. Further, by encouraging American companies to more freely exploit space, the United States would stand the most to gain from the trillion-dollar-or-more economy waiting to be developed in space. All of this can happen for the United States sooner rather than later, but Washington must take the painful first steps toward reform now.

Time is not on America's side. For America to remain a superpower, it must win in space and defeat its competitors—namely China and Russia—at all costs. Winning space will also mean defeating America's stifling space bureaucracy, which so often hinders US space policy. It means fundamentally changing the strategic culture within the Pentagon regarding its views on space, technology development, and staying to the left of boom. After that, US elected officials must embrace a long-term strategy for space defense and development. If Washington continues under its standard operating procedures, then the US will lose the space war—all because of Washington's refusal to accept the reality that the country is threatened from advanced rivals who plan on destroying American satellite constellations. Shoring up these weaknesses would be expensive today, but it would not be as costly as simply waiting for disaster to strike in space. Heeding the calls for reform within this book would be an effective way for never having to live through an actual space war. After all, if the vulnerabilities to American satellites are removed, the Russians, Chinese, North Koreans, and Iranians would have little with which to threaten America.

Whether American leaders have the courage to stand up now remains to be seen. We should all hope and pray that our leaders, regardless of their party, make the correct strategic choice as opposed to the wrong political decision. Time will tell how bad we let things get before making the tough choices—and how much pain we want to endure before making those hard decisions. The future will be won by those nations that want to win more than any other. And space is the future. Whether assessing space in economic, scientific, or military terms, for the United States to not have the leading role in space simply because the situation up there is getting too risky would be akin to the United States Navy mothballing its fleet when submarines or aircraft were invented. What is needed is a clear-eyed understanding of how best to proceed: the United States must preserve its position, enhance that position, and outpace its rivals. There is indeed a new space race. Is America capable of embracing the values that took the country to the Moon, or will it continue insisting upon the hedonistic ideas that mired us in Woodstock?

It's ultimately your choice.

BIBLIOGRAPHY

Aarabi, Kasra. "The Fundamentals of Iran's Islamic Revolution." *Tony Blair Institute for Global Change.* 2019. https://institute.global/insight/co-cxistence/fundamentals-irans-islamic-revolution

Abner, Elihugh M. "Putin's Multipolar World and What It Means for U.S. Strategy." *Association of the United States Army.* 2017. https://www.ausa.org/publications/putins-Multipolar-world

Abrahamson, James A. and Cooper, Henry F. "What Did We Get For Our $30-Billion Investment in SDI/BMD?" The National Institute for Public Policy. 1993.

Adams, Gordon. "The Iron Triangle: The Politics of Defense Contracting." *Foreign Affairs.* 1982. https://www.foreignaffairs./com/reviews/capsule-review/1982–03–01/iron-triangle-politics-defense-contracting

Adelman, Jonathan R. "American Strategic Modernization and the Soviet Struggle." *Air University Review.* 1983. http://www.airpower.maxwell.af.mil/airchronicles/Aureview/1983/nov-dec/Adelman.html

Adib, Denise. "Pop Star Justin Bieber is On the Brink of Superstardom." *ABC News.* 2009. https://abcnews.go.com/GMA/Weekend/teen-pop-star-justin-bieber-discovered-youtube/story?id=9068403

"Advanced MOL Planning." The National Reconnaissance Office. 1969. http://nro.gov/foia/declass/mol/794.pdf

Agencies, "Crowds Chant 'Death to Israel' as Iran Marks 40 Years Since Islamic Revolution." *Times of Israel.* 2019. https://www.timesofisrael.com/iran-marks-40-years-since-islamic-revolution/

Ahmari, Sohrab. "About That New 'Moderate' Iranian Cabinet…" *Wall Street Journal*. 2013. https://www.wsj.com/articles/about-that-new-moderate-iranian-cabinet-1375917151?tesla=y

Ahren, Raphael. "Dershowitz: 'Inept Negotiator' Obama Gave Iran Green Light to Build Nuke." *Times of Israel*. 2015. http://www.timesofisrael.com/dershowitz-inept-negotiator-obama-gave-iran-green-light-to-build-nuke/

Aitoro, Jill. "Brazil Defense Minister: Space Partnership with US Not Dead Yet." *Defense News*. 2017. https://www.defensenews.com/space/2017/12/20/brazil-defense-minister-space-launch-partnership-with-the-us-not-dead-yet/

Aizhu, Chen. "Exclusive: Iran Renews Oil Contracts With China, Taps New Buyers." Reuters. 2015. http://www.reuters.com/article/us-china-iran-oil-idUSKBN0TM0CN20151203

Al-Rodhan, Nayaf R.F. *Meta-Geopolitics of Outer Space: An Analysis of Space Power, Security, and Governance*. New York: Palgrave Macmillan, 2012.

Albright, David. "North Korean Miniaturization." *38 North*. 2013. https://www.38north.org/2013/02/albright021313

Alexander, David. "U.S. Says Will Abide By Mine Ban Treaty Except On Korean Peninsula." Reuters. 2014. http://www.reuters.com/article/us-usa-defense-landmines-idUSKCN0H11U920140923

Allison, Graham and Blackwill, Robert. "Interview: Lee Kuan Yew on the Future of U.S.-China Relations." *Atlantic*. 2013. https://www.theatlantic.com/china/archive/2013/03/interview-lee-kuan-yew-on-the-future-of-us-china-relations/273657/

"America's Far-Off Hypersonic Weapons Freak Out Russia." *War Is Boring*. 2015. https://warisboring.com/america-s-far-off-hypersonic-weapons-freak-out-russia-e95ac401a516#.4ckon8n9u

Ambrosio, Thomas. "The Russo-American Dispute Over the Invasion of Iraq: International Status and the Role of Positional Goods." *Europe-Asia Studies*, 57, no. 8. 2005. http://www.jstor.org/stable/30043987

Amos, Jonathan. "Why India's Mars Mission Is So Cheap—and Thrilling." BBC *News.* 2014. http://www.bbc.com/news/science-environment-29341850

Ananthaswarmy, Anil. "The Quantum Internet is Emerging, One Experiment at a Time." *Scientific American.* 2019. https://www.scientificamerican.com/article/the-quantum-Internet-is-emerging-one-experiment-at-a-time/

Anderlini, Jamil. "UK Move to Join China-Led Bank A Surprise Even to Beijing." *Financial Times.* 2015. https://www.ft.com/content/d33fed8a-d3a1-11e4-a9d3-00144feab7de

Anderson, David and Polan, Shira. "China Made an Artificial Star That's Six Times as Hot as The Sun, and It Could Be the Future of Energy." *Business Insider.* 2018. https://www.businessinsider.com/china-east-experimental-advanced-superconducting-tokamak-nuclear-fusion-reactor-100-million-degrees-2018-12

Andrade, Tonio. *Lost Colony: The Untold Story of China's First Great Victory Over the West.* Princeton: Princeton University Press, 2011.

Antidze, Margarita and Stubbs, Jack. "Before Syria, Russia Struggled to Land Air Strikes on Target." Reuters. 2015. https://www.reuters.com/article/us-mideast-crisis-syria-russia-bombing/before-syria-russia-struggled-to-land-air-strikes-on-target-idUSKCN0SK1WF20151026

Antunes, Rodrigo. "Tritium: A Challenging Fuel for Fusion." *Euro-Fusion.* 2017. https://www.euro-fusion.org/news/2017-3/tritium-a-challenging-fuel-for-fusion/

Anzilotti, Ellie. "The Green New Deal is a Chance to Make Clean Energy Accessible to All." *Fast Company.* 2019. https://www.fastcompany.com/90366185/green-new-deal-100-Percent-clean-energy-will-help-economy

Arango, Tim and Krauss, Clifford. "China Is Reaping Biggest Benefits of Iraq Oil Boom." *New York Times.* 2013. http://www.nytimes.com/2013/06/03/world/middleeast/china-reaps-biggest-benefits-of-iraq-oil-boom.html?_r=0

Aris, Ben and Tkachev, Ivan. "Long Read: 20 Years of Russia's Economy Under Putin in Numbers." *Moscow Times.* 2019. https://www.themoscowtimes.com/2019/08/19/long-read-russias-economy-under-putin-in-numbers-a66924

Arktos. "Alexander Dugin: 'Eurasian Mission' (Arktos, 2014)." YouTube. 2015. https://www.youtube.com/watch?v=xuVcwNUWncQ

"Assessing the Threat from Electromagnetic Pulse (EMP): Executive Report. *Commission to Assess the Threat to the United States from Electromagnetic Pulse (EMP) Attack.* 2018. http://www.firstempcommission.org/uploads/1/1/9/5/119571849/executive_report_on_assessing_the_threat_from_emp_-_final_april2018.pdf

Ashman, Daniel. "North Korean Satellites May Hold Nuclear Bombs." *The Epoch Times.* 2019. https://www.theepochtimes.com/north-korean-satellites-may-hold-nuclear-bombs_2816818.html

Associated Press. "China Lands on the Moon's 'Dark Side' in a Move to Become a Space Power." *Washington Post.* 2019. https://www.washingtonpost.com/lifestyle/kidspost/china-lands-on-the-moons-dark-side-in-a-move-to-become-a-space-power/2019/01/03/aa9ea21a-0f69–11e9–831f-3aa2c2be4cbd_story.html?utm_term=.95803flc4f6b

Associated Press. "India Launches More Than 100 Satellites into Orbit." *The Telegraph.* 2017. https://www.telegraph.co.uk/news/2017/02/15/india-launches-100-satellites-orbit/

Associated Press. "North Korea's New Missile Has Russian Fingerprints 'All Over It.'" *NBC News.* 2019. https://www.nbcnews.com/news/world/north-korea-s-new-missile-has-russian-fingerprints-all-over-n1004151

Aston, Adam. "China's Rare-Earth Monopoly: The Rest of the World is Trying to Find Alternatives to These Crucial Materials." *MIT Technology Review.* 2010. http://www.technologyreview.com/news/421223/chinas-rare-earth-monopoly/

Austin, Greg. "Russia's Eastern Command at Sea." *The Diplomat.* 2015. http://thediplomat.com/2015/03/russias-eastern-command-at-sea/

Axe, David. "The Pentagon's Plan to Put Robot Marines in Space." *The Week.* 2014. https://theweek.com/articles/445664/pentagons-plan-robot-marines-space

Axe, David. "The Real Story of the Secret Space Station." *Daily Beast.* 2015. http://www.thedailybeast.com/articles/2015/10/31/real-story-of-the-secret-space-station.html.

Axe, David. "When It Comes to War In Space, U.S. Has the Edge." Reuters. 2015. http://blogs.reuters.com/great-debate/2015/08/09/the-u-s-military-is-preparing-for-the-real-star-wars/

Axelrod, Tal. "Trump Administration Sanctions Iran's Space Agency." *The Hill.* 2019. https://thehill.com/policy/international/middle-east-north-africa/459764-trump-administration-sanctions-irans-space

Babbin, Jed. "Elon Musk: King of the Crony Capitalists." *American Spectator.* 2018. https://spectator.org/elon-musk-king-of-the-crony-capitalists/

Babbin, Jed. "Shadowboxing with Iran." *American Spectator.* 2019. https://spectator.org/shadowboxing-with-iran/

Badkar, Mamta. "Why A Tiny, Uninhabited Island Chain Is Causing A Huge Row Between China and Japan." *Business Insider.* 2012. http://www.businessinsider.com/history-of-senkaku-diaoyu-dispute-2012-9?page=1

Baker, Peter. *Days of Fire: Bush and Cheney In the White House* (New York: Doubleday, 2013)

Balz, Dan and Atkinson, Rick. "Powell Vows to Isolate Iraqi Army and Kill It." *Washington Post.* 1991. http://www.washingtonpost.com/wp-srv/inatl/Longterm/fogofwar/archive/post012391_3.htm.

Bankston, Keven. "How Sci-Fi Like 'WarGames' Led to a Real Policy During the Reagan Administration." *Slate.* 2018. https://slate.com/technology/2018/10/reagan-wargames-star-wars-science-fiction-policy.html

Barbashin, Anton and Thoburn, Hannah. "Putin's Brain: Alexander Dugin and the Philosophy Behind Putin's Invasion of Crimea." *Foreign Affairs.* 2014 http://www.foreignaffairs.com/articles/russia-fsu/2014–03–31/putins-brain

Barboza, David. "China Passes Japan as Second-Largest Economy." *New York Times.* 2010. http://www.nytimes.com/2010/08/16/business/global/16yuan/.html?pagewanted=all&_r=0

Barkin, Noah. "The U.S. Is Losing Europe in Its Battle with China." *Financial Times.* 2019. https://www.theatlantic.com/international/archive/2019/06/united-states-needs-europe-against-china/590887

Barnatt, Christopher. "Helium-3 Power Generation." *Explaining the Future.* 2016. https://www.explainingthefuture.com/helium3.html

Barnett, Thomas P.M. *The Pentagon's New Map.* New York: Berkley Books, 2003.

Basulto, Dominic. "An Audacious Plan to Mine the Surface of the Moon." *Washington Post.* 2015. http://www/washingtonpost.com/news/Innovations/ wp/2015/03/19/an-audacious-plan-to-mine-the-Surface-of-the-moon/

Baziotis, I.P. "Theoretical Observations of the Ice Filled Craters On Martian Moon Deimos." PhD diss., Aristotle University of Thessaloniki, 2013.

Beck, Jeremy. "China's Helium-3 Program: A Global Game-Changer." *Space Safety Magazine.* 2016. http://www.spacesafetymagazine.com/space-on-earth/everyday/ everyday-life/china-helium-3-program/

Bedard, Paul. "North Korea Reveals Nuclear EMP Attack Plans." *Washington Examiner.* 2018. https://www.washingtonexaminer.com/washington-secrets/ north-korea-reveals-nuclear-emp-attack-plans

Beech, Hannah. "Just Where Exactly Did China Get the South China Sea Nine-Dash Line From?" *Time.* 2016. http://time.com/4412191/ nine-dash-line-9-south-china-sea/

Bechtol, Jr., Bruce E. "North Korea's Illegal Weapons Trade." *Foreign Affairs.* 2018. https://www.foreignaffairs.com/articles/north-korea/2018–06–06/ north-koreas-illegal-weapons-trade

Bender, Jeremy. "China Is Making Moves In Antarctica." *Business Insider.* 2015. http://www.businessinsider.com/china-is-making-moves-in-antarctica-2015–5

Berger, Brian B and Jeff Foust. "SpaceX Falcon 9 Rocket and Amos-6 Satellite Destroyed During Static-Fire Test." *Space News.* 2016. https://spacenews.com/ developing-Explosion-rocks-spacex-falcon-9-pad-at-cape-canaveral/

Berger, Eric. "NASA Rejects Blue Origin's Offer of a Cheaper Upper Stage for the SLS RRocket." *Ars Technica.* 2019. https://arstechnica.com/science/2019/11/ nasa-rejects-blue-oorigins-offer-of-a-cheaper-upper-stage-for-the-sls-rocket/

Berger, Eric. "Russia Embassy Trolls US Launch Industry After New Rocket Engine Sale." *Ars Technica*. 2018. https://arstechnica.com/science/2018/08/russian-embassy-trolls-us-launch-industry-after-new-rocket-engine-sale/

Bergin, Chris. "SpaceX and ULA Go Toe-to-Toe Over EELV Contracts." *NASA Spaceflight*. 2014. https://www.nasaspaceflight.com/2014/03/spacex-and-ula-eelv-contracts

Berman, Ilán. *Implosion: The End of Russia and What It Means for America*. Washington, D.C.: Regnery Publishing, 2013.

Berman, Russell. "Does the U.S. Military Need a Space Corps?" *The Atlantic*. 2017. https://www.theatlantic.com/politics/archive/2017/08/military-space-corps/536124/

Berger, Carl. *History of the Manned Orbiting Laboratory Program (MOL)*. Washington, D.C.: MOL Program Office, Department of the Air Force, 1970.

Berger, Eric. "Japan's HD Photos of the Moon are the Coolest Thing You'll See Today." *Ars Technica*. 2016. https://arstechnica.com/science/2016/10/japans-space-agency-just-released-a-trove-of-jaw-dropping-moon-photos/

Betmann, Maximilian. "A Counterspace Awakening? (Part 1)." *The Space Review*. 2017. https://www.thespacereview.com/article/3247/1

Billings, Lee. "War In Space May Be Closer Than Ever: China, Russia and the U.S. are developing and testing controversial new capabilities in space despite their denial of such work." *Scientific American*. 2015. http://www.scientificamerican.com/article/war-in-space-may-be-closer-than-ever/

Birch, Douglas and Isachenkov, Vladimir. "Crash of US, Russian satellites a threat in Space." *ABC News*. 2015. http://abcnews.go.com/Technology/print?id=6876532

Bisht, Anurag. "What is GLONASS and How is It Different from GPS?" *Beebom*. 2016. https://beebom.com/what-is-glonass-and-how-it-is-different-from-gps/

Blackwill, Robert D. and Harris, Jennifer M. *War By Other Means: Geoeconomics and Statecraft*. Cambridge: The Belknap Press of Harvard University Press, 2016.

Blank, Stephen. "Russia's Proliferation Pathways." *Strategic Insights.* 2008. https://pdfs. semanticscholar.org/61f1/8a380a86513fd2de36c52f8e8b1635b0a38c.pdf

Bodner, Matthew. "Can SpaceX and Blue Origin Best a Decades-Old Russian Rocket Engine Design?" *Technology Review.* 2019. https://www.technologyreview.com/s/613744/spacex-blue-origin-russian-rd180-rocket-engine-design/

Bodner, Matthew. "Energomash Raises Alarm Over U.S. Ban on Russian Rocket Engines." *Space News.* 2018. https://spacenews.com/energomash-raises-alarm-over-u-s-ban-on-russian-rocket-engines/

Bodner, Matthew. "The Long Road to Vostochny: Inside Russia's Newest Launch Facility." *Space News.* 2019. https://spacenews.com/the-long-road-to-vostochny-inside-russias-newest-launch-facility/

Bodner, Matthew. "Putin Approves Roscosmos Merger with Conglomerate." *Space News.* 2015. http://spacenews.com/putin-approves-roscosmos-merger-with-conglomerate/

Boese, Wade. "U.S. Withdraws From ABM Treaty; Global Response Muted." *Arms Control Association.* http://www.armscontrol.org/act/2002_-07–08/abmjul_Aug02

Booker, Christopher. "Iran's Moderate Leader is Nothing of the Kind." *The Telegraph.* 2016. https://www.telegraph.co.uk/news/worldnews/middleeast/iran/12116458/Irans-moderate-leader-is-nothing-of-the-kind.html

Books, Encounter. "Victor Davis Hanson on The Fate of the West, Trump, and The Resistance." YouTube. 2018. http://www.youtube.com/watch?v=Uyui7MMxs8A

Bormann, Natalie and Sheehan, Michael (ed.) *Securing Outer Space.* New York: Routledge, 2009.

Bose, Rakhi. "Flying into Space Could be as Cheap as Buying an SUV: India's 'Space Woman' Susmita Mohanty." *News 18.* 2019. https://www.news18.com/news/buzz/flying-into-space-could-be-as-cheap-as-buying-an-suv-indias-space-woman-susmita-mohanty-2362213.html

Bowermaster, David. "Boeing Probe Intensifies Over Secret Lockheed Papers." *Seattle Times.* 2005. http://www.seattletimes.com/business. boeing-probe-intensifies-over-secret-lockheed-papers/

Bozzato, Fabrizio. "Moon Power: China's Pursuit of Lunar Helium-3." *The Diplomat.* 2014. http://thediplomat.com/2014/06/ moon-power-chinas-pursuit-of-lunar-heium-3/

Bracken, Paul. *The Second Nuclear Age: Strategy, Danger, and the New Power Politics.* New York: Times Books, 2013.

Brandom, Russell. "Trump is Turning Elon Musk into a Crony Capitalist." *The Verge.* 2017. https://www.theverge.com/2017/2/6/14520608/ elon-musk-trump-tesla-business-deal-economic-advisory-team

Braun, Werner von. "Reminisces of German Rocketry." *Journal of the British Interplanetary Society* (ed.) L.J. Carter, vol. 15, no. 3. May-June 1956.

"Brazil." *Encyclopedia Britannica.* 2019. https://www.britannica.com/place/Brazil

Bremmer, Ian. *The End of the Free Market: Who Wins the War Between States and Corporations?* New York: Portfolio, 2010.

Brennan, Elliot. "China's Strengthening Position on Rare Earths." *The Diplomat.* 2015. http://thediplomat.com/2015/07-chinas-strengthening-position-on-rare-earths/

Brennan, Leah. "James Webb Telescope Project Manager, a Columbia Resident, Plans for NASA Launch." *Baltimore Sun.* 2019. https://www.baltimoresun. com/maryland/howard/cng-ho-bill-ochs-james-webb-telescope-nasa-20190724-xnlflaxg4bh3nfgzaooaecgvhi-story.html

Brewin, Bob. "U.S. Army Awarded Contracts to Russian GPS Jammer Vendor." *Computer World.* 2003. https://www.computerworld.com/article/2581651/u-s-army-awarded-contracts-to-russian-gps-jammer-vendor.html

"Brief Overview of China's Cultural Revolution, A." *Encyclopedia Britannica.* Accessed On: 1 November 2019. https://www.britannica.com/story/ chinas-cultural-revolution

Brito, Ricardo and Paraguassu, Lisandra. "Brazil Wants
China to Invest in Its Infrastructure: Vice President." Reuters.
2019. https://www.reuters.com/article/us-brazil-china/
brazil-wants-china-to-invest-in-its-infrastructure-vice-president-idUSKCN1TE2YH

Broder, Jonathan. "What China's New Missiles Mean For the Future of Aircraft
Carriers." *Newsweek.* 2016. http://www.newsweek.com/2016/02/26/china-
Dongfeng-21d-missile-us-aircraft-carrier-427063.html

Broder, Jonathan. "Why the Next Pearl Harbor Could Happen In Space."
Newsweek. 2016. http://www.newsweek.com/2016/05/13/china-us-space-
wars-455284.html

Brown, Alex. "Russia Wants to Limit U.S. Access to Space Station."
Defense One. 2014. http://www.defenseone.com/threats/2014/05/
russia-wants-limit-us-access-space-station/84379

Brown, Harold. "Is SDI Technically Feasible?" *Foreign Affairs.* 1985. https://www.
foreignaffairs.com/articles/1982–02–01/sdi-technically-feasible

"Brown Water Navy in Vietnam." *Naval Historical Foundation.* 2012. https://www.
navyhistory.org/2012/01/brown-water-navy-in-vietnam/

Browne, Ryan. "U.S. Launches Long-Awaited European Missile Defense
Shield." *CNN.* 2016. http://www.cnn.com/2016/05/11/politics/
nato-missile-defense-romania-poland/

"Building a Brown Water Navy." *American Battlefield Trust.* Accessed
On: 29 October 2019. https://www.battlefields.org/learn/articles/
building-brown-water-navy

Bullough, Oliver. "Vladimir Putin: The Rebuilding of 'Soviet' Russia." BBC. 2014.
https://www.bbc.com/news/magazine-26769481

Burns, Robert. "Space Force Would be By Far the Smallest Military Service." *Air
Force Times.* 2019. https://www.airforcetimes.com/news/your-air-force/2019/03/05/
space-force-would-be-by-far-the-smallest-military-service/

Caldicott, Helen and Eisendrath, Craig. *War in Heaven: The Arms Race in Outer
Space.* New York: New Press, 2007.

Cancian, Mark. "Long Wars and Industrial Mobilization: It Won't Be World War II Again." *War on the Rocks*. 2017. https://warontherocks.com/2017/08/long-wars-and-industrial-mobilization-it-wont-be-world-war-ii-again/

Carril, Rodrigo and Duggan, Mark. "The Impact of Industry on Government Procurement: Evidence from Department of Defense Contracting." *The National Bureau of Economic Research*. 2018. https://www.nber.org/papers/w25160

Carroll, Lauren. "Graham: Russia 'Has an Economy the Size of Italy.'" *Politifact*. 2014. https://www.politifact.com/truth-o-meter/statements/2014/jul/27/lindsey-graham/graham-russia-has-economy-size-italy/

Carter, Jamie. "India on Mars? Despite Failed Moon Landing Expect Orbital Spaceflight and Missions to Venus and Mars." *Forbes*. 2019. https://www.forbes.com/sites/jamiecartereurope/2019/09/05/india-on-mars-historic-moon-landing-will-spur-human-spaceflight-and-missions-to-venus-and-mars/#57ae0ad16ac1

Caryl, Christian. "The Democracy Boondoggle in Iraq." *Foreign Policy*. 2013. https://foreignpolicy.com/2013/03/06/the-democracy-boondoggle-in-iraq/

Caton, Jeffrey. "Joint Warfare and Military Dependence on Space." Federation of American Scientists. Accessed On: 29 October 2019. https://fas.org/spp/eprint/LSN3APP2.htm

Cava, Marco della. "'The Ultimate Space Adventure': Virgin Galactic's Space Tourism Plan Demands Courage and $250K." *USA Today*. 2018. https://www.usatoday.com/story/news/2018/12/07/want-travel-space-virgin-galactic-pack-courage-and-250-000/2140284002

Cebul, Daniel. "New Watchdog Report Decries 'Revolving Door' Between the Pentagon and Defense Contractors." *Defense News*. 2018. https://www.defensenews.com/industry/2018/11/06/new-watchdog-report-decries-revolving-door-between-the-pentagon-and-defense-contractors/

Cha, Victor. *The Impossible State: North Korea, Past and Future*. New York: Harper Collins, 2012.

Cha, Victor D. "Kim Jong Un Is No Reformer." *Foreign Policy*. 2012. http://foreignpolicy.com/2012/08/21/kim-jong-un-is-no-reformer/

Chamberlain, Dianne Pfundstein. "NATO's Baltic Tripwire Forces Won't Stop Russia." *The National Interest.* 2016. htttps://nationalinterest.org/blog/the-skeptics/natos-baltic-tripwire-forces-wont-stop-russia-17074

Champion, Marc. "Iran Is Stuck with China to Finance Its Oil Dreams." *Bloomberg.* 2016. https://www.bloomberg.com/news/articles/2016–10–12/tired-of-china-s-grip-iran-confronts-a-harsh-oil-market-reality

Champion, Marc. "Putin is Trump's Brother from Another Motherland." *Bloomberg.* 2018. https://www.bloomberg.com/news/articles/2018–07–09/putin-is-trump-s-brother-from-another-motherland

Champion, Marc and Spiegel, Peter. "Allies React to U.S. Missile U-Turn." *Wall Street Journal.* 2009. http://www.wsj.com/articles/SB125317801774419047

Chandran, Nyshka. "'Serious' Rivalry Still Drives China-Russia Relations Despite Improving Ties." *CNBC.* 2014. https://www.cnbc.com/2018/09/14/china-russia-ties-more-rivalry-than-alliance.html

"Chandrayaan-1: All You Should Know About India's First Lunar Mission." *India Today.* 2015. http://indiatoday.intoday.in/education/story/chandrayaan-1-indias-first-lunar-mission/1/430041.html

"Chandrayaan 3 to Come Soon; ISRO to Prepare for New Moon Mission." *Kerala Kaumudi.* 2019. https://keralakaumudi.com/en/news/news.php?id=154181&u=chandrayaan-3-to-come-soon-%C2%A0isro-to-prepare-for-new-moon-mission

Chang, Gordon G. "Blue Gold: The Coming Water Wars." *World Affairs.* 2013. http://www.worldaffairsjournal.org/article/blue-gold-coming-water-wars

Chang, Gordon G. *The Coming Collapse of China.* New York: Random House, 2001.

Chang, Gordon G. "U.S. Must Put a Ban on Google Helping China Develop a Global Digital Dictatorship." *Daily Beast.* 2019. https://www.thedailybeast.com/google-snubbed-the-pentagonbut-not-the-chinese-military

Chan, DM. "Pocket-Sized Chinese Thruster Has Big Future." *Asia Times*. 2019. https://www.asiatimes.com/2019/09/article/ pocket-sized-chinese-thruster-has-big-future/

Chatzky, Andrew. "Have Sanctions on Russia Changed Putin's Calculus?" *The Council on Foreign Relations*. 2019. https://www.cfr.org/in-brief/ have-sanctions-russia-changed-putins-calculus

Chaudet, Didier, Parmentier, Florent, and Pélopidas, Benoît. *When Empire Meets Nationalism: Power Politics in the US and Russia*. New York: Routledge, 2016.

Chellaney, Brahma. *Water, Peace, and War: Confronting the Global Water Crisis*. Lanham: Rowman & Littlefield, 2013.

Cheney, Dick. "Annual Report to the President and the Congress." *Department of Defense*. 1992. http://history.defense.gov/Portals/70/Documents/ annual_reports/1992_DoD_AR.pdf

Chen, Stephen. "China and Russia Band Together on Controversial Heating Experiments to Modify the Atmosphere." *South China Morning Post*. 2018. https://www.scmp.com/news/china/science/article/2178214/ china-and-russia-band-together-controversial-heating-experiments

Chen, Stephen. "Electric Vehicles in Deep Space: China Hails Its New Ion Thruster for Rockets as the World's Best." *South China Morning Post*. 2016. https://www.scmp.com/tech/science-research/article/1900601/ electric-vehicles-deep-space-china-hails-its-new-ion-thruster

"China Issues S&T Development Guidelines." *GOV.cn*. 2006. http://www.gov.cn/ english/2006–02/09/content_183426.htm

"'China Lobby,' Once Powerful Factor in U.S. Politics, Appears Victim of Lack of Interest." *New York Times*. 1970. https://www.nytimes.com/1970/04/26/archives/ china-lobby-once-powerful-factor-in-us-politics-appears-victim-of.html

"China Military Power: Modernizing a Force to Fight and Win." *Defense Intelligence Agency*. 2019. https://www.dia.mil/Portals/27/Documents/News/Military%20 Power%20Publications/China_Military_Power_FINAL_5MB_20190103.pdf

"China Plans to Make Shenzhen a 'Better Place' Than Hong Kong." *Straits Times*. 2019. https://www.straitstimes.com/asia/east-asia/china-plans-to-make-shenzhen-a-better-place-than-hong-kong

China Power Team. "How Much Trade Transits the South China Sea?" *China Power*. 2017. https://chinapower.csis.org/much-trade-transits-south-china-sea/

"China Unveils New Commercial Carrier Rockets to Compete with India for Global Market." *Economic Times*. 2019. https://economictimes.indiatimes.com/news/science/china-unveils-new-commercial-carrier-rockets-to-compete-with-india-for-global-market/articleshow/71684151.cms

Chin, Josh. "China's Latest Leap Forward Isn't Just Great—It's Quantum." *Wall Street Journal*. 2016. http://www.wsj.com/articles/chinas-latest-leap-forward-Isn't-just-greatits-quantum-14712695555

Chin, Josh. "China Makes Leap Toward 'Unhackable' Quantum Network." *Wall Street Journal*. 2017. https://www.wsj.com/articles/scientists-make-leap-toward-quantum-network.

Chodakiewicz, Marek Jan. "Business as Usual in Belarus: The Alternance Dance." *The Weichert Report*. 2017. https://theweichertreport.com/2017/04/05/business-As-usual-in-belarus-the-alternance-dance/

Cimpanu, Catalin. "Russia Successfully Disconnected from the Internet." *ZD Net*. 2019. https://www.zdnet.com/article/russia-successfully-disconnected-from-the-internet/

C.K., Louis. "Louis CK Everything is Amazing and Nobody is Happy." YouTube. 2016. https://www.youtube.com/watch?v=nUBtKNzoKZ4&list=RDnUBtKNzoKZ4&start_radio=1

Cho, Joohee. "North Korea Nears Completion of Electromagnetic Pulse Bomb." *ABC*. 2011. http://abcnews.go.com/International/electronic-warfare-north-korea-nears-completion-electromagnetic-pulse/story/?id=13081667

Chow, Brian G. "Op-Ed: Nuclear Vulnerability: In-Orbit Bodyguards Would Help Protect NC3 Satellites from Attacks." *Space News.* 2019. https://spacenews. com/op-ed-nuclear-vulnerability-in-orbit-bodyguards-would-help-protect-nc3-satellites-from-attacks/

Chow, Brian G. and Sokoloski, Henry. "The United States Should Follow France's Lead in Space." *Space News.* 2019. https://spacenews.com/the-united-states-should-follow-frances-lead-in-space/

Chow, Denise. "Shuttle Workers Face Big Layoffs as NASA Fleet Retires." *Space.* https://www.space.com/12391-space-shuttle-program-nasa-workers-layoffs.html

Clapper, James R. "Statement for the Record: Worldwide Threat Assessment of the US Intelligence Community." *Director of National Intelligence.* 2016. https://www.armed-services.senate.gov/imo/media/doc/Clapper_02–09–16.pdf

Clark, Colin. "DepSecDef Work Invokes 'Space Control;' Analysts Fear Space War Escalation." *Breaking Defense.* 2015. http://breakingdefense.com/2015/04/depsecdef-work-invokes-space-control-analysts-fear-space-war-escalation/

Clark, Dave. "Taiwan Envoy Hails US Ties After Trump Inauguration." *Yahoo.* 2017. https://www.yahoo.com/news/taiwan-envoy-hails-us-ties-trump-inauguration-014525629.html

Clark, Stephen. "Ariane 5 Rocket, Galileo Satellites On Launch Pad in French Guiana." *Spaceflight Now.* 2018. https://spaceflightnow.com/2018/07/24/ariane-5-rocket-galileo-satellites-on-launch-pad-in-french-guiana/

Clark, Stephen. "Japan Set to Launch Space Station Resupply Mission." *Spaceflight Now.* 2019. https://spaceflightnow.com/2019/09/09/japan-set-to-launch-space-station-resupply-mission/

Cobb, Wendy Whitman. "How SpaceX Lowered Costs and Reduced Barriers to Space." *The Conversation.* 2019. http://theconversation.com/how-spacex-lowered-costs-and-reduced-barriers-to-space-112586

Codevilla, Angelo M. "The Space Force's Value." *Hoover Institution.* 2018. https://www.hoover.org/research/space-forces-value

Cohen, Ariel. "Trump Moves to Protect America from Electromagnetic Pulse Attack." *Forbes.* 2019. https://www.forbes.com/sites/arielcohen/2019/04/05/whitehouse-prepares-to-face-Emp-threat/#22c88976e7e2

Cohen, Zachary. "Missile Threat Alert for Hawaii a False Alarm; Officials Blame Employee Who Pushed 'Wrong Button.'" *CNN.* 2018. https://www.cnn.com/2018/01/13/politics/hawaii-missile-threat-false-alarm/index.html

Cole, Celia. "Overconsumption is Costing Us the Earth and Human Happiness." *The Guardian.* 2010. https://www.theguardian.com/environment/2010.jun/21/overconsumption-environment-relationships-annie-leonard

Coletta, Damon and Pilch, Frances T. (ed.) *Space and Defense Policy.* New York: Routledge, 2009.

Connell, Michael. "Iran's Military Doctrine." *United States Institute of Peace: The Iran Primer.* Accessed On: 28 October 2019. https://iranprimer.usip.org/resource/irans-military-doctrine

Conradi, Peter. *Who Lost Russia? How the World Entered a New Cold War.* London: Oneworld Publications, Ltd., 2018.

Coon, Charli E. "Why President Bush Is Right to Abandon the Kyoto Protocol." *The* Heritage Foundation. 2001. http://www/heritage.org/research/reports/2001/05/President-bush-right-to-abandon-kyoto-protocol

Coram, Robert. *Boyd: The Fighter Pilot Who Changed the Art of War.* Boston: Little, Brown and Company, 2002.

Correll, Diana Stancy. "How the 'Horse Soldiers' Helped Liberate Afghanistan from the Taliban 18 Years Ago." *Military Times.* 2019. https://www.militarytimes.com/news/your-military/2019/10/18/how-the-horse-soldiers-helped-liberate-afghanistan-from-the-taliban-18-years-ago/

Correspondent, Special. "Gaganyaan, Chandrayaan-3 In Mission Mode, Says ISRO." *The Hindu.* 2020. https://www.thehindu.com/sci-tech/science/gaganyaan-chandrayaan-3-in-mission-mode-says-isro/article30449839.ece

"Council Implementing Regulation (EU) No 961/2014." *Official Journal of the European Union.* 2014. http://eur-lex.europa.eu/legal-content/EN/TXT/?uri=CELEX%3A32014R0961

Covault, Craig. "China Readies Military Station—Launch Coincides With Shuttle Phaseout." *Spaceflight Now.* 2009. http://spaceflightnow.com/news/n0903/02chinastation/

Coyer, Paul. "China's Shellacking: The Rule of Law Collides With Strategic Imperative In the South China Sea." *Forbes.* 2016. http://www.forbes.com/sites/paulcoyer/2016/07/16/chinas-shellacking-the-rule-of-law-collides-with-strategic-imperative-in-the-south-china-sea/#6f3695b56b79

Crawford, Neta C. "United States Budgetary Costs of the Post-9/11 Wars Through FY2019: $5.9 Trillion Spent and Obligated." *Brown University Watson Institute of International and Public Affairs.* 2018. https://watson.brown.edu/costsofwar/files/cow/imcc/papers/2018/Crawford_Costs%200f%20War%20Estimates%20Through%20FY2019%20.pdf

Criss, Doug. "Stephen Hawking Says We've Got About 1,000 Years to Find A New Place to Live." *CNN.* 2016. http://www.cnn.com/2016/11/17/health/hawking-humanity-trnd/

Crouch, David. "Swedish Navy Returns to Vast Underground HQ Amid Russia Fears." *The Guardian.* 2019. https://www.theguardian.com/world/2019/sep/30/swedish-navy-returns-to-vast-underground-hq-amid-russia-fears

Crozier, Brian. *The Rise and Fall of the Soviet Empire.* Rocklin: Forum Publishing, 1999.

"Culture and Negotiations: Russian Style." *Skolkovo School of Business.* Accessed on: December 31, 2019. https://embahs.skolkovo.ru/en/emba-hs/blog/culture-and-negotiations-the-russian-style/

Cumings, Bruce. *Korea's Place in the Sun: A Modern History.* New York: W.W. Norton & Company, 2005.

Cuthbertson, Anthony. "China's Baidu Dethrones Google to Take AI Language Crown." *The Independent.* 2019. https://www.independent.co.uk/life-style/gadgets-and-tech/news/ai-china-baidu-artificial-intelligence-google-a9261691.html

Dagger, Richard and Ball, Terrence. "Communism." *Encyclopedia Britannica.* 2008 https://www.britannica.com/topic/communism#ref539199

Daniels, Jeff and Sorkin, Aaron. "America is Not the Greatest Country in the World Anymore." *Speakola.* 2012. https://speakola.com/movie/jeff-daniels-sorkin-newsroom-2012

Daniels, Jeff and Calia, Mike. "Trump: North Korea's 'Reckless Pursuit' of Nuclear Weapons Could Soon Threaten the US." *CNBC.* 2018. https://www.cnbc.com/2018/01/30/trump-.html

Davenport, Christian. "Elon Musk's SpaceX Settles Lawsuit Against Air Force." *Washington Post.* 2015. https://www.washingtonpost.com/business/economy/elon-musks-spacex-to-drop-lawsuit-against-air-force/2015/01/23/c5e8ff80-a34c-11e4–9f89-561284a573f8_story.html

Davenport, Christian. "ULA Chief Accuses Elon Musk's SpaceX of Trying to 'Cut Corners.'" *Washington Post.* 2014. https://www.washingtonpost.com/business/economy/ula-chief-accuses-elon-musks-spacex-of-trying-to-cut-corners/2014/06/18/a7ca0850-f70d-11e3–8aa9-dad2ec039789_story.html

Davey, Brian. "Shale Euphoria: The Boom and Bust of Sub Prime Oil and Natural Gas." *Credo Economics.* 2016. http://www.credoeconomics.com/shale-euphoria-the-boom-and-bust-of-sub-prime-oil-and-natural-gas/

David, Leonard. "China's Anti-Satellite Test: Worrisome Debris Cloud Circles Earth." Space.com. 2013. http://www.space.com/3415-china-anti-satellite-test-wworrisome-debris-cloud-circles-earth.html

David, Leonard. "China Has Big Plans to Explore the Moon and Mars." Space.com. 2014. http://www.space.com/27893-china-space-program-moon-mars.html

David, Leonard. "Declassified: US Military's Secret Cold War Space Project Revealed." *Space.* 2015. https://www.space.com/31470-manned-orbiting-laboratory-military-space-station.html

David, Leonard. "Is Moon Mining Economically Feasible?" Space.com. 2015. http://www.space.com/28189-moon-mining-economic-feasibility.html

David, Leonard. "Mysterious Actions of the Chinese Satellites Have Experts Guessing." Space.com. 2013. http://www.space.com/2270-china-satellite-activities-perplex-experts.html

Davis, Malcom. "China, the US and the Race for Space." *The Strategist.* 2018. https://www.aspistrategist.org.au/china-the-us-and-the-race-for-space/

Davis, Michael E. and Lee, Ricky J. "Twenty Years After: The Moon Agreement and Its Legal Controversies." *Australian International Law Journal.* 1999. http://www.austlii.edu.au/au/journals/AUIntLawJl/1999/4.pdf

Day, Dwayne A. "And the Sky Full of Stars: American Signals Intelligence Satellites and the Vietnam War." *The Space Review.* 2018. http://www.thespacereview.com/article/3430/1

Day, Dwayne. "Blue Suits and Red Ink." *The Space Review.* 2015. http://www.thespacereview.com/article/2858/1

Day, Dwayne. "Blue Suits In Polar Orbits: the MOL Astronauts (Part I)." *The Space Review.* 2015. http://www.thespacereview.com/article/2849/1

Day, Dwayne A. and Kennedy, III, Robert G. "Soviet Star Wars." *Air and Space.* 2010. https://www.airspacemag.com/space/soviet-star-wars-878185

D'Mello, Gwyn. "China Has Grown Cotton on the Far Side of the Moon, and It's the First Country in the World to Do It." *India Times.* 2019. https://www.indiatimes.com/ technology/science-and-future/china-s-lunar-lander-has-successfully-managed-to-grow-cotton-on-the-far-side-of-the-moon-360483.html

de Selding, Peter D. "Australian, U.S. Forces to Share UHF Satellite Capacity." *Space News.* 2010. http://spacenews.com/australian-us-forces-share-uhf-satellite- capacity-agreement-involves-us-mobile-user/

de Selding, Peter D. "Chinese Group to Buy Israel's Spacecom Satellite Operator for $285 Million." *Space News.* 2016. https://spacenews.com/chinese-group-to-buy-israels-spacecom-satellite-operator-for-285-million/

de Selding, Peter B. "New U.S. Space Mining Law's Treaty Compliance Depend on Implementation." *Space News.* 2015. https://spacenews.com/u-s-commercial-space-acts-treaty-compliance-may-depend-on-implementation/

Degaut, Marcos. "Why Washington Doesn't Get Brazil." *The National Interest.* 2016. https://nationalinterest.org/feature/why-washington-doesnt-get-brazil-18473

Demick, Barbara. "A One-Hour Commute to Another World." *L.A. Times.* 2006. http://articles.latimes.com/2006/feb/28/world/fg-commute28/2

DeMint, Jim. "Ex-Im Bank is Crony Capitalism Run Amok." *The Heritage Foundation.* 2015. https://www.heritage.org/trade/commentaryex-im-bank-crony-capitalism-run-amok

Deptula, David A., LaPlante, William A., and Haddick, Robert. "Modernizing U.S. Nuclear Command, Control, and Communications." *The Mitchell Institute for Aerospace Studies.* 2019. http://www.mitchellaerospacepower.org/nc3

DiChristopher, Tom and Schoen, John W. "OPEC States That Wanted Production Cuts Buckle Under the New Oil Order." *CNBC.* 2016. http://www.cnbc.com/2016/06/01/opec-states-that-wanted-production-cuts-buckle-under-the-new-oil-order.html

Dillow, Clay. "Is China's Race to Space a Military Ploy?" *CNBC.* 2016. http://www.cnbc.com/2016/02/18/chinas-space-missions-in-2016-tied-to-military-ambitions.html

Dinerman, Taylor. "Iran's Satellite: A Look at the Implications." *The Space Review.* 2004. http://www.thespacereview.com/article/250/1

Dobriansky, Paula J. "When Trump Supporter Bolsonaro Visits D.C., It Will Reset Brazil's Relationship with U.S." *Miami Herald.* 2019. https://www.miamiherald.com/opinion/op-ed/article227429304.html

Dockrill, Peter. "America Really is In the Midst of a Rising Anxiety Epidemic." *Science Alert.* 2018. https://www.sciencealert.com/americans-are-in-the-midst-of-an-anxiety-epidemic-stress-increase

Dolman, Everett C. "A Debate About Weapons In Space: For U.S. Military Transformation and Weapons In Space." *The SAIS Review of International Affairs.* Vol. 26, No. 1, 2006.

Dolman, Everett C. *Astropolitik: Classical Geopolitics in the Space Age.* New York: Frank Cass Publishers, 2002.

Dolman, Everett Carl. *Pure Strategy: Power and Principles in the Space and Information Age.* New York: Frank Cass Publishers, 2005).

Donnelly, John M. "The Other North Korean Threat: Chemical and Biological Weapons." *Roll Call.* 2018. https://www.rollcall.com/news/politics/the-other-north-korea-threat-chemical-and-biological-weapons

Douhet, Giulio. *Command of the Air.* Washington, D.C.: Office of Air Force History, 1983.

Dueck, Colin. *The Obama Doctrine: American Grand Strategy Today.* New York: Oxford University Press, 2015.

Dueck, Colin. "The Return of Geopolitics." *Real Clear World.* 2013. https://www.realclearworld/com/articles/2013/07/27/the_return_of_geopolitics_105345.html

Dugin, Alexander. *Eurasian Mission: An Introduction to Neo-Eurasianism.* London: Arktos Media, Ltd., 2014.

Dugin, Alexander. *Last War of the World-Island: The Geopolitics of Contemporary Russia.* London: Arktos Media, Ltd., 2015.

Dujmovic, Jurica. "This is What the U.S. Military's 'Space Force' Could Look Like." *Market Watch.* 2018. https://www.marketwatch.com/story/this-is-what-the-us-militarys-space-force-would-look-like-2018–04–18

Editors, History.com. "United States Freezes Japanese Assets." *History.* 2009. https://www.history.com/this-day-in-history-/united-states-freezes-japanese-assets

Eckstein, Megan. "Navy to Field High-Energy Laser Weapon, Laser Dazzler on Ships This Year as Development Continues." *USNI News.* 2019. https://news.usni.org/2019/05/30/navy-t00-field-high-energy-laser-weapon-laser-dazzler-on-ships-this-year-as-development-continues

"EELV: The Right to Compete." *SpaceX.* 2014. http://www.spacex.com/press/2014/04/29/eelv-right-compete

Efron, Shira, Shatz, Howard, J., et al. "The Evolving Israel-China Relationship." *RAND Corporation.* 2019. https://www.rand.org/content/dam/rand/pubs/research_reports/RR2600/RR2641/RAND_RR2641.pdf

Egorov, Boris. "What Weapons Did the Soviet Union Plan to Use in a Space War?" *Russia Beyond.* 2018. https://www.rbth.com/science-and-tech/327998-weapons-soviet-union-space

Eisenhower, Dwight D. "Eisenhower Farewell Address (Full)." YouTube. http://www.youtube.com/watch?v=CWilYW_fBfY

Eisenhower, Dwight D. "Radio and Television Address to the American People On 'Our Future Security.'" *The American Presidency Project.* 1957. http://www.presidency.ucsb.Edu/ws/?pid=10950

Eisenhower, Dwight D. "Military-Industrial Complex Speech, Dwight D. Eisenhower, 1961." *The Avalon Project.* https://avalon.law.yale.edu/20th_century/eisenhower001.asp

Ellyatt, Holly. "Are Russia and China the Best of Friends Now? It's Complicated, Analysts Say." *CNBC.* 2019. https://www.cnbc.com/2019/09/27/russia-and-chinas-relationship-how-deep-does-it-go.html

Embury-Dennis, Tom. "Vladimir Putin Says Russia's Borders 'Do Not End Anywhere.'" *The Independent.* 2016. http://www.independent.co.uk/news/world/europe/putin-Russia-border-do-not-end-anywhere-comments-quote-eu-us-tensions-a7438686.html

End, Rae Botsford. "China's Lunar Mission Chang'e 5-T1 Sets Stage For Future Possibilities." *Spaceflight Insider.* 2015. http://www.spaceflightinsider/com/missions/chinas-lunar-mission-sets-stage-for-future/

Episkopos, Mark. "Russia Has 1 'Weapon' NATO Can't Easily Defeat." *The National Interest.* 2019. https://nationalinterest.org/blog/buzz/ russia-has-1-weapon-nato-cant-easily-defeat-55387

Erwin, Sandra. "Commercial Satellite Roaming Possible with Existing Military Terminals, Experiment Shows." *Space News.* 2019. https://spacenews.com/ commercial-satellite-roaming-possible-with-existing-military-terminals-experiment-shows/

Erwin, Sandra. "U.S. Army 'Space Week' to Shine Light on Why Soldiers Care About Satellites." *Space News.* 2017. https://spacenews. com/u-s-army-space-week-to-shine-light-on-why-soldiers-care-about-satellites/

Eschner, Kat. "How Industrial Espionage Started America's Cotton Revolution." *Smithsonian Magazine.* 2017. https://www.smithsonianmag.com/smart-news/ how-industrial-espionage-started-americas-cotton-revolution-180967608

Escobar, Pepe. "Trump Will Try to Smash the China-Russia-Iran Triangle . . . Here's Why He Will Fail." *South China Morning Post.* 2017. https://www.scmp/com/week-asia/opinion/article/2064005/ trump-will-try-smash-china-russia-iran-triangle-heres-why-he-will

Etherington, Darrell. "NSLComm's First Spring-Loaded Expanding Antenna Satellite is Headed to Space." *Tech Crunch.* 2019. https://techcrunch.com/2019/07/03/ nslcomms-first-spring-loaded-expanding-antenna-satellite-is-headed-to-space/

"F-35 Jet Fighters to Take Integrated Avionics to a Whole New Level." *Military & Aerospace Electronics.* 2003. https://www.militaryaerospace.com/computers/ article/16709227/f35-jet-fighters-to-take-integrated-avionics-to-a-whole-new-level

"Factbox-Strait of Hormuz: The World's Most Important Oil Artery." Reuters. 2019.https://www.reuters.com/article/usa-iran-oil-strait/ factbox-strait-of-hormuz-the-worlds-most-important-oil-artery-idUSL5N2254EM

"The Failed Attempt to Avert War with Japan, 1941." *Association for Diplomatic Studies and Training.* 2013. https://adst.org/2013/11/the-failed-attempts-to-avert-war-with-japan-1941/?fbclid=IwAR3kkbL1PAuPe400nood-eDdepz8EkpbfnWFJg wtbEtnyI08tRuBKC-Iac8

Farber, Madeline. "India Just Launched 20 Satellites In a Single Mission." *Fortune Magazine.* 2016. http://fortune.com/2016/06/22/india-20-satellites-single-mission/

Federico, Giulia. "OSPEK: After ISS—The Next Mir?" *Space Safety Magazine.* 2013. http://www.spacesafetymagazine.com/space-exploration/international-space-station/iss-mir/

Feldscher, Jacqueline and Bender, Bryan. "FAA Put on Notice That It Needs to Catch Up with Spaceflight Boom." *Politico.* 2019. https://www.politico.com/newsletters/politico-space/2019/11/08/faa-put-on-notice-that-it-needs-to-catch-up-with-spaceflight-boom-487620

Ferguson, Niall. *Civilization: Is the West History?* Directed by Pennink, Adrian. London, 2011.

Fernholz, Tim. "The US is Worried That a Russian Satellite is Really a Weapon." *QZ.* 2018. https://qz.com/1361037/the-us-is-worried-that-a-russian-satellite-is-really-a-weapon/

Feulner, Edwin J. "The Strategic Defense Initiative at 34." Heritage Foundation. 2017. https://www.heritage.org/missile-defense/commentary/the-strategic-defense-initiative-34

"50 Years Ago: NASA Benefits from Manned Orbiting Laboratory Cancellation." *NASA.* 2019. https://www.nasa.gov/feature/50-years-ago-nasa-benefits-from-mol-cancellation

"First Mover," *Investopedia.* Accessed on: 15 September 2018. https://www.investopedia.com/terms/f/firstmover.asp

Fisher, Jr., Richard D. "China's Close Call." *Wall Street Journal.* 2008. http://www.wsj.com/news/articles/SB12239460905385099?mg=ren064-wsj&url=http%3A%2F%2Fdefault%2Farticle%2FSB122539460905385099.html

"5 Tensions Between Cybersecurity and Other Public Policy Concerns" in *At the Nexus of Cybersecurity and Public Policy: Some Basic Concepts and Issues.* National Research Council (Washington, D.C.: The National Academies Press, 2014). https://www.nap.edu/read/18749/chapter/7

Floyd, Paul. "The Militarization of Space," *STRATFOR Media Center* video, 2:16, 2015. http://www.stratfor.com/video.militarization-space

Fluxspiel. "2012—Eric Schmidt and Peter Thiel." YouTube. 2015. https://www.youtube.com/watch?v=PsXFwy6gG_4

Forest, Dave. "This Region is China's Next Target for Resource Deals." *Oil Price.* 2017. https://oilprice.com/Metals/Commodities/This-Region-Is-Chinas-Next-Target-For-Resource-Deals.html

Forster, Katie. "Japan is Launching an Asteroid Mining Space Program." *Business Insider.* 2014. https://www.businessinsider.com/japan-is-launching-an-asteroid-mining-space-program-2014–9

Foust, Jeff. "Commerce Department Seeks to Increase American Space Industry's Global Competitiveness." *Space News.* 2019. https://spacenews.com/commerce-department-seeks-to-increase-american-space-industrys-global-competitiveness/

Foust, Jeff. "House Passes Commercial Space Bill." *Space.* 2015. https://www.space.com/31147-house-passes-commercial-space-bill.html

Foust, Jeff. "India Confirms Plans for Second Lunar Lander Mission." *Space News.* 2020. https://spacenews.com/india-confirms-plans-for-second-lunar-lander-mission/?fbclid=IwAR3ikFkoMWW1wkjqAxMjc ctVa97-wkQ9B1Qut6mLFkQjRp-NLPTV70Unc7s

Foust, Jeff. "Key House Appropriator Remains Skeptical About Artemis." *Space News.* 2019. https://spacenews.com/key-house-appropriator-remains-skeptical-about-artemis/

Foust, Jeff. "NASA and JAXA Reaffirm Intent to Cooperate in Lunar Exploration." *Space News.* 2019. https://spacenews.com/nasa-and-jaxa-reaffirm-intent-to-cooperate-in-lunar-exploration/

Foust, Jeff. "Russian Official Sounds Skeptical Note About Gateway." *Space News.* 2018. https://spacenews.com/russian-official-sounds-skeptical-note-about-gateway/

Frantzman, Seth J. "Space: Israel's Final Frontier." *The Jerusalem Post.* 2019. https://www.jpost.com/Magazine/Israels-final-frontier-567448

Fravel, M. Taylor. "Why India Did Not 'Win' the Standoff with China." *War on the Rocks.* 2017. https://warontherocks.com/2017/09/why-india-did-not-win-the-standoff-with-china/

Freedberg, Jr., Sydney J. "China's Fear of US May Tempt Them to Preempt: Sinologists." *Breaking Defense.* 2013. http://www.breakingdefense.com/2013/10/chinas-Fear-of-us-may-tempt-them-to-preempt-sinologists/

Freedberg, Jr., Sydney J. "McCain Compromises On RD-180s: Are Rocket Wars Over?" *Breaking Defense.* 2016. http://breakingdefense.com/2016/06/mccain-Compromises-on-rd-180s-in-ndaa-are-russian-rocket-wars-finally-over/

Freedberg, Jr., Sydney J. "US 'Gets Its Ass Handed to It' in Wargames: Here's a $24 Billion Fix. 2019. https://breakingdefense.com/2019/03/us-gets-its-ass-handed-to-it-in-wargames-heres-a-24-billion-fix/

Freedman, Lawrence. *Deterrence.* Malden: Polity Press, 2004.

Friedman, George. *Flash Points: The Emerging Crisis in Europe.* New York: Doubleday, 2015.

Friedman, George. *The Next 100 Years: A Forecast for the 21ˢᵗ Century.* New York: Doubleday, 2009.

Friedman, Uri. "The Word That Derailed the Trump-Kim Summit." *The Atlantic.* 2018. https://www.theatlantic.com/international/archive/2018/05/libya-trump-kim/561158/

Fu-Lee, Kai. *AI Superpowers: China, Silicon Valley, and the New World Order* (New York: Houghton Mifflin, 2019).

Fukuyama, Francis. *The End of History and the Last Man.* New York: Free Press, 1992.

Gabbatt, Adam. "Barack Obama Sends Letter to Kim Jong-il." *The Guardian.* 2009. https://www.theguardian.com/world/2009/dec/16/obama-letter-kim-jong-il

Gaddis, John Lewis. *The Cold War: A New History.* New York: Penguin Press, 2006.

Gaddis, John Lewis. *Surprise, Security and the American Experience.* Cambridge: Harvard University Press, 2004.

Gady, Franz-Stefan. "India to Test First Missile Capable of Hitting China." *The Diplomat.* 2016. https://thediplomat.com/2016/12/india-to-test-fire-nuclear-missile-capable-of-hitting-china/

Gaier, Rodrigo Viga. "Brazil's Bolsonaro Taps Astronaut, Courts Judge for Cabinet." Reuters. 2018. https://www.reuters.com/article/us-brazil-politics/brazils-bolsonaro-taps-astronaut-courts-judge-for-cabinet-idUSKCN1N52Q3

Gallagher, Sean. "Trump's Missile Defense Strategy: Build Wall in Space, Make Allies Pay for It." *Ars Technica.* 2019. https://arstechnica.com/tech-policy/2019/01/new-dod-missile-defense-strategy-star-wars-ii-the-wrath-of-trump/

Galeon, Dom. "China Claims They Have Actually Created an EM Drive." *Futurism.* 2017. https://futurism.com/china-claims-they-have-actually-created-an-em-drive

Gallman, Philip G. *Green Alternatives and National Energy Strategy: The Facts Behind the Headlines.* Baltimore: The Johns Hopkins University Press, 2011.

Gangale, Thomas. *The Development of Outer Space: Sovereignty and Property Rights in International Space Law.* Santa Barbara: ABC-CLIO, Inc, 2009.

Garretson, Peter. "Guess What Could Be Totally Missing From the New U.S. President's Intel Briefing." *War Is Boring.* 2016. https://warisboring.com/guess-what-could-be-totally-missing-from-the-new-u-s-presidents-intel-briefing-9ce4881643e5#.a5c9aqxae

Garun, Natt. "Israeli Team Will No Longer Send a Second Spacecraft to the Moon." *The Verge.* 2019. https://www.theverge.com/2019/6/25/18758794/spaceil-beresheet-2-moon-spacecraft-new-objective-challenge

Gambrell, John. "US Says Iran Took Mine Off Tanker; Iran Denies Involvement." *Military Times.* 2019. https://www.militarytimes.com/news/your-military/2019/06/14/us-says-iran-removed-unexploded-mine-from-oil-tanker/

Gannon, Megan. "China's President Xi Wants More Military Use of Space: Report." (2014). http://www.space.com/25517-china-space-technology.html

Garza, Alejandro De La. "How Historians Are Reckoning with the Former Nazi Who Launched America's Space Program." *Time.* 2019. https://time.com/5627637/nasa-Nazi-von-braun/

Gavin, James M. *War & Peace In the Space Age.* New York: Harper & Brothers, 1958.

Gay, Louise. "Climate Change: Temperatures to Increase 6C By End of Century." *Telegraph.* 2009. http://www.telegraph.co.uk/news/earth/environment/Climatechange/6593389/Climate-change-temperatures-to-increase-6C-by-end-of-century.html

Gearty, Robert. "Kim Jong Un Threatens to Renew Testing Nuclear Weapons, Long-Range Missiles." *Fox News.* 1 January 2020. https://www.foxnews.com/world/kim-jong-un-testing-nuclear-weapons-long-range-missiles

Gertz, Bill. "Inside the Ring: Pentagon's Air-Sea Battle." *Washington Times.* 2011. http://www.washingtontimes.com/news/2011/nov/9/inside-the-ring-584291397/?page=all.

Gertz, Bill. "PLA On Cyberwarfare Buildup." *Washington Times.* 2016. http://www.washingtontimes.com/news/2016/feb/17/inside-the-ring-china-plans-cyberwarfare-force-to-/

Gertz, Bill. "U.S. Opposes China-Russia Space Arms Treaty." *Washington Times.* 2015. https://www.washingtontimes.com/news/2015/dec/2/inside-the-ring-us-opposes-china-russia-space-arms/

Ghoshal, Debalina. "North Korea's Toxic Space Program." *The Gatestone Institute.* 2018. https://www.gatestoneinstitute.org/13157/north-korea-space-program

Gill, Timothy M. "Why the Power Elite Continues to Dominate American Politics." *The Washington Post.* 2018. https://www.washingtonpost.com/outlook/2018/12/24/why-power-elite-continues-dominate-american-politics/

Gilpin, Robert. *The Political Economy of International Relations.* Newark: Princeton University Press, 1987.

Glaser, Bonie S. "Chinese Perceptions of U.S. Decline and Power." *The Jamestown Foundation.* 2009. http://www.jamestown.org/programs/chinabrief/single/?tx_ttnews%5Btt_news%5D=35241&cHash=db. .#.V-gZhzKZN7N

Glass, Andrew. "George W. Bush Receives Bin Laden Memo: Aug. 6, 2001." *Politico.* 2009. https://www.politico.com/story/2009/08/george-w-bush-receives-bin-laden-memo-aug-6-2001–025834

Goble, Paul. "Plan to Settle Two Million Central Asians in Russian Far East Triggers Anger There." *Window on Eurasia—New Series.* 2016. http://windowoneurasia2.blogspot.com/2016/07/plan-to-settle-two-million-central.html

Goble, Paul. "Russians in Far East 'Ethnicizing' Distinctive Regional Identities, Khabarovsk Scholar Says." *Window on Eurasia—New Series.* 2016. http://windowoneurasia2.blogspot.com/2016/07/russians-in-far-east-ethnicizing.Html

Goble, Paul. "There is an 'Operation Trust' in Belarus—But Not What Regnum Editor Describes." *Window on Eurasia—New Series.* 2017. http://windowoneurasia2.blogspot.com/2017/03/there-is-operation-trust-in-belarus-but.html

Goldman, David P. "The Digital Age Produces Binary Outcomes." *American Affairs Journal.* Spring 2017, Volume I, Number 1. https://americanaffairsjournal.org/2017/02/digital-Age-produces-binary-outcomes/

Goldman, David P. "Dumb and Dumber: When Neocons and Obama Liberals Agree." *Tablet Magazine.* 2013. https://www.tabletmag.com/jewish-news-and-politics/132459/dumb-and-dumber

Goldman, David P. "The Norm is NOT Democracy—The Norm is Extinction." PJ Media. 2018. https://pjmedia.com/spengler/norm-is-not-democracy-norm-extinction/

Goldman, David P. "US Ban Won't Derail Huawei's European 5G Rollout." *Asia Times.* 2019. https://www.asiatimes.com/2019/05/article/us-ban-wont-derail-huaweis-european-5g-rollout/

Goldstein, Joseph. "Kenyans Say Chinese Investment Brings Racism and Discrimination." *The New York Times.* 2018. https://www.nytimes.com/2018/10/15/world/africa/kenya-china-racism.html

Gosnold, "Space Situational Awareness from the Paris Airshow." *Satellite Observation.* 2019. https://satelliteobservation.net/2019/06/20/space-situational-awareness-news-from-the-paris-airshow/

Goswami, Namrata. "Waking Up to China's Space Dream." *The Diplomat.* 2018. https://thediplomat.com/2018/10/waking-up-to-chinas-space-dream/

Goward, Dana A. "Russia, China Alliance on Navigation Satellites Threatens GPS." *National Defense.* 2019. https://www.nationaldefensemagazine.org/articles/2019/8/23/viewpoint-russia-china-alliance-on-navigation-satellites-threatens-gps

Graham, William R. "North Korea Nuclear EMP Attack: An Existential Threat." *38 North.* 2017. https://www.38north.org/2017/06/wgraham060217

Grammaticas, Damian. "China Establishes 'Air-Defence Zone' Over East China Sea." BBC. 2013. http://www.bbc.com/new/world-asia-25062525

Grammaticas, Damian. "Chinese Colonialism?" BBC *News.* 2012. http://www.bbc.com/news/world-asia-18901656

Grammaticas, Damian. "Inside North Korea's Space Centre." BBC *News.* 2012. http://www.bbc.com/news/world-asia-china-17684617

Griffin, Jennifer. "Military Eyeing Former Cold War Mountain Bunker as 'Shield' Against EMP Attack?" *FOX News.* 2015. http://www.foxnews.com/politics/2015/05/05/military-eyeing-former-cold-war-mountain-bunker-as-shield-against-emp-attack.html

Griffith University, "Quantum 'Spooky Action at a Distance,' Becoming Practical." *Phys.org.* 2018. https://phys.org/news/2018–01-quantum-spooky-action-distance.html

Grossman, David. "Actually, About that Whole 'We're Going to the Moon in 2024 Thing . . .'" *Popular Mechanics.* 2019. https://www.popularmechanics.com/space/moon-mars/a29492598/congress-nasa-2024-moon-return/

Grossman, Lisa. "US Probe Enters Mars Orbit, Indian Craft Close Behind." *New Scientist.* 2014. https://www.newscientist.com/article/dn26244-us-probe-enters-mars-orbit-indian-craft-close-behind/

Grow, Brian, Grey, Stephen, and Amin, Roman. "In Murky Deal with Russia, Big Profit for a Tiny Florida Firm." Reuters. 2014. https://www.reuters.com/investigates/special-report/comrade-capitalism-rocket-men/

Grush, Loren. "Japanese Spacecraft Successfully Touches Down on an Asteroid, Grabbing a Sample of Dust" *Inverse.* 2019. https://www.theverge.com/2019/2/21/182345782/jaxa-hayabusa-2-ryugu-asteroid-sample-return-mission

Grush, Loren. "Lawmaker Who Helps Fund NASA Questions the Agency's Need to Get to the Moon by 2024." *The Verge.* 2019. https://www.theverge.com/2019/10/16/20917171/nasa-artemis-program-moon-jose-serrano-jim-bridenstine-budget-2024

Gruss, Mike. "Air Force: Atlas 5 Will be Grounded If RD-180 is Found to Violate U.S. Sanctions." *Space News.* 2016. http://spacenews.com/one-of-several-u-s-senators-moscow-barred-from-russia-in-retaliation-for-the-2014-sanctions/

GuruFocus. "Climate Change Spurs Growing Investment In Water Stocks." *Forbes.* 2016. http://www.forbes.com/sites/gurufocus/2016/04/05/gurus-invest-in-water-stocks-as-risk-to-resource-grows/#2dc4d2161652

Grier, Peter. "Are North Korea's Nuclear Weapons Small Enough to Fit a Ballistic Missile." *The Christian Science Monitor.* 2013. https://www.csmonitor.com/USA/Politics/DC-Decoder/2013/0408/Are-North-Korea-s-nuclear-weapons-small-enough-to-fit-a-ballistic-missile

Groll, Elias. "North Korean Missiles Just Keep Getting Better." *Foreign Policy.* 2019. https://foreignpolicy.com/2019/10/03/north-korean-missiles-just-keep-getting-better/

Grygiel, Jakub J. and Mitchell, A. Wess. *The Unquiet Frontier: Rising Rivals, Vulnerable Allies, and the Crisis of American Power.* Princeton: Princeton University Press, 2016.

Gwynne, Peter. "The Cooling World." *Newsweek*. 1975.

Hadhazy, Adam. "Artificially Intelligent Rockets Could Slash Launch Costs." *Space*. 2011. https://www.space.com/11181-rocket-launches-artificial-intelligence-japan.html

Hadley, Greg. "'Rocket Man': Trump Uses Belittling New Nickname for Kim Jong Un in UN Speech." *McClatchy DC*. 2017. https://www.mcclatchydc.com/latest-news/article174107001.html

Haines, John R. "Schrödinger's Nuke: How Iran's Nuclear Weapons Program Exists—and Doesn't Exist—at the Same Time." *Foreign Policy Research Institute*. 2015. https://www.fpri.org/article/2015/05/schrodingers-nuke-how-irans-nuclear-weapons-program-exists-and-doesnt-exist-at-the-same-time/

Hallonquist, Al. "The MOL-Men Come into the Light." *Air and Space Magazine*. 2015. https://www.airspacemag.com/daily-planet/mol-men-come-light-180957353

Hambling, David. "EmDrive: China's Radical New Space Drive." *Wired Magazine*. 2013. http://www.wired.co.uk/article/emdrive-and-cold-fusion

Hambling, David. "Why DARPA is Betting a Million Bucks on an 'Impossible' Space Drive." *Popular Mechanics*. 2018. https://www.popularmechanics.com/space/rockets/a24219132/darpa-emdrive/

Hammes, T.X. "Offshore Control is the Answer." *Proceedings*. 2012. https://www.usni.org/magazines/proceedings/2012/December/offshore-control-answer

Harding, Luke and Traynor, Ian. "Obama Abandons Missile Defence Shield In Europe." *The Guardian*. 2009. https://www.theguardian.com/world/2009/sep/17/missile-defence-shield-barack-obama

Harney, Alexandra, Sevastopulo, Demetri, and Alden, Edward. "Top General Warns US Over Attack." *Financial Times*. 2005. https://www.ft.com/content/28cfe55a-f4a7-11d9-9dd1-00000e2511c8

Hartung, William D. "Pentagon Socialism: The Pentagon's Plan to Dominate the Economy." *Salon*. 2018. https://www.salon.com/2018/11/12/pentagomn-socialism_partner/

Harris, Francis. "Beijing Secretly Fires Lasers to Disable US Satellites." *The Telegraph*. 2006. http://www.telegraph.co.uk/news/worldnews.1529864/Beijing-secretly-fires-lasers-to-disable-US-satellites.html

Hayward, Steven F. "Bureaucracy in America Now Goes All the Way Down." *Forbes*. 2014. https://www.forbes.com/sites/stevenhayward/2014/01/20/bureaucracy-in-america-now-goes-all-the-way-down/#1bb9d72db1c0

Graham-Harrison, Emma, Luhn, Alec, Walker, Shaun, Walker, Shaun, and Rice-Oxley, Mark. "China and Russia: The World's New Superpower Axis?" *The Guardian*. https://www.theguardian.com/world/2015/jul/07/china-russia-superpower-axis

Harrison, Todd. "The Future of MILSATCOM." *The Center for Strategic and Budgetary Assessments*. (2013). Pp. 4–46.

Hays, Brooks. "New U.S. Space Mining Law May Violate International Treaty." *UPI*. 2015. http://www.upi.com/Science_News/2015/11/27/New-US-space-mining-law-may-violate-international-treaty/8751448634436/

Hendrickx, Bar. "Russia's Secret Satellite Builder." *The Space Review*. 2019. http://www.thespacereview.com/article/3709/1

Henry, Caleb. "Brazil to Order Second Dual Civil-Military Communications Satellite." *Space News*. 2019. https://spacenews.com/brazil-to-order-second-dual-civil-military-communications-satellite/

Henry, Caleb. "Spacecom Begins Service with a Borrowed Satellite Rebranded Amos-7." *Space News*. 2017. https://spacenews.com/spacecom-begins-service-with-a-borrowed-satellite-rebranded-amos-7/

Herschensohn, Bruce. *Taiwan: The Threatened Democracy*. Los Angeles: World Ahead Publishing, Ltd., 2006.

Herszenhorn, David M. "What Is Putin's 'New Russia'?" *The New York Times*. 2014. http://www.nytimes.com/2014/04/19/world/europe/what-is-putins-new-russia.html?_r=0

Hewitt, John. "China is Going to Strip Mine the Moon for Helium-3 Fusion Fuel." *Extreme Tech.* 2015. http://www.extremetech.com/ extreme/197784-china-is-going-to-mine-the-moon-for-helium-3-fusion-fuel

Hey, Nigel. *The Star Wars Enigma: Behind the Scenes of the Cold War Race for Missile Defense.* Washington, D.C.: Potomac Books, Inc., 2006.

Hill, Fiona. "Energy Empire: Oil, Gas and Russia's Revival." *The Foreign Policy Centre.* 2004. https://www.brookings.edu/wp-content/uploads/2016/06/20040930. pdf

Hill, Jeffrey. "Startup Space Spotlight: Israel's NSLComm is Supercharging Smallsat Bandwidth." *Satellite Today.* 2018. https://www.satellitetoday.com/ business/2018/05/29/startup-israels-nslcom-is-supercharging-smallsat-bandwidth/

Hirst, Tomas. "Russia Threatens to Cut Off Gas Supplies to Europe By Friday." *BusinessInsider Australia.* 2015. http://www.businessinsider.com.au/ russia-gazprom-risks-another-gas-standoff-with-ukraine-2015–2

Hitchens, Christopher. "Imperialism: Superpower Dominance, Malignant and Benign." *Slate.* 2002. https://slate.com/news-and-politics-/2002/12/american-imperialism-then-and-now.html

Hitchens, Theresa. "Pentagon Eyes Military Space Station." *Breaking Defense.* 2019. https://breakingdefense.com/2019/07/pentagon-eyes-military-space-station/

Hoare, Callum. "WW3: Why US Feared Soviet Nuclear Missile Attack from SPACE After Secret Launch." *Express.* 2019. https://www.express.co.uk/news/ world/1128487/ww3-nasa-sputnik-us-soviet-union-nuclear-space-war-eisenhower-khrushchev-spt

Hollings, Alex. "No Space Marines: Pentagon Document Shows Space Force Plans to Draw Troops from Army, Navy, and Air Force." *SOFREP.* 2018. https://sofrep. com/news/no-space-marines-pentagon-document-shows-space-force-plans-to-draw-troops-from-army-navy-and-air-force/

Hong, Brendon. "China's Looming Land Grab in Outer Space." *Daily Beast.* 2018. https://www.thedailybeast.com/chinas-looming-land-grab-in-outer-space?ref=scroll

"Hong Kong," *2019 Index of Economic Freedom.* Accessed On: 11 November 2019. https://www.heritage.org/index/country/hongkong

Hopkirk, Peter. *The Great Game: The Struggle for Empire in Central Asia.* New York: Kodansha, USA, 1994.

Houser, Kristin. "Americans Don't Really Care About Reaching Mars, Says Poll." *Futurism.* 2019. https://futurism.com/the-byte/americans-dont-care-mars-poll

Howard, Brian Clark. "Aral Sea's Eastern Basin Is Dry For First Time in 600 Years." *National Geographic.* 2014. http://news.nationalgeographic.com/news/2014/10/141001-aral-sea-shrinking-drought-water-environment

Howell, Elizabeth. "Baikonur Cosmodrome: Russian Launch Complex." Space.com. 2016. http://www.space.com/33947-baikonur-cosmodrome.html

Howell, Elizabeth. "India Looks Beyond Moon to Mars, Venus and Astronaut Missions." *Space.* 2019. https://www.space.com/india-moon-mars-venus-exploration-plans-Beyond-chandrayaan-2.html

Howell, Elizabeth. "Japan's Brainy Epsilon Rocket Launching on 1st Test Flight Tuesday." Space.com. 2016. http://www.space.com/22530-japan-epsilon-smart-rocket-test-launch.html

Howell, Elizabeth. "JAXA: Japan's Aerospace Exploration Agency." Space.com. 2016. http://www.space.com/22672-japan-aerospace-exploration-agency.html

"How Donald Trump Thinks About Foreign Policy." *Vox.com.* YouTube video. 2016. https://www.youtube.com/watch?v=xpExmP7ZAyQ

Hsiung, James C, *The Xi Jinping Era: His Comprehensive Strategy Toward the China Dream.* New York: CN Times Books, 2015.

Huang, Cary. "US, Japan, India, Australia . . . is Quad the First Step to an Asian Nato?" *The South China Morning Post.* 2017. https://www.scmp.com/week-asia/opinion/article/2121474/us_japan-india-australia-quad-first-step-asian-nato

Hughes, Jr., Wayne P. "Build a Green-Water Fleet." *Proceedings.* 2018. https://www.usni.org/magazines/proceedings/2018/June-build-green-water-fleet

Hui, Mun Dong. "North Korean Propaganda Promotes EMP Attacks Using Nuclear Weapons." *Daily NK.* 2018. https://www.dailynk.com/english/north-korean-propaganda-Promotes-emp-attacks-using-nuclear-weapons/

Huizhong, Wu. "India's First Homegrown 'Space Shuttle' Test Successful." *CNN.* May. http://www.cnn.com/2016/05/23/asia/india-space-shuttle-success/

Huntington, Samuel P. "The Clash of Civilizations?" *Foreign Affairs.* 1993. https://www.foreignaffairs.com/articles/united-states-1993–06–01/clash-civilizations

Hutchinson, Benji. "Left of Boom—Defeating the Threat Among Us." *NEC Today.* 2017. https://nectoday.com/left-of-boom-defeating-the-threat-among-us/#

Huth, Paul K. *Extended Deterrence and the Prevention of War.* New Haven: Yale University Press, 1988.

"India to Overtake Japan to Become 3rd Largest Economy in 2025." *The Economic Times.* 2019. https://economictimes.indiatimes.com/news/economy/indicators/india-to-overtake-japan-to-become-3rd-largest-economy-in-2025/articleshow/70193869.cms

India Today Web Desk. "Remembering Chandrayaan-1 Launch: When India Touched Moon 11 Years Ago." *India Today.* 2019. https://www.indiatoday.in/science/story/remembering-chandrayaan-1-launch-when-india-touched-moon-11-years-ago-1611686–2019–10–22

Insinna, Valerie. "Rep. Mike Turner on Why He's Softened on Space Force, and the Importance of an East Coast Missile Defense Site." *Defense News.* 2019. https://www.defensenews.com/space/2019/04/08/rep-mike-turner-on-why-hes-softened-on-space-force-and-the-importance-of-an-east-coast-missile-defense-site/

"Intel's New 3D Chip Technology May Help Prolong Moore's Law." *MIT Technology Review.* 2018. https://www.technologyreview.com/f/612587/intels-new-3d-chip-technology-may-help-prolong-moores-law/

Ionescu, Imanuela. "Brazil-Russia Military-Technical Cooperation." *Army University Press.* 2018. https://www.armyupress.army.mil/journals/Military-Review/English-Edition-Archives/November-December-2018/Ionescu-Brazil-Russia/

"Iran." *Nuclear Threat Initiative.* 2018. https://www.nit.org/learn/countries/iran/nuclear/

"Iran: Nuclear Intentions and Capabilities." *Office of the Director of National Intelligence.* 2007. https://www.dni.gov/files/documents/Newsrooms/Reports%20and%20Pubs/20071203_release.pdf

Iritani, Evelyn. "Great Idea but Don't Quote Him: Deng Xiaoping's Famous One-Liner Started China On the Way to Capitalism. The Only Problem is There's No Proof He Actually Said It." *The Los Angeles Times.* 2004. http://latimes.com/2004/sep/09/business/fi-deng9

Isaacson, Walter. "How America Risks Losing Its Innovation Edge." *Time.* 2019. https://time.com/longform/america-innovation/

"Israel's Wars." *Israel Ministry of Foreign Affairs.* 2013. https://mfa.gov.il/MFA/AboutIsrael/History/Pages/Israel-Wars.aspx

Irving, Michael. "Global Quantum Internet Dawns, Thanks to China's Micius Satellite." *New Atlas.* 2018. https://newatlas.com/micius-quantum-internet-encryption/53102/

Jabari, Lawahez and Givetash, Linda. "Israel Claims It Thwarted Iranian Drone Attack with Syrian Airstrikes." *NBC News.* 2019. https://www.nbcnews.com/news/world/israel-Claims-it-thwarted-iranian-drone-attack-syria-airstrikes-n1046151

Jaffe, Greg. "Rumsfeld Aims to Elevate Role of Special Forces." *The Wall Street Journal.* 2006. https://www.wsj.com/articles/SB114020280689677176

Jakhar, Partik. "How China's GPS 'Rival' Beidou is Plotting to Go Global." BBC. 2018. https://www.bbc.com/news/technology-45471959

Jekielek, Jan. "Exclusive: On Brazil Joining NATO and Defending the Soul of the West—Foreign Minister Ernesto Anaujo." *The Epoch Times.* 2019. https://www.theepochtimes.com/exclusive-on-brazil-joining-nato-and-defending-the-soul-of-the-west-foreign-minister-ernesto-araujo_2848997.html

Jian, Ma. "Xi Jinping—A Son of the Cultural Revolution." *The Japan Times.* 2016. https://www.japantimes.co.jp/opinion/2016/05/15/commentary/world-commentary/xi-jinping-son-cultural-revolution/#.Xbu7vy-ZPyU

Johnson-Freese, Joan. *Heavenly Ambitions: America's Quest to Dominate Space.* Philadelphia: University of Pennsylvania Press, 2009.

Johnson-Freese, Joan. "Stopping the Slide Towards A War in Space: The Sky's Not Falling, Part 2." *Breaking Defense.* 2016. https://breakingdefense.com/2016/12/stopping-the-slide-towards-a-war-in-space-the-skys-not-falling-part-2/

Johnson, Paul. *Modern Times: The World From the Twenties to the Nineties.* New York: Harper Perennial Classics, 1991.

Johnston, Ryan. "Report: Obama Admin Planted Cyber 'Bombs' Inside Russian Infrastructure." *Cyber Scoop.* 2017. https://www.cyberscoop.com/report-obama-planted-cyber-bombs-inside-russian-infrastructure/

Jones, Andrew. "China, Russia to Cooperate on Lunar Orbiter, Landing Missions." *Space News.* 2019. https://spacenews.com/china-russia-to-cooperate-on-lunar-orbiter-landing-missions/

Jones, Andrew. "Chinese Commercial Launch Companies are Preparing Second-Generation Rockets." *Space News.* 2019. https://spacenews.com/chinese-commercial-launch-companies-are-preparing-second-generation-rockets/

Jones, Andrew. "New China-Brazil Earth Resources Satellite in Launch in H2 2019." *GB Times.* 2018. https://gbtimes.com/new-china-brazil-earth-resources-satellite-in-launch-in-h2–2019

Jones, Andrew. "Successful Long March 5 Launch Opens Way for China's Major Space Plans." *Space News.* 2019. https://spacenews.com/successful-long-march-5-launch-opens-way-for-chinas-major-space-plans/

Kadivar, Mohsen. "Islam in the Modern World: Routinizing the Iranian Revolution." *Kadivar.com.* 2013. http://en.kadivar.com/2013/11/22/routinizing-the-iranian-revolution/

Kang, Tae-Jun. "South Korea's Quest To Be A Major Space Power." *The Diplomat.* 2015. http://thediplomat.com/2015/03/south-koreas-quest-to-be-a-major-space-power/

Kagan, Frederick W. "It's Mad to Forgo Missile Defense." *American Enterprise Institute.* 2015. https://www.aei.org/publication/its-mad-to-forgo-missile-defense/

Kagan, Robert. "The Ambivalent Superpower." *Politico.* 2014. https://www.politico.com/magazine/story/2014/02/united-states-ambivalent-superpower-103860

Kagan, Robert. "Not Fade Away: The Myth of American Decline." *The New Republic.* 2012. https://newrepublic.com/article/99521/america-world-power-declinism

Kagan, Robert. *Of Paradise and Power: America and Europe In the New World Order.* New York: Alfred A. Knopf, 2003.

Kaplan, Fred. *Dark Territory: The Secret History of Cyberwar.* New York: Simon & Schuster, 2016.

Katsner, Jens. "Taiwan Lowers Its Military Sights." *Asia Times.* 2011. http://www.atimes.com/atimes/China/MH02Ad02.html

Kelly, Lorelei and Bjarnason, Robert. "Our 'Modern' Congress Doesn't Understand 21st Century Technology." *Tech Crunch.* 2018. https://techcrunch.com/2018/05/06/our-modern-congress-doesnt-understand-21st-century-technology/

Kennan, George F. "The Long Telegram." *The National Security Archive of George Washington University.* 1946. http://nsarchive.gwu.edu/coldwar/documents/episode-1/kennan.htm

Kennedy, John F. "John F. Kennedy Moon Speech—Rice Stadium." *NASA.* 1962. https://er.jsc.nasa.gov/seh/ricetalk.htm

Kennedy, Merrit. "Trump Created the Space Force. Here's What It Will Actually Do." *NPR.* 2019. https://www.npr.org/2019/12/21/790492010/trump-created-the-space-force-heres-what-it-will-do

Kenton, Will. "Export-Import Bank of the United States." *Investopedia.* 2018. https://www.investopedia.com/e/ex-im-bank.asp

Kerr, Paul. "Iran, North Korea Deepen Missile Cooperation." *Arms Control Association.* 2007. https://www.armscontrol.org/act/2007–01/iran-nuclear-briefs/iran-north-korea-deepen-missile-cooperation

Kessler, Sarah. "Obama Warned Trump That North Korea Would Be His Biggest Problem." *Quartz.* 2017. https://qz.com/924760/donald-trumps-biggest-problem-will-be-north-Koreas-latest-icbm-test-stacks-up-idUSKBN1DT01F

Keohane, Robert O. *After Hegemony: Cooperation and Discord in the World Economy.* Princeton: Princeton University Press, 2005.

Kim, Jack and Kim, Christine. "North Korea Nuclear Threat Means South Must Not Delay Anti-Missile System." Reuters. 2017. https://www.yahoo.com/news/north-korea-nuclear-threat-means-south-must-not-023734948.html

Kissinger, Henry. *Diplomacy.* New York: Simon & Schuster, 1994.

Klare, Michael. "Which Is More Likely: Oil Wars or Water Wars?" *Big Think.* 2008. http://bigthink.com/videos/which-is-more-likely-oil-wars-or-water-wars

Kleiman, Matthew J. *The Little Book of Space Law.* Illinois: ABA Publishing, 2013.

Klein, John J. *Space Warfare: Strategy, Principles, and Policy.* New York: Routledge Publishing Group, 2006.

Koda, Yoji. "China's Blue Water Navy Strategy and Its Implications." *Center for a New American Security.* 2017. https://www.cnas.org/publications/reports/chinas-blue-water-navy-strategy-and-its-implications

Koebler, Jason. "Iran's 'Space Program' May Be a Cover for Developing Better Ballistic Missiles." *Vice.com.* 2015. http://motherboard.vice.com/read/irans-space-Program-may-be-a-cover-for-developing-better-ballistic-missiles

Konviser, Bruce I. "U.S. Missiles in E. Europe Opposed by Locals, Russia." *The Washington Post.* 2007. http://www.washingtonpost.com/wp-dyn/content/article/2007/01/27/AR2007012701370.html

Kotkin, Stephen. *Stalin, Volume I: Paradoxes of Power: 1878–1928.* New York: Penguin Press, 2014.

Kramer, Andrew E. "Gazprom of Russia to Double Prices for Georgia." *The New York Times.* 2006. http://www.nytimes.com/2006/12/22/business/worldbusiness/22iht-Gazprom.3992669.html?_r=0

Kramer, Katie. "Build the Economy Here on Earth By Exploring Space: Tyson." *CNBC*. 2015. http://www.cnbc.com/2015/05/01/build-the-economy-here-on-earth-by-exploring-space-tyson.html

Kramer, Katie. "Neil deGrasse Tyson Says Space Ventures Will Spawn First Trillionaire." *NBC News*. 2015. http://www.nbcnews.com/science/space/Neil-degrasse-tyson-says-space-ventures-will-spawn-first-trillionaire-n352271.

Kramer, Miriam. "MAVEN: NASA's Orbiter Mission to Mars—Mission Details." *Space*. 2018. https://www.space.com/23617-nasa-maven-mars-mission.html

Kramer, Miriam. "Philae Spacecraft to Drill into Comet As Battery Life Dwindles." Space.com. 2014. http://www.space.com/27774-philae-comet-landing-Drill-battery.html.

Krauss, Lawrence. "The Shuttle Was a Dud But Space Is Still Our Destiny." *The Wall Street Journal*. 2011. http://www.wsj.com/articles/SB10001424052702304911104576445934254203762

Krauthammer, Charles. "The Unipolar Moment." *Foreign Affairs*. 1991. https://www.foreignaffairs.com/articles/1991–02–01/unipolar-moment

Kravchenko, Stepan, Meyer, Henry, and Talev, Margaret. "U.S. Strikes Killed Scores of Russian Fighters in Syria, Sources Say." *Bloomberg*. 2018. https://www.bloomberg.com/news/articles/2018–02–13/u-s-strikes-said-to-kill-scores-of-russian-fighters-in-syria

Krugman, Paul. "Enemies of the Sun." *The New York Times*. 2015. https://www.nytimes/com/2015/10/05/opinion/paul-krugman-enemies-of-the-sun.html

Kumar, Chethan. "India's Next Moon Shot Will Be Bigger, In Pact with Japan." *The Times of India*. 2019. https://timesofindia.indiatimes.com/india/indias-next-moon-shot-will-be-bigger-in-pact-with-japan/articleshow/71030437.cms

Kurlantzick, Joshua. *State Capitalism: How the Return of Statism is Transforming the World*. New York: Oxford University Press, 2016.

Kyodo, Jiji. "Japan's Lunar Probe Discovers Moon Cave, which May Be Optimal Base for Exploration." *Japan Times.* 2017. https://www.japantimes.co.jpnews/2017/10/18/national/science-health/japans-lunar-orbiter-discovers-moon-cave-potentially-suitable-use-shelter/#.XbhPFS-ZPyU

Lake, Eli. "Iran's Nuclear Program Helped by China, Russia." *The Washington Times.* 2011. https://www.washingtontimes.com/news/2011/jul/5/irans-nuclear-program-helped-by-china-russia/

Lambakis, Steven. *On the Edge of Earth: The Future of American Space Power.* Lexington: University of Kentucky Press, 2001.

Lardinois, Frederic. "IBM Will Soon Launch a 53-Quibit Quantum Computer." *Tech Crunch.* 2019. https://techcrunch.com/2019/09/18/ibm-will-soon-launch-a-53-quibit-quantum-Computer/

Last, Jonathan V. *What to Expect When No One's Expecting: America's Coming Demographic Disaster.* New York: Encounter Books, 2013.

Lauder, John A. "Russian Proliferation to Iran's Weapons of Mass Destruction and Missile Programs." *Iran Watch.* 2000. https://www.iranwatch.org/library/government/united-states-/executive-branch/central-intelligence-agency/Russian-proliferation-irans-weapons-mass-destruction-and-missile

Laurelle, Marlène. *Russian Eurasianism: An Ideology of Empire.* Washington, D.C.: Woodrow Wilson Center Press, 2008.

Laxman, Srinivas. "China's Unmanned Moon Mission to Bring Back Lunar Soil to Earth." *Asian Scientist.* 2012. http://www.asianscientist.com/2012/03/topnews/china-unmanned-moon-mission-to-bring-back-lunar-soil-2012

Lebrow, Richard Ned. "Was Khrushchev Bluffing in Cuba?" *Bulletin of the Atomic Scientists.* 1988. https://www.tandfonline.com/doi/abs/10.1080/00963402.1988.11456136

Leebaert, Derek. *To Dare and To Conquer: Special Operations and the Destiny of Nations, from Achilles to Al-Qaeda.* New York: Little, Brown, 2006.

Leebaert, Derek. *The Fifty-Year Wound: How America's Cold War Victory Shapes Our World.* New York: Little, Brown and Company, 2002.

Lehmann, John. "CIA Bungle: Agents Tracked Hijackers but Told No One." *New York Post.* 2002. https://nypost.com/2002/06/03cia-bungle-agents-tracked-hijackers-but-told-no-one/

Lele, Ajey. "Commentary: GSAT-7: India's Strategic Satellite." *Space News.* 2013. https://spacenews.com/37142gsat-7-indias-strategic-satellite/

Leone, Dario. "That Time Serbia Shot Down an American F-117 Stealth Fighter." *National Interest.* 2019. https://nationalinterest.org/blog/buzz/time-serbia-shot-down-american-f-117-stealth-fighter-70546

Leskin, Paige. "American Kids Want to Be Famous on YouTube, and Kids in China Want to Go to Space: Survey." *Business Insider.* 2019. https://www.businessinsider.com/American-kids-youtube-star-astronauts-survey-2019-7?fhclid=1wAR35XokQkveZAkPTDE2eXNFimpbOEWTrc4tEk1YUUqZ8KWxKz_DX18QRkBY

Letzter, Rafi. "China's Quantum-Key Network, the Largest Ever, Is Officially Online." *Live Science.* 2018. https://www.livescience.com/61474-micius-china-quantum-key-intercontinental.html

Lewis, John S. *Mining the Sky: Untold Riches From the Asteroids, Comets, and Planets.* New York: Basic Books, 1996.

Liang, Qiao and Xiangsui, Wang. *Unrestricted Warfare: China's Master Plan to Destroy America.* Panama City: Pan American Publishing Company, 1999.

Liberatore, Stacy. "NASA Chief Jim Bridenstine Doubles Down on Claim Pluto is a Planet, Citing Its Moons, Oceans and Organic Compounds." *Daily Mail.* 2019. https://www.dailymail.co.uk/sciencetech/article-7653569/NASA-chief-Jim-Bridenstine-says-Pluto-planet-AGAIN.html?ito=social-facebook&__twitter_impression=true&fbclid=1wAR3-0Alq5Q2UPfdMatPdFAA01IFeDw3yALTpneK_UyCOaNPMWUJ6fSLEQvU

Lima, Mario Sergio and Said, Flavia. "Bolsonaro Takes the Reins in Brazil in Nationalist Surge." *Bloomberg.* 2019. https://www.bloomberg.com/news/articles/2019-01-01/former-army-captain-takes-brazil-s-reins-in-surge-of-nationalism

Lincoln, Taylor. "Sleighted: Accounting Tricks Create False Impression That Small Businesses are Getting Their Share of Federal Procurement Money, and the Political Factors That Might Be at Play." *Public Citizen*. 2015. http://www.citizen. org/documents/Small-business-contracting-report.pdf

Lin, Jeffrey and Singer, P.W. "China's Latest Quantum Radar Could Help Detect Stealth Planes, Missiles." *Popular Science*. 2018. https://www.popsci.com/ china-quantum-radar-detects-stealth-planes-missiles

Lin, Jeffrey, Singer, P.W., and Costello, John. "China's Quantum Satellite Could Change Cryptography Forever." *Popular Science*. 2016. http://www.popscie.com/ chinas-quantum-satellite-could-change-cryptography-forever

Lin, Jeffrey and Singer, P.W. "Is China's Space Laser for Real?" *Popular Science*. 2018. https://www.popsci.com/china-space-laser#page-2

Lipp, James. "On Methods for Studying the Psychological Effects of Unconventional Weapons." *RAND Corporation*. http://www.dtic.mil/dtic/tr/ fulltext/u2/108425.pdf

Livni, Ephrat. "Female Astronauts Schooled Trump from Outer Space." *Quartz*. 2019. https://qz.com/1731692/female-astronauts-school-trump-from-outer-space/

Lockie, Alex. "If North Korea Wants a Reliable ICBM, It's Going to Have to Fire One Towards Somebody." *Business Insider*. 2017. https://www.businessinsider. com/north-korea-icbm-reliable-overfly-2017–8

Logsdon, David and Wolford, Reeve. "Op-Ed: U.S. Brazil Should Act Now to Forge a Relationship in Space." *Space News*. 2018. https://spacenews.com/ op-ed-u-s-brazil-should-act-now-to-forge-a-partnership-in-space/

Logsdon, John M. *After Apollo? Richard Nixon and the American Space Program*. New York: Palgrave Macmillan, 2015.

Longfellow, Henry Wadsworth. "The Midnight Ride of Paul Revere." *National Center*. 1860. https://nationalcenter.org/PaulRevere%27sRide.html

"Long-Range Plan: Implementing USSPACECOM Vision for 2020." Petersen AFB, CO: US Space Command, Director of Plans, 1998.

Lopez, Linette. "It's Time to Start Worrying About What Russia's Been Up to In Latin America." *Business Insider.* 2015. http://www.businessinsider.com/its-time-to-start-worrying-about-russias-been-up-to-in-latin-america-2015–3

Loris, Nicolas. "Four Big Problems With the Obama Administration's Climate Change Regulation." Heritage Foundation. 2015. http://www/heritage.org/research/Reports/2015/08/four-big-problems-with-the-obama-administrations-climate-Change-regulations

Lukin, Artyom. "Putin's Silk Road Gamble." *The Washington Post.* 2018. https://www.washingtonpost.com/news/theworldpost/wp/2018/02/08/putin-china/

Lupton, David E. *On Space Warfare: A Space Power Doctrine.* Maxwell Air Force Base: Air University Press, 1988.

Luttwak, Edward. *The Rise of China vs. The Logic of Strategy.* Cambridge: The Belknap Press of Harvard University Press, 2013.

Lyons, Richard D. "Failures Mark Russian Space Program." *The New York Times.* 1973. https://www.nytimes.com/1973/09/26/archives/failures-mark-russian-space-program-mechanical-failure.html

MacDonald, Bruce W. "China, Space Weapons, and U.S. Security." *Council Special Report,* no. 38 (2008) : 3–38.

Mack, Eric. "How NASA was Born 60 Years Ago from Panic Over a 'Second Moon.'" *C-NET.* 2018. https://www.cnet.com/news/how-nasa-got-its-start-60-years-ago-sputnik-eisenhower/

Mackinder, Halford. *Democratic Ideals and Reality: A Study in the Politics of Reconstruction.* New York: Henry Holt, 1919.

Mackinder, Halford. "The Geographical Pivot of History." *The Geographical Journal, Vol. 23, No. 4* (1904) : 421–437.

Madhumathi D.S. "Chandrayaan-2: A Rapid Dive to 15 Minutes of Terror." *The Hindu.* 2019. https://www.thehindu.com/sci-tech/science/chandrayaan-2-a-rapid-dive-to-15-minutes-of-terror/article29369002.ece

Majaski, Christina. "Brazil, Russia, India, and China (BRIC." *Investopedia.* 2019. https://www.investopedia.com/terms/b/bric.asp

Majumdar, Dave. "5 North Korean Weapons South Korea Should Fear." *The National Interest.* 2016. http://nationalinterest.org/feature/5-north-korean-weapons-south-korea-should-fear-14825

Majumdar, Dave. "Get Ready, America: Russia and China Have Space Weapons." *The National Interest.* 2016. http://nationalinterest.org/blog/the-buzz/get-ready-america-russia-china-have-space-weapons-15027

Majumdar, Dave. "Is Russia Preparing For War?" *The National Interest.* 2016. http://nationalinterest.org/blog/the-buzz/russia-preparing-war-17716

Makovsky, David. "The Silent Strike." *The New Yorker.* 2012. https://www.newyorker.com/magazine/2012/09/17/the-silent-strike

Mankoff, Jeffrey. *Russian Foreign Policy: The Return of Great Power Politics.* New York: Rowman & Littlefield Publishers, 2009.

Mansharamani, Vikram. "China Wants to Mine the Moon for 'Space Gold.'" PBS. 2016. https://www.pbs.org/newshour/economy/china-wants-to-mine-the-moon-for-space-gold

Mansoor, Raja. "Pakistan is Losing the space race." *The Diplomat.* 2018. https://thediplomat.com/2018/02/pakistan-is-losing-the-space-race/

Maranzani, Barbara. "8 Things You Didn't Know About Catherine the Great." *History Channel.* 2012. http://www.history.com/news/8-things-you-didn't-know-about-catherine-the-great

"Maritime Chokepoints: The Backbone of International Trade." *Ship-Technology.* 2017. https://www.ship-technology.com/features/featuremaritime-chokepoints-the-backbone-of-international-trade-5939317

Marquand, Robert. "Dutch Still Wincing at Bush-Era 'Invasion of The Hague Act.'" *CSMonitor.* 2009. http://www.csmonitor.com/World/Europe/2009/0213/p05s01-woeu.html

Marshall, Andrew. "Russia Warns NATO of a 'Cold Peace.'" *The Independent*. 2014. https://www.independent.co.uk/news/russia-warns-nato-of-a-cold-peace-1386966.html

Massicot, Dara. "Anticipating a New Russian Military Doctrine in 2020: What It Might Contain and Why It Matters." *War on the Rocks*. 2019. https://warontherocks.com/2019/09/anticipating-a-new-russian-military-doctrine-in-2020-what-it-might-contain-and-why-it-matters/

Matignon, Louis de Gouyon. "The Soviet Almaz Military Space Station." *Space Legal Issues*. 2019. https://www.spacelegalissues.com/the-soviet-almaz-military-space-station/

Mazzucato, Mariana. *The Entrepreneurial State: Debunking Public vs. Private Sector Myths*. New York: Anthem Press, 2015.

McBride, Edward. "A Ravenous Dragon." *The Economist*. 2008. http://www.economist.com/node/10795714

McCaney, Kevin. "Navy Cranks Up the Power on Laser Weapon." *Defense Systems*. 2015. https://defensesystems.com/articles/2016/06/28/navy-150-kilowatt-laser-weapon-test.aspx

McCurry, Justin. "Discovery of 50km Cave Raises Hopes for Human Colonisation of Moon." *The Guardian*. 2017. https://www.theguardian.com/science/2017/oct/19/lunar-cave-discovery-raises-hopes-for-human-colonisation-of-moon

McDougall, Walter A. *Promised Land, Crusader State: The American Encounter With the World Since 1776*. New York: Houghton Mifflin Company, 1997.

McFadden, Christopher. "23 Great NASA Spin-off Technologies." *Interesting Engineering*. 2018. https://interestingengineering.com/23-great-nasa-spin-off-technologies

McFate, Sean. *The New Rules of War: Victory in the Age of Durable Disorder*. New York: William Murrow, 2019.

McFaul, Michael. "Michael McFaul on Vladimir Putin and Russia." *Uncommon Knowledge*. YouTube video. 2014. https://www.youtube.com/watch?v=iHgp9fLUzpE

McGregor, Grady. "China Is Launching Its 5G Network Ahead of Schedule and On a Spectrum the U.S. Can't Yet Match." *Fortune.* 2019. https://fortune.com/2019/10/31/china-5g-rollout-spectrum/

McIntyre, Jamie. "The Last Korean Meltdown." *The Daily Beast.* 2010. http://www.thedailybeast.com/articles/2010/11/24/the-last-korean-meltdown-bill-clinton-on-the-brink-of-war.html

McKenzie, Sheena. "Germany's Far-Right Makes Big Gains in State Elections." *CNN.* 2019. https://www.cnn.com/2019/09/02/europe/saxony-brandenburg-germany-state-election-results-grm-intl/index.html

McKie, Robin. "How France's Space Ambitions Took Off in French Guiana." *The Guardian.* 2011. https://www.theguardian.com/world/2011/oct/22/france-space-french-guiana

McLoughlin, Bill. "Iran vs. US: Tehran Could Hit America with a 'Dirty Bomb Terror Attack' Warns Expert," *The Daily Express.* 2019. https://www.express.co.uk/news/world/1143301/Iran-news-terror-oil-tanker-attack-US-drone-middle-east-latest

McNeil, Maj. Samuel L. "Proposed Tenets of Space Power: Six Enduring Truths." *Air & Space Power Journal,* 2004.

McNeill, Jena Baker and Weitz, Richard. "Electromagnetic Pulse (EMP) Attack: A Preventable Homeland Security Catastrophe." Heritage Foundation. 2008. http://www.heritage.org/research/reports/2008/10/electromagnetic-pulse-emp-attack-a-preventable-homeland-security-catastrophe

Mead, Walter Russell. "It's Kim Jong-un's World' We're Just Living In It." *The American Interest.* 2016. http://www.the-america-interest-com/2016/09/09/its-kim-jong-uns-world-were-just-living-in-it/

Mead, Walter Russell. *Special Providence: American Foreign Policy and How It Changed the World.* New York: Routledge, 2002.

Mearshimer, John J. *The Tragedy of Great Power Politics.* New York: W.W. Norton & Company, 2001.

Mehta, Aaron. "Here are the Biggest Weaknesses in America's Defense Sector." *Defense One*. 2019. https://www.defensenews.com/pentagon/2019/06/27/ here-are-the-biggest-weaknesses-in-americas-defense-sector/

Messier, Doug. "China Ready to Cooperate on Orbital Space Station." *Parabolic Arc*. 2019. https://www.parabolicarc.com/2019/01/15/ china-ready-cooperate-russia-orbital-space-station/

Meycrs, Lawrence. "Politics Really is Downstream from Culture." *Breitbart*. 2011. https://www.breitbart.com/entertainment/2011/08/22/ politics-really-is-downstream-from-culture/

Milani, Livia Peres. "Brazil's Space Program: Finally Taking Off?" *Wilson Center*. 2019. https://www.wilsoncenter.org/blog-post/brazils-space-program-finally-taking

Miles, Aaron. "Escalation Dominance in America's Oldest New Nuclear Strategy." *War on The Rocks*. 2018. https://warontherockes.com/2018/09/ escalation-dominance-in-americas-oldest-new-nuclear-strategy/

Millard, Peter. "Why Billions of Barrels of Oil Go Untapped in Brazil." *Bloomberg*. 2019. https://www.bloomberg.com/news/articles/2019–03–03/ why-billions-of-barrels-of-oil-go-untapped-in-brazil-quicktake

Millis, Walter. *The Forrestal Diaries*. New York: Viking Publishing, 1951.

Minick, Wendell. "Analysis: China Builds Amphibious Capabilities For Taiwan Scenario." *Defense News*. 2016. http://www.defensenews. com/story/defense/international/asia-pacific/2016/05/14/ analysis-china-builds-amphibious-capabilities-taiwan-scenario/84372014

"Missile Wars: Missile Defense, 1944–2002." *PBS Frontline*. 2002. https://www.pbs. org/wgbh/pages/frontline/shows/missile/etc/cron.html

Mizokami, Kyle. "The Lovely Little Town That Would Have been Absolutely Screwed By World War III: This Small German City Was Ground Zero for a Cold War Turned Hot." *War Is Boring*. 2013. http://warisboring.com/the-lovely-little-town-that-would-have-been-absolutely-screwed-by-world-war-iii-933229e2ea52#.9fvz9sm7c

Mizokami, Kyle. "North Korea Claims That It Tested the 'H-Bomb of Justice.'" *Popular Mechanics.* 2016. http://www/popularmechanics.com/military/weapons/news/a18848/north-korea-h-bomb-of-justice/

Mizokami, Kyle. "North Korean Video Graphically Threatens to Turn Seoul Into a 'Sea of Flames.'" *Popular Mechanics.* 2016. http://www.popularmechanics.com/ military/weapons/news/a20306/north-korea-threatens-to-turn-seoul-into-a-sea-of-flames/

Mizokami, Kyle. "Trump's Space Force Isn't the Only Military Space Program: Here's What China and Russia Are Up To." *Jalopnik.* 2019. https://foxtrotalpha.jalopnik.com/as-trump-s-space-force-ramps-up-what-are-russia-and-ch-1832772367

Mola, Roger A. "Does Missile Defense Actually Work?" *Air & Space Magazine.* 2013. http://www.airspacemag.com/military-aviation/does-missile-defense-actually-work-18626187/

Moltz, James Clay. *Asia's space race: National Motivations, Regional Rivalries, and International Risks.* New York: Columbia University Press, 2012.

Moltz, James Clay. *Crowded Orbits: Conflict and Cooperation in Space.* New York: Columbia University Press, 2014.

Moltz, James Clay. *The Politics of Space Security: Strategic Restraint and the Pursuit of the National Interest.* Stanford: Stanford University Press, 2008.

Moltz, James Clay. "Russia and China: Strategic Choices in Space." *Space and Defense Policy.* New York: Routledge, 2009.

Moore, Charles. "Barack Obama Must Do More Than Manage America's Decline." *The Telegraph.* 2012. http://www.telegraph.co.uk/news/worldnews/us-election/9349286/President-Barack-Obama-must-do-more-than-manage-Americas-decline.html

Moore, Mike. *The Twilight War: The Folly of U.S. Space Dominance.* Oakland: Independent Institute Press, 2008).

Moore, Stephen. "No, Bill Clinton Didn't Balance the Budget." *CATO Institute.* 1998. http://www.cato.org/publications/commentary/no-bill-clinton-didnt-balance-budget.

Moravcsik, Andrew. "Power of Connection: Why the Russia-Europe Gas Trade is Strangely Untouched by Politics." *Nature.* 2019. https://www.nature.com/articles/d41586–019-03694-y

Morrison, Spencer P. "America: China's Mercantile Resource Colony." *American Greatness.* 2018. https://amgreatness.com/2018/12/11/america-chinas-mercantile-resource-colony/

Morris, David Z. "Is SpaceX Undercutting the Competition More Than Anyone Thought?" *Fortune.* 2017. https://fortune.com/2017/06/17/spacex-launch-cost-competition/

Mortier, Jan and Finnis, Benjamin. "China Leads Race to the Moon: States are Quietly Preparing to Secure Fuel for the Fourth Generation of Nuclear Weapons, and China Is Winning." *The Diplomat.* 2015. http://thediplomat.com/2015/01/china-leads-race-to-the-moon/

Mortimer, Caroline. "Ukraine Crisis: Why is Crimea So Important to Russia? The Peninsula is a Crucial Point of Contention—and History—Between the Two Countries." *The Independent.* 2014. http://www.independent.co.uk/news/world/europe/ukraine-crisis-why-is-crimea-so-important-to-russia-9166447.html

Mosher, Dave. "Video Shows a Top-Secret Chinese Space Mission Failing in Mid-Flight—China's Second Rocket Loss of the Year." *Business Insider.* 2019. https://www.businessinsider.com/china-long-march-rocket-launch-failure-weibo-video-pictures-2019–5

Mowthorpe, Matthew. *The Militarization and Weaponization of Space.* Lanham: Lexington Books, 2004.

Mullen, Jethro and Armstrong, Paul. "North Korea Carries Out Controversial Rocket Launch." *CNN.* 2012. https://www.cnn.com/2012/12/11/world/asia/north-korea-rocket-launch/index.html

Munro, Kelsey. "China Cabinet: Black Swans, Grey Rhinos, an Elephant in the Room." *The Interpreter.* 2019. https://www.lowyinstitute.org/the-interpreter/china-cabinet-Black-swans-grey-rhinos-elephant-room

Murgia, Madhumita and Waters, Richard. "Google Claims to Have Reached Quantum Supremacy." *Financial Times.* 2019. https://www.ft.com/content/b9bb4e54-dbc1-11e9–8f9b-77216ebe1f17

Myers, Meghann. "The Space Force is Officially the Sixth Branch. Here's What That Means." *The Air Force Times.* 2019. https://www.airforcetimes.com/news/your-military/2019/12/21/the-space-force-is-officially-the-sixth-branch-heres-what-that-means/

"Mysterious Russian Space Object Could Be the Return of Istrebitel Sputnikove—the 'Satellite Killer.'" *National Post.* 2014. https://nationapost.com/news/mysterious-russian-space-object-could-be-the-return-of-istrebitel-sputnikov-the-satellite-killer

"NASA Confirms Evidence That Liquid Water Flows On Today's Mars." *NASA.* 2015. https://www.nasa.gov/press-release/nasa-confirms-evidence-that-liquid-water-flows-on-today-s-mars

"NASA's Strategic Direction and the Need for a National Consensus." *The National Academies of Sciences, Engineering, and Medicine.* 2012. https://www.nap.edu/read/18248/chapter/2

Nardon, Laurence. "Developed Space Programs." *The Politics of Space: A Survey,* ed. Eligar Sadeh. London: Routledge, 2011.

"National Space Policy of the United States of America." *The Office of the President of the United States.* 2010. https://www.whitehouse.gov/sites/default/files/national_space_policy_6–28–10.pdf

Nau, Henry R. *Conservative Internationalism: Armed Diplomacy Under Jefferson, Polk, Truman, and Reagan.* Princeton: Princeton University Press, 2013.

Nesbit, Jeff. "Europa's Water Plumes Are a Big Deal." *U.S. News & World Report.* 2016. http://www.usnews.com/news/articles/2016–09–27/europa-has-water-plumes-and-that-is-a-very-big-deal

"The New Colonialists." *The Economist.* 2008. https://www.economist.com/leaders/2008/03/13/the-new-colonialists

"New WGS-11 Satellite to Offer Greater Coverage, Efficiency Than Predecessors." *Aerotech News.* 2019. https://www.aerotechnews.com/blog/2019/12/27/ new-wgs-11-satellite-to-offer-greater-coverage-efficiency-than-predecessors/

Nixon, Richard. "Conversation 7, Tape 464." *Richard Nixon Presidential Library.* 1971.

Nofi, Albert A. "Recent Trends in Thinking About Warfare." *CAN.* 2006. https:// www.cna.org/CNA_files/PDF/D0014875.A1.pdf

Nordhaus, William D. "Schumpeterian Profits in the American Economy: Theory and Measurement." *National Bureau of Economic Research.* April 2004, pp. 34–35. https://www.nber.org/papers/w10433.pdf

"North Korea Nuclear Timeline Fact." *CNN.* 2016. http://www.cnn. com/2013/10/29/world/asia/north-korea-nuclear-timeline—fast-facts/

Nowakowski, Tomasz. "Tiangong-2 Arrives at Chinese Launch Center for September Liftoff." *Spaceflight Insider.* 2016. http://www. spaceflightinsider.com/organizations/china-national-space-administration/ tiangong-2-arrives-chinese-launch-center-september-liftoff/

Oba, Mintaro. "Iran-North Korea Relationship Reflects Failed US Policies." *Atlantic Council.* 2019. https://www.atlanticcouncil.org/blogs/iransource/ iran-north-korea-relationship-Reflects-failed-us-policies/

Oberdorfer, Don and Carlin, Robert. *The Two Koreas: A Contemporary History.* New York: Basic Books, 2014.

Oberg, Jim. *Space Power Theory.* Washington, D.C.: Government Printing Office, 1999.

O'Dwyer, Gerard. "Norway Accuses Russia of Jamming Its Military Systems." *Defense News.* 2019. https://www.defensenews.com/global/europe/2019/03/08/ norway-alleges-signals-jamming-of-its-military-systems-by-russia/

Ohanian, Lee. "Seattle Schools Propose to Teach That Math Education is Racist— Will California Be Far Behind?" *Hoover Institution.* 2019. https://www.hoover.org/ research/scattle-schools-propose-teach-math-education-racist-will-california-be-far-behindseattle

"Old School Air Force Can't Handle 'Space Corps' Challenge?" *Military*. 2017. https://www.military.com/defensetech/2017/10/09/ old-school-air-force-cant-handle-space-corps-challenge

Oliphant, Roland. "How Vladimir Putin's Military Firepower Compares to the West: As the Russian President Prepares to Unveil His New Tanks at Saturday's Victory Day Parade, Here is a Look at How Russian Forces Measure Up to Those of Britain and the US." *The Telegraph*. 2015. http://www.telegraph.co.uk/news/ worldnews/vladimir-putin/11586021/How-Putins-military-firepower-compares-to-the-West.html

O'Neill, Jim. "Building Better Global Economic BRICs." Goldman Sachs. 2001. https://www.goldmansachs.com/insights/archive/archive-pdfs/build-better-brics.pdf

Oppenheim, Maya. "Donald Trump Shaken by 'Scary' Intelligence Briefings: 'We Have Some Big Enemies Out There." *The Independent*. 2017. https://www. independent.co.uk/news/world/americas/donald-trump-scary-intelligence-briefings-interview-big-enemies-out-there-a7534031.html

Opsal, Ryan. "China Is Cozying Up to Latin America For Its Oil." *Business Insider*. 2014. http://www.businessinsider.com/china-and-latin-america-oil-trade-2015–6

"Options for Deploying Missile Defenses in Europe." The Congressional Budget Office. 2009. http://www.cbo.gov/sites/default/files/111th-congress-2009–2010/ reports/02–27-missiledefense.pdf

Orlova, Karina. "The Siloviki Coup in Russia." *The American-Interest*. 2016. https://www.the-american-interest.com/2016/09/21/the-siloviki-coup-in-russia/

Ortel, Janka. "Germany Chooses China Over the West." *Foreign Policy*. 2019. https://foreignpolicy.com/2019/10/21/ Germany-merkel-chooses-china-over-united-states-eu-huawei/

Osborn, Kris. "Stealth Strike! U.S. Air Force Wants to Test F-35 Laser Weapons by 2021." *The National Interest*. 2019. https://nationalinterest.org/blog/buzz/ stealth-strike-us-air-force-wants-test-f-35-laser-weapons-2021–88716

"OSPEK." *Russian Space Web*. http://www.russianspaceweb.com/ospek.html

"Outer Space Treaty of 1967." *National Aeronautics and Space Administration.* Accessed On: 22 June 2019. https://history.nasa.gov/1967treaty.html

Parker, Laura. "What You Need to Know About the World's Water Wars." *National Geographic.* 2016. http://news.nationalgeographic.com/2016/07/world-aquifiers-water-wars/

Park, Han S. *North Korea: The Politics of Unconventional Wisdom.* Boulder: Lynne-Reiner, 2002.

Pasandideh, Shahryar. "Iran's Space Program Won't Get Off the Ground While Under Sanctions." *World Politics Review.* 2019. https://www.worldpoliticsreview.com/articles/28167/iran-s-space-program-won-t-get-off-the-ground-while-under-sanctions

Patel, Neel V. "It's Time For America to Start Worrying About Losing the space race." *Inverse.* 2016. https://www.inverse.com/article/12535-it-s-time-for-america-to-start-worrying-about-losing-the-space-race

Patranobis, Sutirho. "ISRO Record: Chinese State Media Says Beijing Can Learn Lessons from India." *Hindustan Times.* 2017. https://www.hindustantimes.com/india-news/isro-launch-china-could-learn-some-lessons-from-india-on-space-commerce-says-state-media/story-hUXI9C9ReS0KmTMOnwZYEM.htm.

Pawlikowski, Ellen; Loverro, Doug, and Cristler, Tom. "Space: Disruptive Challenges, New Opportunities, and New Strategies." *Strategic Studies Quarterly* (2014): 27–54.

Pawlyk, Oriana. "Air Force to Congress: 'No Space Corps.'" *Military.* 2017. https://www.military.com/daily-news/2017/06/22/air-force-congress-no-space-corps.html

Pawlyk, Oriana. "It's Official: Trump Announces Space Force as 6th Military Branch." *Military.com.* 2018. https://www.military.com/daily-news/2018/06/18/its-official-Trump-announces-space-force-6th-military-branch.html.

Pearce, Fred. "China Spends Big On Nuclear Fusion as French Plan Falls Behind." *New Scientist.* 2015. https://www.newscientist.com/article/dn27944-china-spends-big-on-nuclear-fusion-as-french-project-falls-behind/

Perlez, Jane. "American and Chinese Ships Nearly Collided in South China Sea." *New York Times.* 2013. http://www.nytimes.com/2013/12/15/world/asia/chinese-and-american-ships-nearly-collide-in-south-china-sea.html

Peston, Robert. "Russia 'Planned Wall Street Bear Raid.'" BBC *News.* 2014. http://www.Bbc.com/news/business-26609548

Peterson, Matthew J. "Thank God Trump Isn't a Foreign Policy Expert." *American Greatness.* 2018. https://amgreatness.com/2018/06/12/thank-god-trump-isnt-a-foreign-policy-expert/

Peters, Ralph. "Constant Conflict." *Parameters.* 1997. http://strategicstudiesinstitute.army.mil/pubs/parameters/Articles/2010winter/Peters.pdf

Peters, Ralph. "Stability, America's Enemy." *Lines of Fire: A Renegade Writes on Strategy, Intelligence, and Security.* Mechanicsburg: Stackpole Books, 2011.

Phillips, Tom. "China's Tiangong-1 Space Station 'Out of Control' and Will Crash to Earth." *Guardian.* 2016. http://www.theguardian.com/science/2016/sep/21/chinas-tiangong-1-space-station-out-of-control-crash-to-earth

Pickrell, Ryan. "The US Has Been Getting 'Its Ass Handed to It' In War Games Simulating Fights Against Russia and China." *Business Insider.* 2019. https://www.businessinsider.com/the-us-apparently-gets-its-ass-handed-to-it-in-war-games-2019–3

Pillsbury, Michael. *The Hundred-Year Marathon: China's Secret Strategy to Replace America as the Global Superpower.* New York: Henry Holt & Co., 2015.

Pinto, Nolan. "4 Astronauts Identified for Gaganyaan Mission: ISRO Chief K Sivan." *India Today.* 2020. https://www.indiatoday.in/science/story/gaganyaan-4-Astronauts-identified-isro-k-sivan-1633024–2020–01–01

Pipes, Richard. "Why the Soviet Union Thinks It Could Fight & Win a Nuclear War." *Commentary.* 1977. https://www.commentarymagazine.com/articles/why-the-soviet-union-thinks-it-could-fight-win-a-nuclear-war

"PLA Colonels on 'Unrestricted Warfare': Part I, A November 1999 Report from the U.S. Embassy Beijing." Federation of American Scientists. Accessed on: 5 November 2019. https://fas.org/nuke/guide/china/doctrine/unresw1.htm

Ponniah, Kevin. "How Pragmatic Finland Deals with Its Russian Neighbour." BBC. 2017. https://www.bbc.com/news/world-europe-40731415

Posen, Barry. *Restraint: A New Foundation for U.S. Grand Strategy.* Ithaca: Cornell University Press, 2014.

Preston, Bob, Johnson, Dana J., et al. *Space Weapons, Earth Wars.* Santa Monica: RAND Corporation, 2002.

Prisco, Jacopo. "Two Abandoned Soviet Space Shuttles Left in the Kazakh Desert." *CNN.* 2017. https://www.cnn.com/style/article/baikonur-buran-soviet-space-shuttle/index.html

"Putin: Soviet Collapse a Genuine Tragedy: In Annual Speech, Russian Leader Cites Chechen Violence as Result." *NBC News.* 2005. http://www.nbcnews.com/id/7632057/ns/world_news/t/putin-soviet-collapse-genuine-tragedy/#.V1-SVVfE5Hg

Putin, Vladimir. "Putin on Novorossiya." *DirectLine.* YouTube Video. 2014. http://www.youtube.com/watch?v=C1_eF6_EMwM

Pyle, Rod. "Probing the Mysteries of Europa, Jupiter's Cracked and Crinkled Moon." *CalTech.* 2015. https://www.caltech.edu/news/probing-mysteries-europa-jupiters-cracked-and-crinkled-moon-48593

Quora, "Why Didn't the Space Shuttle Program Continue?" *Forbes.* 2018. https://www.forbes.com/sites/quora/2018/06/07/why-didnt-the-space-shuttle-program-continue/#2034801690ae

Rabinovitch, Ari. "Israel Uses Military to Join Commercial space race." Reuters. 2015. https://www.reuters.com/article/us-israel-space/israel-uses-military-expertise-to-join-commercial-space-race-idUSKBN0M016Y20150304

Rachman, Gideon. "Think Again: American Decline: This Time It's For Real." *Foreign Policy.* 2011. http://foreignpolicy.com/2011/01/03/think-again-american-decline/

Ramzy, Austin. "Leaders of China and Taiwan Talk of Peace Across the Strait." *New York Times.* 2015. http://www.nytimes.com/2015/11/08/world/asia/presidents-china-taiwan-meet-shake-hands-singapore.html

Ramzy, Austin. "Taiwan's New Leader Faces A Weak Economy and China's Might." *New York Times.* 2016. http://www.nytimes.com/2016/01/18/world/asia/tsai-Ing-wen-taiwan-president.html

Rathi, Akshat. "In Search of Clean Energy, Investments in Nuclear-Fusion Startups are Heating Up." *Quartz.* 2018. https://qz.com/1402282/in-search-of-clean-energy-investments-in-nuclear-fusion-startups-are-heating-up/

ReaganFoundation. "National Security: President Reagans Address on Defense and National Security 3/23/83/" 2019. https://www.youtube.co/watch?v=ApTnYwh5KvE

Reagan, Ronald. "Encroaching Control." YouTube *video.* 1961. https://www.youtube.com/watch?v=SDouNtnR_IA

"Rebuilding America's Defenses." *The Project For the New American Century.* 2000. http://www.informationclearinghouse.info/pdf/RebuildingAmericasDefenses.pdf

Reevell, Patrick. "Russia's Crumbling Baikonur Spaceport is Earth's Only Launch Pad for Manned Flights." *ABC News.* 2018. https://abcnews.go.com/international/russias-crumbling-baikonur-spaceport-earths-launchpad-manned-flights/story?id59677789

"Remarks by Alireza Jafarzadeh on New Information on Top Secret Projects of the Iranian Regime's Nuclear Program." *Iran Watch.* 2002. https://www.iranwatch.org/library/ncri-new-information-top-secret-nuclear-projects-8–14–02

Rempfer, Kyle. "Air Force May Investigate SpaceX CEO for Smoking Pot, While Canada Greenlights Its Use for Troops." *Air Force Times.* 2018. https://www.airforcetimes.com/news/your-air-force/2018/09/11/air-force-may-investigate-spacex-ceo-for-smoking-pot-while-canada-greenlights-its-use-for-troops/

Rempfer, Kyle. "Why This US General Says Russian Wagner Mercenaries in Africa 'Concern Me Greatly.'" *Military Times.* 2019. https://www.militarytimes.com/news/your-military/2019/04/04/why-this-us-general-says-russian-wagner-mercenaries-in-africa-concern-me-greatly/

"Report of the Commission to Assess United States National Security Space Management and Organization." *Department of Defense.* 2001. http://www.dod.gov/pubs/20010111.html

"Reusability: The Key to Making Human Life Multi-Planetary." *SpaceX.* 2015. https://www.spacex.com/news/2013/03/31/reusability-key-making-human-life-multi-planetary

Reynolds, Emily. "Astronauts 'Step' Inside NASA's Beam Habitat For the First Time." *Wired.* 2016. http://www.wired.co.uk/article/nasa-inflatable-habitat-bigelow-isis

"Revenue Growth is the Top Metric to Measure the Success of Innovation." KPMG. 2018 https://home.kpmg/sk/en/home/media/press-releases/2018/04/revenue-growth-is-top-metric-to-measure-innovation-success.html

Rinke, Andreas and Hansen, Holger. "With or Without Huawei? German Coalition Delays Decision on 5G." Reuters. 2019. https://www.reuters.com/article/us-germany-china-huawei/with-or-without-huawei-german-coalition-delays-decision-on-5g-rollout-idUSKBN1YL22Z

Roblin, Sebastien. "Russia's Cold War Super Weapon (Put Lasers on Everything It Can)." *The National Interest.* 2017. https://nationalinterest.org/blog/the-buzz/russias-cold-war-super-weapon-put-lasers-everything-it-can-21553

Roblin, Sebastien. "Study This Picture: This is Why Japan Attacked Pearl Harbor (And Dragged America Into World War II)." *The National Interest.* 2018. https://nationalinterest.org/blog/buzz/study-picture-why-japan-attacked-pearl-harbor-and-dragged-america-world-war-ii-37712

Roblin, Sebastien, "Quantum Radars Could Unstealth the F-22, F-35 and J-20 (Or Not)." *National Interest.* 2018. https://nationalinterest.org/blog/the-buzz/quantum-radars-could-Unstealth-the-f-22-f-35-j-20-or-not-25772

Rogan, Tom. "Be Glad Iran's Satellite Launch Failed." *The Washington Examiner.* 2019. https://www.washingtonexaminer.com/opinion/ be-glad-irans-satellite-launch-failed

"Roger Shawyer Explaining the Basic Science Behind #EmDrive." YouTube. 2016. https://www.youtube.com/watch?v=wBtk6xWDrwY

Rohrlich, Justin. "Why Can't Startup Companies Get US Government Contracts?" *Vice News.* 2015. https://www.vice.com/en_us/article/d3978k/ why-cant-startup-companies-get-us-government-contracts

Ronel, Asaf. "Jerusalem, We Have a Problem: Why Israel's NASA Isn't Taking Off." *Haaretz.* 2018. https://www.haaretz.com/israel-news/. premium-israel-s-nasa-struggles-to-lift-off-amid-budgetary-issues-1/6471276

"Rosatom to Build Four Nuclear Power Plants." *Power Engineering International.* 2018. https://www.powerengineeringint.com/2018/06/11/ rosatom-to-build-four-nuclear-power-plants-in-china/

Rosenbaum, Eric and Russo, Donovan. "China Plans a Solar Power Play in Space That NASA Abandoned Decades Ago." *CNBC.* 2019. https://www.cnbc. com/2019/03/15/china-plans-a-solar-power-play-in-space-that-nasa-abandoned- long-ago.html

Rosenberg, David. "How Israel is Turning Its High-Tech into Global Political Power." *Fathom.* 2018. http://fathomjournal.org/ how-israel-is-turning-its-high-tech-into-global-political-power

Rosen, James. "Why is Russia Using 'Dirty Bombs' In Syria?" *Miami Herald.* 2015. https://www.miamiherald.com/news/nation-world/world/article37968804.html

Roset, Claudia. "The Audacity of Silence On Possible Iran-North Korea Nuclear Ties." *Forbes.* 2016. http://www.forbes.com/sites/claudiarosett/2016/12/15/ the-audacity-of-silence-on-possible-iran-north-korea-nuclear-ties/#f41fc8c8a8e

Roth, Andrew. "Iran Announces Delivery of Russian S-300 Missile Defense System." *Washington Post.* 2016. https://www.washingtonpost.com/world/iran- announces-delivery-of-russian-s-300-missile-defense-system/2016/05/10/944afa2e- 16ae-11e6–971a-dad9ab18869_story.html?utm_term=.756339c2822d

Rouge, Joseph D. "Space-Based Solar Power as an Opportunity for Strategic Security." *Report to the Director, National Security Space Office Interim Assessment.* 2007. https://space.nss.org/space-based-solar-power-as-an-opportunity-for-strategic-security/

Rubin, Michael. "The Genesis of Iran's Space Program." *American Enterprise Institute.* 2012. http://www.aei.org/publication/the-genesis-of-irans-space-program/

"Russia, China Sign Space Exploration Deal." *Moscow Times.* 2018. https://www.themoscowtimes.com/2018/06/08/Russia-China-Sign-Space-Exploration-Deal-a61736

"Russia Could Build Orbital Assembly After 2020—Energiacorporation." *Interfax.* 2009. http://www.webcitation.org/5j8D34fZx

"Russia Halts Gas Supplies to Ukraine After Talks Breakdown." BBC *News.* 2015. http://www.bbc.com/news/world-europe-33341322

"Russia, US Competing for Space Partnership With Brazil." *VOA News.* 2015. http://www.voanews.com/a/russia-united-states-compete-for-space-partnership-with-brazil/2824135.html

"Russian/Soviet Doctrine." Federation of American Scientists. 2000. https://fas.org/nuke/guide/russia/doctrine/intro.htm

"Russia's Natural Population Declines for 4th Straight Year—Audit Chamber." *MoscowTimes.* 2019. https://www.themoscowtimes.com/2019/11/07/russias-natural-population-declines-for-4th-straight-year-audit-chamber-a68066

Rutenberg, Jim. "In Farewell, Rumsfeld Warns Weakness Is 'Provocative.'" *New York Times.* 2006. http://www.nytimes.com/2006/12/16/washington/16prexyHtml?_r=0.

Sadeh, Eligar (ed.) *The Politics of Space: A Survey.* New York: Routledge, 2011.

Sadeh, Eligar (ed.) *Space Strategy In the 21ˢᵗ Century: Theory and Policy.* New York: Routledge, 2013.

Safire, William. "On Language: The Crackdown Watch." *New York Times.* 1991. http://www.nytimes.com/1991/01/13/magazine/on-language-the-crackdown-watch.html

"Samuel Slater." PBS.com. https://www.pbs.org/wgbh/theymadeamerica/whomade/slater_hi.html

Sanger, David E. "From Iran to Syria, Trump's 'America First' Approach Faces Its First Tests." *New York Times.* 2016. http://www.nytimes.com/2016/11/18/us/politics/from-iran-to-syria-trumps-america-first-approach-faces-its-first-tests.html?_r=0

Sanger, David E. and Broad, William J. "How U.S. Intelligence Agencies Underestimated North Korea." *New York Times.* 2018. https://www.nytimes.com/2018/01/06/world-asia/north-korea-nuclear-missile-intelligence.html

"Say Cheese! Russia Snaps Photos of Top Secret US Spysats: Washington Has Become Extremely Paranoid About a Series of Rare Images of Its Lacrosse Spy Satellites (#2, 3, 4, and 5), Which Were Recently Released By Russia." Sputnik *News.* 2015. http://sputniknews.com/world/20150430/1021559093.html#ixzz3Yn1whqQZ

Schachtman, Noah. "With New Mini-Satellites, Special Ops Takes Its Manhunts into Space." *Wired.* 2013. https://www.wired.com/2013/05/special-ops-mini-sats-manhunts/

Scheer, Steve. "After SpaceX Blast, Israeli Satellite Firm Struggles to Keep Deal With the Chinese. *Haaretz.* 2016. http://www.haaretz.com/israel-news/business/1.740144

Schelling, Thomas C. *Arms and Influence.* New Haven: Yale University Press, 2008.

Schiavenza, Matt. "China Economy Surpasses US In Purchasing Power, But Americans Don't Need to Worry." *International Business Times.* 2014. http://www.ibtimes.com/china-economy-surpasses-us-purchasing-power-americans-dont-need-to-worry-1701804

Schmitt, Harrison. "Mining the Moon." *Popular Mechanics.* 2004. http://www.popularmechanics.com/space/moon-mars/a235/1283056

Schmitt, Harrison H. "Summary Testimony." *Subcommittee on Science, Technology, And Space of the Senate Commerce, Science, and Space Committee.* 2003. http://www.globalsecurity.or/space/library/congress/2003_h/031106.doc

Schneider, Mark B. "The Russian Nuclear Threat." *Real Clear Defense.* 2019. https://www.realcleardefense.com/articles/2019/05/28/the_russian_nuclear_threat_114457.html

Schneider, Mark B. "The Russian Nuclear Weapons Buildup and the Future of the New START Treaty." *Real Clear Defense.* 2016. https://www.realcleardefense.com/articles/2016/11/02/the_russian_nuclear_weapons_buildup_110294.html

Sciutto, Jim. "Exclusive: China Warns U.S. Surveillance Plane." *CNN.* 2015. http://edition.cnn.com/2015/05/20/politics/south-china-sea-navy-flight/index.html

Sciutto, Jim. "U.S. Military Prepares For the Next Frontier: Space War." *CNN.* 2016. http://www.cnn.com/2016/11/28/politics/space-war-us-military-preparations/

Seddon, Max and Farchy, Jack. "Russians Rally to the Brexit Flag in Britain's EU Referendum: Cameron Remarks on Putin Bemused Moscow but Russian Nationals in the UK Have a Stake in the Outcome." *Financial Times.* 2016. http://www.ft.com/cms/s/0/ba26flac-20fa-11e6–9d4d-c11765124d.html#axzz4IN8ukWIb

Seidel, Jamie. "China's Claim It Has 'Quantum' Radar May Leave $17 Billion F-35 Naked." *News.com.au.* 2017. https://www.news.com.au/technology/innovation/inventions/chinas-Claim-it-has-quantum-radar-may-leave-17-billion-f35-naked/news-story/207ac01ff3107d21a9f36e54b6f0fbab

Seligman, Lara. "Russian Jamming Poses a Growing Threat to U.S. Troops in Syria." *Foreign Policy.* 2018. https://foreignpolicy.com/2018/07/30/russian-jamming-poses-a-growing-threat-to-u-s-troops-in-syria/

Shanker, Thom and Landler, Mark. "Putin Says U.S. is Undermining Global Stability." *New York Times.* 2007. https://www.nytimes.com/2007/02/11/world/europe/11munich.html?Mtrref=www.google.com&gwh=2A9EBFA12D122979579919DIE784175E&gwt=pay&assetType=REGIWALL

Shear, Eric. "Conspiracy Theories Regarding Amos-6 Falcon 9 Explosion Not Based on Physics, Reality." *Spaceflight Insider.* 2016. https://www.spaceflightinsider. com/organizations/space-exploration-technologies/conspiracy-theories-regarding-amos-6-falcon-9-explosion-not-based-on-physics-reality/

Shellenberger, Michael. "The Reason Renewables Can't Power Modern Civilization is Because They Were Never Meant To." *Forbes.* 2019. https://www.forbes.com/ sites/michaelshellenberger/2019/05/06/the-reason-renewables-cant-power-modern-civilization-is-because-they-were-never-meant-to/#738de505ea2b

Shenkar, Oded. *Copycats: How Smart Companies Use Imitation to Gain a Strategic Edge.* Boston: Harvard Business School Publishing, 2010.

Shilpak, David A. and Johnson, Michael W. "Reinforcing Deterrence on NATO's EasternFlank." *RAND Corporation.* 2016. https://www.rand.org/pubs/research_reports/RR1253.html.

Shieber, Jonathan. "China Nears Completion of Its GPS Competitor, Increasing the Potential for Internet Balkanization," *Tech Crunch,* 2019. https://techcrunch. com/2019/12/28/china-nears-completion-of-its-gps-competitor-increasing-the-potential-for-internet-balkanization/

Shim, Elizabeth. "North Korea's 'Electronic Bomb,' Technology of Russian Origin, Experts Say." *UPI.* 2017. https://www.upi.com/Top_News/World-News/2017/09/14/North-Koreas-electronic-bomb-technology-of-Russian-origin-experts-say/6401505408979

Shlapak, David A. and Johnson, Michael W. "Outnumbered, Outranged, and Outgunned: How Russia Defeats NATO." *War on the Rocks.* 2016. https://warontherocks.com/2016/04/ outnumbered-outranged-and-outgunned-how-russia-defeats-nato/

Shlapak, David A. and Johnson, Michael W. "Reinforcing Deterrence on NATO's Eastern Flank." *RAND Corporation.* 2016. https://www.rand.org/pubs/research_reports/RR1253.html

Showalter, Dennis E. "Derek Leebaert, To Dare and To Conquer: Special Operations and the Destiny of Nations, from Achilles to Al-Qaeda." *Michigan War Studies Review.* 2006. http://www.miwsr.com/2006/downloads/20061201.pdf

Siddiqui, Zeba. "Factbokx: India and Pakistan—Nuclear Arsenals and Strategies." Reuters. 2019. https://www.reuters.com/article/us-india-kashmir-pakistan-nuclear-factbo/factbox-india-and-pakistan-nuclear-arsenals-and-strategies-idUSKCN1QI405

Siegel, Ethan. "NASA's Impossible Space Engine, the EmDrive, Passes Peer Review (But That Doesn't Mean It Works)." *Forbes.* 2016. http://www.forbes.com/startwithabang/2016/09/02/nasas-impossible-space-engine-the-emdrive-passes-peer-review/#3ca9a346692c

Silver, Calen. "Top 20 Economies in the World." *Investopedia.* 2019. https://www.investopedia.com/insights/worlds-top-economies/

Silveira, Evanildo de. "Um Ministro em Óbita." *Apublica.* 2019. https://apublica.org/2019/03/um-ministro-em-orbita/

Singer, P.W. and Cole, August. *Ghost Fleet: A Novel of the Next World War.* New York: Houghton Mifflin Harcourt Publishing Company, 2015.

Sinitskaya, Anastasia. "Chinese Engine Tested in Orbit." *Asgardia: The Space Nation.* 2019. https://asgardia.space/en/news/Chinese-Ion-Engine-Tested-in-Orbit

Sixsmith, Martin. *Russia: A 1,000-Year Chronicle of the Wild East.* New York: The Overlook Press, 2012.

Sloat, Sarah. "Russia and China Teaming Up to Heat the Atmosphere is 'Not That Exciting.'" *Inverse.* 2018. https://www.inverse.com/article/51972-ionosphere-russia-china-experiments

Smedley, Tim. "Goodbye Nuclear Power: Germany's Renewable Energy Revolution." *The Guardian.* 2013. https://www.theguardian.com/sustainable-business/nuclear-power-germany-renewable-energy

Smithberger, Mandy. "Brass Parachutes: The Problem of the Pentagon Revolving Door." *POGO.* 2018. https://www.pogo.org/report/2015/01/23/c5e8ff80-a34c-11e4-9f89-561284a573f8_story.html

Smith, Belinda. "Women Astronauts May Be Better Suited to Space Than Men—But Not By Much." *AU News.* 2019. https://www.abc.net/au/news/science/2019–04–08/ why-women-may-be-slightly-better-suited-to-space-living-than-men/10941616

Smith, David. *Monkey Mind: A Memoir of Anxiety.* New York: Simon and Schuster, 2013.

Smith, David. "Workers Claim Abuse as China Adds Zimbabwe to Its Scramble for Africa." *The Guardian.* 2012. https://www.theguardian.com/world/2012/jan/02/ china-zimbabwe-Workers-abuse

Smith, Richard K. *The Airships Akron & Macon: Flying Aircraft Carriers of the United States Navy.* Annapolis: Naval Institute Press, 2012.

Smith, Marcia S. "McCain, James Trade Barbs Over RD-180 Engines." *Space Policy Online.* 2016. http://ww.spacepolicyonline.com/news/ mccain-james-trade-barbs-over-rd-180-engines

Snowden, Scott. "China Plans to Build the World's First Solar Power Station in Space." *Forbes.* 2019. https://www.forbes.com/sites/scottsnowden/2019/03/05/ china-plans-to-Build-the-worlds-first-solar-power-station-in-space/#53c361d45c94

Sokolovskiy, V.D. *Soviet Military Strategy.* Ed. Harriet Fast Scott. New York: Crane, Russak, & Company, Inc., 1968.

Solomon, Lewis D. *The Privatization of Space Exploration: Business, Technology, Law and Policy.* New Jersey: Transaction Publishers, 2012

Sonne, Paul and Fassihi, Farnaz. "In Skies Over Iran, a Battle for Control of Satellite TV. *The Wall Street Journal.* 2011. https://www.wsj.com/articles/SB10001 424052970203501304577088380199787036

Southerland, Daniel. "Communist Influence in Peace Movement: Threat or Red Herring?" *Christian Science Monitor.* 1982. https://www.csmonitor. com/1982/1129/112936.html

"South American Powers Ranked by Military Strength." *Global Firepower.* Accessed On: 30 October 2019. https://www.globalfirepower.com/countries-listing-south-america.asp

"Space: Investing in the Final Frontier." *Morgan Stanley.* 2019. https://www.
morganstanley.com/ideas/investing-in-space

"Space Threat 2018: North Korea Assessment." *Center for Strategic and International
Studies: Aerospace.* 2018. https://aerospace.csis.org/space-threat-2018-north-korea/

Spring, Baker. "U.S. National Security Policy Toward Space: The Debate
That Should End But Won't." *Space News.* 2005. http://spacenews.com/
us-national-security-policy-Toward-space-debate-should-end-wont/

Spudis, Paul D. "China is Now Positioned to Dominate the Moon: What are
they up to?" *AirSpaceMag.com.* 2014. http://www.airspacemag.com/daily-planet/
China-now-positioned-dominate-moon-180953267/?no-ist.

Standish, Reid. "China's Path Forward is Getting Bumpy." *The Atlantic.*
2019. https://www.theatlantic.com/international/archive/2019/10/
china-belt-road-initiative-problems-kazakhstan/597853

Staff. "Bolsonaro Says Brazil 'Liberated from Socialism' at Inaugural
Ceremony." *France 24.* 2019. https://www.france24.com/
en/20190101-live-jair-bolsonaro-inauguration-oath-president-brazil

Staff. "Chandrayaan-2: Was India's Moon Mission Actually a Success?" BBC. 2019.
https://www.bbc.com/news/world-asia-india-49875897

Staff. "China, Brazil to Launch New Earth Resource Satellite Next Year." *Xinhua.*
2018 http://www.xinhuanet.com/english/2018–11/22/c_137624776.htm

Staff. "Iran News: U.S. Says Marines Used in Tanker Attacks Bear 'Striking
Resemblance' to Weapons Touted by Iran." *CBS News.* 2019. https://www.
cbsnews.com/news/iran-news-us-shows-limpet-mine-parts-case-against-iran-in-
tanker-attacks-today-2019–06–19/

Staff. "Japan Launches Epsilon Rocket Carrying Seven Satellites, Including One
Supposed to Generate Fake Meteor Shower." *Japan Times.* 2019. https://www.
japantimes.co.jp/news/2019/01/18/national/japan-launches-epsilon-rocket-
carrying-seven-satellites-including-one-supposed-generate-fake-meteor-shower/#.
XbfZbi-ZPyU

Staff. "Kazakhstan Mulls Ending Russia's Cosmodrome Lease." *Moscow Times.* 2012. https://www.themoscowtimes.com/2012/12/10/kazakhstan-mulls-ending-russias-cosmodrome-lease-a20035

Staff. "Man Helps Young Girl into Manhole in Hawaii." *CNN.* 2018. https://www.cnn.com/videos/us/2018/01/14/Hawaii-residents-false-missile-alert-girl-manhole-sot.com

Staff. North Korea Nuclear Tests: What Did They Achieve?" BBC. 2017. https://www.bbc.com/news/world-asia-17823706

Staff. "Pentagon Concerned About North Korea Jamming GPS Signals, Officials Say." *FoxNews.* https://www.foxnews.com/us/pentagon-concerned-about-north-korea-jamming-gps-signals-officials-say

Staff. "Toyota Motor Corp. and Japan Space Agency to Develop Manned Lunar Rover." The Hindu. YouTube. 2019. https://www.youtube.com/watch?v=-OsEdVS-73Y

Staff. "Why Americans Lost Interest in Putting Men on the Moon." BBC. 2014. https://www.bbc.com/news/av/world-us-canada-28450386why-americans-lost-interest-in-putting-men-on-the-moon

Staff Writer. "IRGC's Salami: Iran Today is Stronger Than Its Enemies." *Al Arabiya English.* http://english.alarabiya.net/en/News/middle-east/2019/10/25/IRGC-s-Salami-Iran-today-is-stronger-than-its-enemies.html

Stalcup, Travis C. "U.S. in Space: Superiority, Not Dominance: Trying for Dominance in Space is Counterproductive. The U.S. Should Settle for a More Modest Goal." *The Diplomat.* 2014. http://www.thediplomat.com/2014/01/u-s-in-Space-superiority-not-dominance/?allpages=yes.

Stanglin, Doug. "Vladimir Putin Blames Lenin for Soviet Collapse." *USA Today.* 2016. http://www.usatoday.com/story/news/2016/01/21/vladimir-putin-blames-lenin-soviet-collapse/79116132/

Stanway, David. "China Targets Nuclear Fusion Power Generation by 2040." Reuters. 2019. https://www.reuters.com/article/us-china-nuclearpower-fusion/china-targets-nuclear-fusion-power-generation-by-2040-idUSKCN1R00NB

Starr, Barbara. "Sources: Rumsfeld Calls for Special Ops Covert Action." *CNN*. 2002. https://www.cnn.com/2002/US/08/02/rumsfeld.memo/index.html

Stavridis, James. "The U.S. Needs a Cyber Force More Than a Space Force." *Bloomberg*. 2018. https://www.bloomberg.com/opinion/articles/2018–08–14/u-s-needs-a-space-force-and-a-cyber-force

Steil, Benn. "Russia's Clash with the West is About Geography, Not Ideology." *Foreign Policy*. 2018. https://foreignpolicy.com/2018/02/12/russias-clash-with-the-west-is-about-geography-not-ideology/

Stelter, Brian, Kaku, Michio, and Mirkinson, Jack. "CNN's Reliable Sources." YouTube Video. 2014. https://www.youtube.com/watch?v=P9PR362V-hY

Stephens, Brett. *America In Retreat: The New Isolationism and the Coming Global Disorder*. New York: Sentinel, 2014.

Steury, Donald P. "How the CIA Missed Stalin's Bomb: Dissecting Soviet Analysis, 1946–50." *CIA*. 2007. https://www.cia.gov/library/center-for-the-study-of-intelligence/csi-publications/csi-studies/v0149n01/html_files/stalins_bomb_3.html

Stewart, Phil. "U.S. Studying India Anti-Satellite Weapons Test, Warns of Space Debris." Reuters. 2019. https://www.reuters.com/article/us-india-satellite-usa/u-s-studying-india-anti-satellite-weapons-test-warns-of-space-debris-idUSKCN1R825Z

Stockton, Nick. "How to Track North Korea's New Satellite As It Streaks Through Space." *Wired.com*. 2016. http://www.wired.com/2016/02/tracking-north-korea-satellite/

Stockton, Nick. "Maybe the West's Water Wars Aren't As Bad As You Think." *Wired*. 2016. https://www.wired.com/2016/09/maybe-californias-water-wars-arent-bad-think/

Stone, Madeline and Webb, Kevin. "9 Early YouTube Stars Who Are Still Wildly Popular After More Than a Decade." *Business Insider*. 2019. https://www.businessinsider.com/9-early-youtube-stars-still-famous-after-a-decade-of-fame-2015–6

Strange, Joseph L. and Iron, Richard. "Center of Gravity: What Clausewitz Really Meant." *Joint Forces Quarterly*. 2004. https://theforge.defence.gov.au/sites/default/files/Adfwtc06_strange_and_iron_-_clausewitz.pdf

Strassler, Robert B. *The Landmark Thucydides: A Comprehensive Guide to the Peloponnesian War*. New York: Touchstone Books, 1996.

"Strategic Defense: How Much Will It Really Cost?" *The* Heritage Foundation. 1987. http://www.heritage.org/research/reports/1987/10/strategic-defense-how-much-will-it-really-cost

Strout, Nathan. "Russian Satellite Creeps Up to Intelsat Satellite—Again." *C4ISRNet*. 2019. https://www.c4isrnet.com/battlefield-tech/2019/09/03/russian-satellite-creeps-up-to-intelsat-satellite-again/

Stuckey, Alex. "A Few Good Men." *The Houston Chronicle*. 2019. https://www.houston chronicle.com/local/space/mission-moon-article/NASA-is-stalled-Apollo-era-legends-13788864.php

"Study Investigates How Men and Women Adapt Differently to Spaceflight." *NASA*. 2014. https://www.nasa.gov/content/men-women-spaceflight-adaptation

Stuff, Strategy. "The Strategy of Eurasianism." YouTube. 2019. https://www.youtube.com/watch?v=5Z98moTOa7Y

Subramanian, Courtney and Fritze, John. "Trump Calls NASA Astronauts Jessica Meir and Christina Koch to Congratulate Them on First All-Female Spacewalk." *USA Today*. 2019. https://www.usatoday.com/story/news/politics/2019/10/18/trump-calls-nasa-female-astro-jessica-meir-and-christina-koch-congratulate-them-first-all-female-spa/4022523002/

Sullivan, Mark. "VC Peter Thiel: You Can Either Invest in 'Bits' or 'Atoms." *Venture Beat*. 2014. https://venturebeat.com/2014/09/08/vc-peter-thiel-you-can-either-invest-in-bits-or-atoms/

Sweet, Cassandra. "Tariffs Boost Solar-Panel Makers in U.S." *The Wall Street Journal*. 2014. http://www.wsj.com/articles/tariffs-boost-solar-panel-makers-in-u-s-1406329365

Takeyh, Ray. "The Nuclear Deal is Iran's Legal Path to the Bomb." *Politico.* 2017. https://www.politico.com/magazine/story/2017/09/22iran-nuclear-deal-bomb-215636

Talmadge, Eric. "Possible Peace Declaration Looms Large Over Trump-Kim Summit." *AP.* 2018. https://www.apnews.com/758766909000c45d2a77213v0d020235a

Tangredi, Sam J. *Anti-Access Warfare: Countering A2/AD Strategies.* Annapolis: Naval Institute Press, 2013.

Tate, Karl. "A Village in Orbit: Inside NASA's Space Colony Concepts (Infographic)." Space.com. 2014. http://www.space.com/22228-space-station-colony-concepts-explained-infographic.html

Tate, Karl. "How Quantum Entanglement Works (Infographic)." *LiveScience.* 2013. http://www.livescience.com/28550-how-quantum-entanglement-works-infographic,html?_ga=1.129607052.1803845722.1472708474

Taverney, Thomas D. "Resilient, Disaggregated, and Mixed Constellations," *The Space Review.* 2011. http://www.thespacereview.com/article/1918/1

Tavernise, Sabrina. "As Fewer Americans Serve, Growing Gap Is Found Between Civilians and Military." *New York Times.* 2011. http://www.nytimes.com/2011/11/25/us/civilian-military-gap-grows-as-fewer-americans-serve.html_r=0

Teitel, Amy Shira. "The Manned Orbiting Laboratory the Air Force Failed to Launch." *Popular Science.* 2015. http://www.popsci.com/manned-orbiting-laboratory-air-force-failed-to-launch

Tellis, Ashley J. "Does China Threaten the United States in Space?" *Carnegie Endowment for Peace.* http://carnegieendowment.org/2014/01/28/does-China-threaten-united-states-in-space

"Ten 'Big Facts' About India." BBC. 2014. https://www.bbc.com/news/world-asia-25881705

Tetsuro, Kosaka. "As North Korea's Missile Arsenal Grows, So Does Its Nuclear Threat." *Nikkei Asian Review.* 2019. https://asia.nikkei.com/Spotlight/Comment/As-North-Korea-s-missile-arsenal-grows-so-does-its-nuclear-threat

The Hechinger Report, "Ranking Countries By the Worst Students: More Than a Quarter of American 15-Year-Old Students Are Low-Performing In at Least One Subject." 2016. http://www.usnews.com/news/articles/2016–02–16/ranking-countries-by-the-worst-students

Thiel, Peter. "Good for Google, Bad for America." *New York Times.* 2019. https://www.nytimes.com/2019/08/01/opinion/peter-thiel-google.html

Thiessen, Mark and Yamaguchi, Mari. "75 Years Later, 'Forgotten' WWII Battle on Alaskan Island Haunts Soldiers." *Army Times.* 2018. https://www.armytimes.com/veterans/2018/05/27/75-years-later-forgotten-wwii-battle-in-alaska-haunts-soldiers/

Thompson, Loren. "U.S. Growing Dependent on Russia for Satellite Propulsion Systems." *Forbes.* 2018. https://www.forbes.com/sites/lorenthompson/2018/09/14/u-s-satellite-makers-turn-to-foreign-sources-for-in-space-propulsion-despite-buy-american-push/#65a60ad51590

Tibon, Amir and Harel, Amos. "U.S. Senate Warns Israel Over Deepening Ties with China, Citing 'Serious Security Concerns.'" *Haaretz.* 2019. https://www.haaretz.com/us-news/.premium-u-s-senate-condemns-deepening-israel-china-ties-cites-serious-security-concerns-1.7368680

Tiezzi, Shannon. "China's Response to the US Cyber Espionage Charges." *The Diplomat.* 2014. http://thediplomat.com/2014/05/chinas-response-to-the-us-cyber-espionage-charges/

Timmons, Heather. "The BRICs Era is Over, Even at Goldman Sachs." *QZ.* 2015. https://qz.com/544410/the-brics-era-is-over-even-at-goldman-sachs/

Tirone, Jonathan. "Nuclear Fusion." *Washington Post.* 2019. https://www.washingtonpost.com/business/energy/nuclear-fusion/2019/06/20/c6bd5682-938d-11e9–956a-88c291ab5c38_story.html

Torchia, Christopher. "9/11 Conspiracy Theories Rife In Muslim World." *Washington Post.* 2010. http://www.washingtonpost.com/wp-dyn/content/ article/2010/10/02/AR2010100200663.html

Toll, Ian W. "A Reluctant Enemy." *New York Times.* 2011. https://www.nytimes. com/2011/12/07/opinion/a-reluctant-enemy.html

Totten, Michael J. "Dispatches: Vladimir Putin's Next Move." *World Affairs.* 2014. http://www.worldaffairsjournal.org/blog/michael-j-totten/ vladimir-putins-next-move

Toucas, Boris. "Russia's Design in the Black Sea: Extending the Buffer Zone." *Center for Strategic and International Studies.* 2017. https://www.csis.org/analysis/ russias-design-black-sea-extending-buffer-zone

"Tracking Solar Flares." *Solar Center at Stanford University.* Accessed on: January 1, 2020. http://solar-center.stanford.edu/SID/activities/ionosphere/html

"Treaty on the Principles Governing the Activities of States in the Exploration and Use of Outer Space, Including the Moon and Other Celestial Bodies." *United Nations Office for Outer Space Affairs.* Accessed On: 1 November 2019. https:// www.unoosa.org/oosa/en/ourwork/spacelaw/treaties/introouterspacetreaty.html

Trevithick, Joseph. "Revealed—The Pentagon's Scheme for A Spy Outpost in Space." *War Is Boring.* 2015. https://warisboring.com/ the-u-s-military-had-plans-for-a-spy-outpost-in-space/

Trevithick, Joseph. "Russia Says It Used Precision Guided Munitions In Strikes On Syrian Rebel Drone Makers." *The Drive.* 2018. https://www.thedrive.com/the-war-zone/23376/russia-says-it-used-precision-guided-munitions-in-strikes-on-syrian-rebel-drone-makers

Trump, Donald J. "Trump to UN: 'Rocket Man' on a Suicide Mission." CNN, YouTube. https://www.youtube.com/watch?v=8mstcJDHaGE

Trump, Donald J. *Twitter.* 7 June 2019. 1:38 PM. https://twitter.com/ realdonaldtrump/status/1137051097955102729?lang=en

Tselichtchev, Ivan. "Chinese in the Russian Far East: A Geopolitical Time Bomb?" *South China Morning Post.* https://www.scmp.com/week-asia/geopolitics/article/2100228/chinese-russian-far-east-geopolitical-time-bomb

Tucker, Patrick. "America's Top Threats in Space are Lasers and Nukes." *Defense One.* 2014. http://www.defenseone.com/threats/2014/07/americas-top-Threats-space-are-lasers-and-nukes/89519/print/

"Turkey's Downing of Russian Warplane Explained." *Caspian Report.* YouTube Video. 2015. https://www.youtube.com/watch?v=qkfGrbayVxE

Tyler, Patrick E. "Russia and China Sign 'Friendship' Pact." *New York Times.* 2001. http://www.nytimes.com/2001/07/17/world/russia-and-china-sign-friendship-pact.html

"U.N. Convention on the Law of the Sea: Living Resources Provisions." *Every CRS Report.* 2004. https://www.everycrsreport.com/reports/RL32185.html

U.S. Air Force, *Counter-Space Operations, Air Doctrine Document 2–2.1.* 2004. Pp. 1–66.

U.S. Air Force, *National Security Space Strategy: Unclassified Summary.* 2011. http://fas.org/irp/eprint/nsss.pdf

"U.S. and China Lead Tech Innovation and Disruption, Even as Innovation Spreads Globally: KPMG Report." *KPMG.* 2017. https://home.kpmg.com/us/en/home/media/press-releases/2017/03/us-and-china-lead-tech-innovation-and-disruption-even-as-innovation-spreads-globall-kpmg-report.html

Umland, Andreas. "Stalin's Russocentrism in Historical and International Context." *The Journal of Nationalism and Ethnicity,* vol. 39, Issue 1. 2011.

Ungureanu, Horia. "EmDrive Engine Can Theoretically Bring Us to Mars in 70 Days: What's Next After Peer Review?" *Tech Times.* 2016. https://www.techtimes.com/articles/176203/20160903/emdrive-engine-can-theoretically-bring-us-to-mars-in-70-days-whats-next-after-peer-review.htm

"United Nations Treaties and Principles on Outer Space." *The United Nations.* http://www.unoosa.org/pdf/publications/STSPACE11E.pdf

Vance, Ashlee. "This City in Serbia Could Become the Next Silicon Valley." Bloomberg. YouTube. 2016. https://www.youtube.com/watch?v=66MJEzkU8HE

Klinger-Vidra, Robyn. "Building the Venture Capital State." *American Affairs Journal.* Fall 2018, Vol. 2, No. 3. https://americanaffairsjournal.org/2018/08/building-the-venture-capital-state/

Vizard, Frank. "Safeguarding GPS." *Scientific American.* 2003. https://www.scientificamerican.com/article/safeguarding-gps/

Wall, Mike. "China Launches Pioneering 'Hack-Proof' Quantum-Communications Satellite." Space.com. 2016. http://www.space.com/33760-china-launches-quantum-communications-satellite.html

Wall, Mike. "Japanese Spacecraft Successfully Snags Sample of Asteroid Ryugu." *Space.* 2019. https://www.space.com/japanese-asteroid-probe-lands-ryugu.html

Wall, Mike. "NASA's Shuttle Program Cost $209 Billion—Was it Worth It?" Space.com. 2011. https://www.space.com/12166-space-shuttle-program-cost-promises-209-billion.html

Wall, Mike. "New Space Mining Legislation Is 'History In the Making.'" Space.com. 2015. http://www.space.com/31177-space-mining-commercial-spaceflight-congress.html

Wall, Mike. "North Korea Launches Satellite to Space." Space.com. 2016. http://www.space.com/31860-north-korea-satellite-launch.html

Wall, Mike. "Space Tourist: How a U.S. Millionaire Bought a Ticket Into Orbit." Space.com. 2011. http://www.space.com/11492-space-tourism-pioneer-dennis-tito.html

Wall, Mike. "Toyota, Japan to Launch Huge Moon Rover for Astronauts in 2029." *Space.* 2019. https://www.space.com/toyota-japan-moon-rover-2029-timeline.html

Wall, Mike. "Water Ice Common on Asteroids, Discovery Suggests." Space.com. 2015. http://www.space.com/9292-water-ice-common-asteroids-discovery-suggests.html

Wall, Mike. "Yes, NASA's New Megarocket Will Be More Powerful Than the Saturn V" Space.com. 2016. http://www.space.com/33691-space-launch-system-most-powerful-rocket.html

Walker, Joshua W. and Azuma, Hidetoshi. "Japan's Plan to Bring the United States and Russia Together: The G-7 and Abe's Eurasian Adventures." *Foreign Affairs.* 2016. http://www.foreignaffairs.com/articles/japan/2016–05–17/japans-plan-bring-united-states-and-russia-together

Wang, Brian. "China Says Tests of Propellentless EMDrive on Tiangong-2 Space Station Were Successful." *Next Big Future.* 2016. https://www.nextbigfuture.com/2016/12/china-says-Tests-of-propellentless.html

Wang, Zheng. *Never Forget National Humiliation: Historical Memory in Chinese Politics and Foreign Relations.* New York: Columbia University Press, 2012.

Watson, Kathryn. "Trump Says Future Belongs to 'Patriots,' Not 'Globalists,' in U.N. General Assembly Speech." *CBS News.* 2019. https://www.cbsnews.com/live-news/trump-un-speech-future-belongs-to-patriots-not-globalists-united-nations-general-assembly-today/

Weatherby, Leif. "Politics is Downstream from Culture, Part I: Right Turn to Narrative." *The Hedgehog Review.* 2017. https://hedgehogreview.com/blog/infernal-machine/posts/politics-is-downstream-from-culture-part-1-right-turn-to-narrative

Weeden, Brian and Samson, Victoria. "Op-Ed: India's ASAT Test is Wake-Up Call for Norms of Behavior in Space." *Space News.* 2019. https://spacenews.com/op-ed-indias-asat-test-is-wake-up-call-for-norms-of-behavior-in-space/

Weichert, Brandon J. "A Catwalk in Space." *American Greatness.* 2019. https://amgreatness.com/2019/10/20/a-catwalk-in-space/

Weichert, Brandon J. "America Should Not Abandon the Arctic." *American Greatness.* 2019. https://amgreatness.com/2019/11/17/america-should-not-abandon-the-arctic/

Weichert, Brandon J. "The Arms Race in Space Is Here." *The American Spectator.* 2019. https://spectator.org/the-arms-race-in-space-is-here/

Weichert, Brandon J. "AOC Isn't Even a Good Environmentalist." *The American Spectator.* 2019. https://spectator.org/aoc-isnt-even-a-good-environmentalist/

Weichert, Brandon J. "The Art of Economic Warfare: China Gains the Upper Hand By Playing a Different Game." *American Greatness.* 7 December 2016.

http://amgreatness.com/2016/12/07/
art-economic-warfare-china-gains-upper-hand-playing-different-game/

Weichert, Brandon J. "Brazil Does Not Need to Join NATO." *The American Spectator.* 2019. https://spectator.org/brazil-does-not-need-to-join-nato/

Weichert, Brandon J. "The Case for an American Space Corps." *American Greatness.* 2017. https://amgreatness.com/2017/12/03/the-case-for-an-american-space-corps/

Weichert, Brandon J. "The Case for Space Dominance." *American Greatness.* 27 December 2016. http://amgreatness.com/2016/12/27/case-space-dominance/

Weichert, Brandon J. "China's Exhortation to Study (and Conquer)." *The American Spectator.* 2018. https://spectator.org/
chinas-exhortation-to-study-and-conquer/

Weichert, Brandon J. "Deregulate Space." *American Greatness.* 2017. https://
amgreatness.com/2017/10/18/deregulate-space/

Weichert, Brandon J. "How to Create Your Own Enemies." *The American Spectator.* 2019. https://spectator.org/how-to-create-your-own-enemies/

Weichert, Brandon J. "Ignoring Iran Will Not Help." *The Weichert Report.* 2019. https://theweichertreport.com/2019/06/16/ignoring-iran-will-not-help/

Weichert, Brandon J. "Iran Keeps Asking for It." *The American Spectator.* 2019. https://spectator.org/iran-keeps-asking-for-it/

Weichert, Brandon J. "Musings in Singapore." *The Weichert Report.* 2018. https://
theweichertreport.com/2018/06/11/musings-in-singapore/

Weichert, Brandon J. "Nationalism is the Key Ingredient for Space Exploration." *American Greatness.* 2019. https://amgreatness.com/2019/08/10.
nationalism-is-the-key-ingredient-for-space-exploration/

Weichert, Brandon J. "NATO Must Pay." *The American Spectator.* 2018. https://spectator.org/nato-must-pay/

Weichert, Brandon J. "North Korea's Nuclear Program and the Cone of Uncertainty." *The Weichert Report.* 2017. https://theweichertreport.com/2017/12/07/north-koreas-nuclear-program-and-the-cone-of-uncertainty/

Weichert, Brandon J. "Preparing for a Russian 'Space Pearl Harbor.'" *Orbis.* 23 August 2019. https://www.fpri.org/article/2019/08/preparing-for-a-russian-space-pearl-harbor/

Weichert, Brandon J. "Russia and America Must Build a Lunar Space Station." *Space News.* 2017. https://spacenews.com/op-ed-russia-and-america-must-build-a-lunar-space-station/

Weichert, Brandon J. "Russia is Not Going to Change." *The New English Review.* 2018. https://www.newenglishreview.org/custpage.cfm?frm=189386&sec_id=189386

Weichert, Brandon J. "Russia's Preemptive Nuclear War Doctrine." *The Weichert Report.* 2016. https://theweichertreport.com/2016/10/14/russias-preemptive-nuclear-war-doctrine/

Weichert, Brandon J. "The True Ambitions of Russian Foreign Policy Today." *The Institute of World Politics.* 2017. https://www.iwp.edu/events/the-true-ambitions-of-russian-foreign-policy-today/

Weichert, Brandon J. "Space Weapons Trump Peace with Russia." *The American Spectator.* 2018. https://spectator.org/space-weapons-trump-peace-with-russia/

Weichert, Brandon J. "Trump is Winning the Little Cold War with Iran." *The American Spectator.* 2019. https://spectator.org/trump-is-winning-the-little-cold-war-with-iran/

Weichert, Brandon J. "Trump's Most Unhelpful Moon Tweet." *Space News.* 2019. https://spacenews.com/trumps-most-unhelpful-moon-tweet/

Weichert, Brandon J. "VLOG: The Geopolitics of the Moon." *The Weichert Report.* 2017. https://theweichertreport.com/2017/01/16/vlog-the-geopolitics-of-the-moon/

Weichert, Brandon J. "Washington is Still Not Getting Space Force Right." *American Greatness.* 2019. https://amgreatness.com/2019/04/20/washington-is-still-not-getting-space-force-right/

Weinberger, Sharon. "The Very Real Plans to Put Marines in Space." *Popular Mechanics.* 2010. http://www.popularmechanics.com/military/a5539/Plans-for-marines-in-space/

Weisfield, Daniel. "Peter Thiel at Yale: We Wanted Flying Cars, Instead We Got 140 Characters." *Yale School of Management.* 2013. https://som.yale.edu/blog/peter-thiel-at-yale-we-wanted-flying-cars-instead-we-got-140-characters

Weitering, Hanneke. "India Successfully Launches RISAT-2B Earth-Observation Satellite." *Space.* 2019. https://www.space.com/india-risat-2b-earth-satellite-launch-success.html

Weitering, Hanneke. "Israeli Moon Lander Suffered Engine Glitch Before Crash." *Space.* 2019. https://www.space.com/beresheet-moon-crash-engine-glitch.html

Weitering, Hanneke. "Pluto's 'Heart' Hints at Deep, Underground Ocean." Space.com. 2016. http://www.space.com/34179-plutos-heart-hints-at-deep-underground-ocean.html

Wen, Quan, Yang, Liwei, et al. "Impacts of Orbital Elements of Space-Based Laser Station on Small Scale Space Debris Removal." *Optik,* vol. 154. 2018. Pp. 83–92.

Werner, Debra. "NSLComm Reports 90 Percent Success on First Cubesat Mission." *Space News.* 2019. https://spacenews.com/nslcomm-mission-one/

Wenzl, Ron. "Star Trek and Vietnam: When Klingons Were Stand-Ins for the Soviets." *History.* 2018. https://www.history.com/news/star-trek-series-episodes-vietnam-war

Wenz, John. "5 Big, Bold Soviet Space Missions That Never Were." *Popular Mechanics.* 2015. https://www.popularmechanics.com/space/a16037/failed-soviet-space-missions/

"What Are Satellites Used For?" *Union of Concerned Scientists.* 2015. https://www.ucsusa.org/resources/what-are-satellites-used

"What is Machine Learning?" *Expert System.* 2017. https://expertsystem.com/machine-learning-definition/

"Why Are Science and Technology Critical to America's Prosperity in the 21st Century?" in *Rising Above the Storm: Energizing and Employing America For a Brighter Economic Future.* Washington, D.C.: The National Academies Press, 2007. Accessed on: December 30, 2019. https://www.nap.edu/read/11463/chapter/4

Wieslander, Anna. "What Makes an Ally? Sweden and Finland as NATO Partners." *Atlantic Council.* 2019. https://www.atlanticcouncil.org/blogs/new-atlanticist/what-Makes-an-ally-sweden-and-finland-as-nato-partners/

WildFilmsIndia, "Troops at Indo-China Border Face Off Against Each Other." YouTube. 2017. https://www.youtube.com/watch?v=84b1PSO!AaU

Wille, Dennis. "What Should We Call Members of a Space Force?" *Slate.* 2019. https://slate.com/technology/2019/02/space-force-members-names-sentinels-troopers.html

Willetts, David. "Forces' Star Wars: Special Forces Soldiers Could Soon Be Deployed in SPACE to Strike Anywhere on Earth in Minutes." *The Sun.* 2017. https://www.thesun.co.uk/news/4015794/special-forces-soldiers-could-soon-be-deployed-in-space-to-strike-anywhere-on-earth-in-minutes/

Williamson, Ray A. "The Growth of Global Space Capabilities: What's Happening and Why It Matters." Secure the World Foundation. 2009. https://swfound.org/media/36582/swf-testimony%20to%20the%20us%20congress%2019%20nov%202009.pdf

Williams, Lauren C. "Navy Declares EMS a Full-Fledged Warfighting Domain." *Defense Systems.* 2018. https://defensesystems.com/articles/2018/10/24/navy-electronic-warfare-williams-aspx

Winfrey, David. "The Last Spacemen: MOL and What Might Have Been." *The Space Review.* 2015. http://www.thespacereview.com/article/2866/1

Wire Staff, CNN. "Obama, Russian President Sign Arms Treaty." *CNN.* 2010. http://www.cnn.com/2010/POLITICS/04/08/obama.russia.treaty/

Wirtz, James J. "Limited Nuclear War Reconsidered." *On Limited Nuclear War In the 21ˢᵗ Century* (ed.) Jeffrey A. Larsen and Kerry M. Kartchner. Stanford: Stanford University Press, 2014.

Woody, Christopher. "Russia Reportedly Warned Mattis It Could Use Nuclear Weapons in Europe, and It Made Him See Moscow as an 'Existential Threat' to the US." *Business Insider.* 2018. https://www.businessinsider.com/russia-warned-mattis-it-could-use-tactical-nuclear-weapons-baltic-war-2018–9

Wohlstetter, Roberta. *Pearl Harbor: Warning and Decision.* Stanford: Stanford University Press, 1962.

Wolf, Jim and Alexander, David. "U.S. Successfully Tests Airborne Laser on Missile." Reuters. 2010. http://www.reuters.com/article/usa-arms-laser-idUSN1111660620100212

Woolsey, James and Pry, Peter. "A Shariah-Approved Nuclear Attack." *The Washington Times.* 2015. https://www.washingtontimes.com/news/2015/aug/18/jams-woolsey-peter-pry-emp-a-sharaiah-approved-nucl/

Woolsey, James and Pry, Vincent. "How North Korea Could Kill 90 Percent of Americans." *The Hill.* 2017. https://thehill.com/blogs/pundits-blog/defense/326094-how-north-korea-could-kill-up-to-90-percent-of-americans-at-any

Woolsey, R. James and Pry, Peter Vincent. "The Miniaturization Myth: Obama and 'Experts' Wrongly Measure North Korea's Nuclear Intentions." *The Washington Times.* 2016. http://www.washingtontimes.com/news/2016/apr/24/r-james-woolsey-peter-vincent-pry-obama-wrong-on-n/

Woolsey, James R., Graham, Graham, William R., et al. "Understanding Nuclear Missile Threats from North Korea and Iran." *The National Review.* 2016. http://www.nationalreview.com/article/431206/iran-north-korea-nuclear

"World War II in Colour." Episode 8. Jonathan Martin, 2010. *Military History.* https://www.imdb.com/title/tt2101277

Wuthnow, Joel, Limaye, Satu, and Samaranayake, Nilanthi. "Doklam, One Year Later: China's Long Game in the Himalayas." *War on the Rocks.* 2018. https//warontherocks.com/2018/06/doklam-one-year-later-chinas-long-game-in-the-himalayas/

Wu, Nicholas. "French President Emmanuel Macron Announces Creation of French Space Force." *USA Today.* 2019. https://www.usatoday.com/story/news/world/2019/07/13/french-space-force-macron-announces-creation-space-force-command/1723998001

Xenakis, John J. "World View: China Builds Navy Designed to Overwhelm the US Navy." *Breitbart.* 2015. http://www.breitbart.com/national-security/2015/05/05/world-view-china-builds-navy-designed-to-overwhelm-the-us-navy/

Yarow, Jay. "Peter Thiel: Progress Ended After We Landed on the Moon and 'Hippies Took Over the Country.'" *Business Insider.* 2014. https://www.businessinsider.com/peter-thiel-on-progress-2014–10

Yufan, Hao. *Sino-American Relations: Challenges Ahead.* Oxford: Routledge Publishing, 2010.

Zaborskiy, Victor. "Policy and Strategic Considerations of the Russian Space Program." (ed.) Eligar Sadeh. *Space Strategy in the 21ˢᵗ Century: Theory and Policy.* London: Routledge, 2013.

Zak, Anatoly. "The Almaz Military Space Station Program." *Russian Space Web.* http://www.russianspaceweb.com/almaz.html

Zak, Anatoly. "Did the Soviets Actually Build a Better Space Shuttle?" *Popular Mechanics.* 2013. https://www.popularmechanics.com/space/rockets/a9763/did-the-soviets-actually-build-a-better-space-shuttle-16176311

Zak, Anatoly. "Energia-M: The Swan Song of the Soviet Space Program." *Russian Space Web.* 2014. http://www.russianspaceweb.com/energia_m.html

Zak, Anatoly. "Here Is the Soviet Union's Secret Space Cannon" *Popular Mechanics.* 2015. http://www.popularmechanics.com/military/weapons/a18187/here-is-the-soviet-unions-secret-space-cannon

Zak, Anatoly. "The Hidden History of the Soviet Satellite Killer." *Popular Mechanics.* 2013. https://www.popularmechanics.com/space/satellites/a9620/the-hidden-history-of-the-soviet-satellite-killer-16108970/

Zak, Anatoly. "Russia's Space Agency Might Break Up with the U.S. to Get with China." *Popular Mechanics.* 2018. https://www.popularmechanics.com/space/moon-mars/a19159719/roscosmos-china-collaboration/

Zeihan, Peter. "The Russian Land Grab." YouTube. 2017. https://www.youtube.com/watch?v=rkuhWA9GdCo

Zhen, Liu. "Russia Offers Rocket Engine Tech as China's Long March 5 Struggles to Get Off the Ground." *South China Morning Post.* 2019. https://www.scmp.com/news/china/military/article/3024766/russia-offers-rocket-engine-tech-chinas-long-march-5-struggles

Zhou, Cissy and Bermingham, Finbarr. "China Steps Up Efforts to Close Failed Zombie Companies by 2020, But Faces Harsh Economic Reality." *The South China Morning Post.* 2019. https://www.scmp.com/economy/china-economy/article/2185186/china-steps-efforts-close-failed-zombie-companies-2020-faces

Ziarnick, Brent. "The Age of the Great Battlestars." *The Space Review.* 2009. http://www.thespacereview.com/article/1438/1

Ziarnick, Brent. "Defense Department Status Quo is a Weak Argument Against Space Force." *The Hill.* 2019. https://thehill.com/opinion/national-security/463558-defense-department-status-quo-is-a-weak-argument-against-space

Ziarnick, Brent D. "Why the Space Corps Needs to Use Naval Rank." *The Space Review.* 2019. http://www.thespacereview.com/article/3761/1

Zolfagharifard, Ellie. "Russia's Plan to Conquer the Moon: Nation Will Send 12 Cosmonauts to Lunar Surface Ahead of Creating A Permanent Base By 2020." *Daily Mail.* 2016. http://www.dailymail.co.uk/sciencetech/article-3653942/Russia-s-plan-to-conquer-moon-Nation-send-12-cosmonauts-lunar-surface-ahead-creating-permanent-base-2030.html

Zubrin, Robert. *Entering Space: Creating a Spacefaring Civilization.* New York: Penguin Putnam, Inc., 1999.

Zubrin, Robert. "Victory from Space: In 20th-century wars, the key was air power. In 21st-century wars, it will be space power." *National Review.* 2014. http://www.nationalreview.com/node/403920/print

INDEX